SCIENCE REORGANIZED
SCIENTIFIC SOCIETIES IN
THE EIGHTEENTH CENTURY

Science Reorganized

SCIENTIFIC SOCIETIES IN
THE EIGHTEENTH CENTURY

James E. McClellan III

COLUMBIA UNIVERSITY PRESS
NEW YORK   1985

Library of Congress Cataloging in Publication Data

McClellan, James E. III, 1946–
Science reorganized.

Bibliography: p.
Includes index.
1. Science—Societies, etc.—History. 2. Learned
institutions and societies—History. I. Title.
Q10.M38 1985   506   84-22993
ISBN 0-231-05996-5
ISBN 0-231-05997-3 (pbk.)

Columbia University Press
New York   Guildford, Surrey

Printed in the United States of America

Clothbound editions of Columbia University Press books are
Smyth-sewn and printed on permanent and durable acid-free paper

Contents

Illustrations and Figures

Illustrations

Du reste, c'est ainsi que les sciences et la littérature vivaient au XVIIIe siècle. Elles aimaient la noblesse dont elles étaient aimées, et elles flattaient la fortune qui était fière de ses attentions. La république des lettres n'était en réalité que l'expression de toutes les délicatesses de la monarchie. Elle vivait de sa table et de sa bourse sans craindre d'être humiliée. Elle était heureuse, je n'ose dire qu'elle fut reconnaissante.

-Chaillou des Barres (1851)

About This Book

ACADEMIES and societies of science have had their historians at least since Thomas Sprat and his *History of the Royal Society* which appeared in 1667. The fact that Sprat's *History* was commissioned by and just after the foundation of the Royal Society of London in 1662 is testimony to an impulse on the part of these institutions to commemorate their reputations and place in the world through the writing of their own histories. Indeed, from Sprat's day, official institutional histories have been a mainstay of the historiography of academies and societies, and the tradition continues in any number of memorial, anniversary, and like volumes which appear regularly. An unofficial antiquarianism complements this official literature of scientific societies.

With the rise of academic scholarship in the nineteenth century a new style of institutional history appeared—more analytical and disinterested histories written by professional scholars. Adolf von Harnack's enduring *Geschichte der königlich preussischen Akademie der Wissenschaften zu Berlin* (1900) ranks as the exemplar of this type of endeavor. Since then, as our biography reveals, many excellent studies of individual academies and societies written by professional historians have appeared. Scores of competent special studies treating aspects of science and scientific societies in the eighteenth century augment these histories of individual institutions.

As this corpus of work has grown, so has the need and desire for a more comprehensive study—a synthesis—of the phenomenon of eighteenth-century scientific societies. Even though it just touches on national academies and societies of science like the Royal Society of London and the Paris Academy of Science, Edward Conradi's 1905 article on "Learned Societies and Academies in Early Times" may well be considered the first effort at such a synthetic treatment.[1] J. P. Mahaffy's slightly personal "On the Origins of Learned Academies in Modern Europe" (1913) is another early study. Michele Maylender's encyclopedic survey of Italian academies (1926–1930), Daniel Mornet's pioneering study of the French provincial academies (1933), and Arthur Almhult's short article on the Swedish academies (1937) were early efforts to treat national subgroups of societies. In 1931 Alex von Harnack offered what is still a reputable overview of all scientific societies from the seventeenth to the nineteenth centuries in his "Die Akademien der Wissenschaften." This was followed the next year by Bernard Fäy's less successful "Learned Societies in Europe and America in the Eighteenth Century." An excellent short review of the collective growth of eighteenth-century scientific academies and societies by Douglas McKie, "Scientific Societies to the End of the Eighteenth Century," appeared in 1948.[2] Still, as evidenced by Henri Roddier's (1959) "Pour une histoire des académies" or even more by Robert Schofield's 1963 article, "Histories of Scientific Societies: Needs and Opportunities for Research," by the 1960s the history—particularly the collective history—of eighteenth-century academies and societies had yet to be fully explored.[3]

In the two decades since Schofield's call for further work a great deal has been accomplished. The Russian historian Iudif' Khaimovna Kopelevich produced her informative and workmanlike overview of eighteenth-century academies, *Vozniknovenie nauchnykh akademii*, in 1974.[4] Recent German scholarship devoted to eighteenth-century scientific societies has been especially rich. Among others, one needs to mention especially Ludwig Hammermayer. His "Akademiebewegung und Wissenschaftsorganisation"—a most thorough survey of the growth and spread of learned and scientific societies in the eighteenth century—appeared in 1976, and it is only the most relevant instance of his contributions to this historiography. Similarly, in France a major lacuna was filled with the appearance in 1978 of Daniel Roche's comprehensive treatment of eighteenth-century French provincial academies: *Le siècle des lumières en province: Académies et académiciens provinciaux, 1680–1789*.

This book is intended as a general survey of the history of scientific academies and societies in the eighteenth century and a further contri-

[xii]

bution in English to this substantial literature. If this work adds anything not already in the literature, it is a more detailed sense of the ways in which, through their concrete actions and interactions, academies and societies collectively reshaped organized science and scientific communications on the international level in the eighteenth century.

As a synthesis, this book covers a lot of territory. Chronologically, its central focus is on the nearly one hundred and fifty years between the 1660s and the 1790s, and at the extreme it takes into account changes from the early seventeenth century into the first decades of the nineteenth. Over one hundred scientific and related societies are touched on, some in detail, and more than that number are pointed to along the way. Social and scientific circumstances in a dozen or more different countries are considered, and orders of politics and culture ranging from great capitals to provincial outposts are discussed. Somewhat against the grain of contemporary scholarship, this book is anything but a narrow and specialized study.

Such wide scope is not without its problems, and this work is offered somewhat hesitantly on that account. I am acutely aware that it is impossible to cover thoroughly and with full scholarly rigor every facet of the history of academies and societies of science from the 1660s to the French Revolution. But the subject demands such inclusive treatment. Indeed, the guiding ideas in the pages that follow are that the scientific societies of the eighteenth century constitute a coherent subject important in understanding the development of science since the Scientific Revolution and that only by considering the scientific societies collectively as a single phenomenon can the proper perspectives be brought to bear on their impact and history.

Thus, more than is usually the case with historical writing, this book depends on a large body of preexisting research, and is possible only because the considerable amount of quality historical work just referred to has already been done. The many scholarly and intellectual debts owed by this book are repaid, adequately I hope, in the notes. Still, some are so great that separate mention must be made. As is obvious in the notes, Daniel Roche's work on the academies of provincial France and Ludwig Hammermayer's on the academic movement are major pillars on which this study stands. Similarly, without Roger Hahn's *Anatomy Of A Scientific Institution: The Paris Academy of Sciences,*

1666–1803 this work could not have been attempted. Charles C. Gillispie's *Science and Polity in France at the End of the Old Regime*, among his other works, has likewise proved invaluable.

The list of others in the community of historians past and present whose work is a sine qua non of this could easily be extended: Sten Lindroth (the Swedish Academy), Hans Hubrig (the patriotic and economic societies), Dorothy Stimson and Sir Henry Lyons (Royal Society of London), Alexander Vucinich (science in Russia), Brooke Hindle (science in America), David Kronick (the scientific press), Harry Woolf (the transits of Venus), W. Knowles Middleton (the *Accademia del Cimento*), Harcourt Brown (scientific organizations in seventeenth-century France), John Heilbron (eighteenth-century physical science communities), Karl Hufbauer (the German chemical community), and many more.

But the list would not be complete without mentioning Martha Ornstein's *The Rôle of Scientific Societies in the Seventeenth Century* (1913). My work is not intended as a mechanical sequel to that classic.[5] Her book is about new institutional alternatives to university-based science and the emergence of experimental science in the seventeenth century. This book is about the assimilation of science by government, the penetration of science into society, and the professionalization of science in the eighteenth century. Other differences of substance and approach separate these two works and their authors. Yet Ornstein's modest but enduring volume has proved not only a valuable source but a model for emulation.

Otherwise, this book has been fairly long in the making, and it has accumulated large debts. Charles C. Gillispie is the original promoter of this work, and through to the present both the work and the author have continued to benefit from his scholarship and counsel. Harold Dorn also exerted a major personal and intellectual influence on the book's coming-to-be. The debt owed him for his commitment to critical thinking and clear writing is large. To Daniel Roche and the late Sten Lindroth go thanks for personally introducing me to the academies of their respective countries. Karl Hufbauer read the work in manuscript at a late stage, and his several suggestions have made it a better book than it otherwise would have been. Craig B. Waff was instrumental in getting the book published, and I thank him in addition for his several

[xiv]

substantive contributions as a fellow historian of science and the eighteenth century. To other friends and colleagues with whom I have discussed this history wholly or in part over the years, I have to apologize for just a general note of thanks and for perhaps taking more than I gave.

For permissions to consult and to publish archival material I thank the Royal Society of London, the Académie des Sciences of the Institut de France, the Swedish Royal Academy of Sciences in Stockholm, the Deutsche Staatsbibliothek and the Akademie der Wissenschaften der DDR in Berlin, the Académie des Sciences, Arts et Belles-Lettres of Dijon, the Archives Départementales of the Hérault (Montpellier), the Bibliothèque Municipale of Montpellier, and the American Philosophical Society in Philadelphia.

The illustrations are not tied tightly to the text, but I intend a point with them. The illustration from Sprat's *History of the Royal Society* is by permission of the Royal Society of London. The remaining illustrations are from originals held by the American Philosophical Society. I thank these institutions for cooperating in allowing reproduction here.

A small host of kindnesses is to be repaid to the people of the abovementioned institutions for their assistance and welcome. I thank particularly N. H. Robinson and the staff of the Library of the Royal Society of London for meeting my research needs so effectively and so pleasantly. Similarly, I am grateful to the personnel of the archives of the Paris Academy of Sciences who aided me during my various visits. Many friends and associates at the American Philosophical Society Library have long been of assistance; the book owes fairly recent debts to Hildegard Stephans and Roy E. Goodman, for several aids and courtesies. I look back to ancient visits to the library of the Swedish Academy of Sciences and to the archives of the Academy of Sciences of the German Democratic Republic, and I thank, respectively, Dr. Wilhelm Odelberg and Dr. Christa Kirsten, for their aid then and since. Many anonymous reference desk librarians and stack people in various locales are owed a word of thanks, but those of the New York Public Library cannot be passed entirely without separate mention.

Any book owes a lot to the house that publishes it, and I thank Columbia University Press for the kind treatment afforded me and my products. I also need to acknowledge my own institution, Stevens Institute of Technology, whose institutional resources greatly supported this endeavor. Particular thanks are owed the Humanities Department of Stevens for material and emotional support and to the staff of Williams Library.

I have learned that beyond a certain point authors get too much credit for their books. Donna Walsh contributed a great deal through her expert copyediting. Ken Venezio designed the book, and Doris China prepared the figures that accompany the text. Diana Witt prepared the index, and Jim Cramer helped out in proofreading. Joan McQuary, Theresa Yuhas, and others in various departments of Columbia University Press lent their considerable expertise to making my book a reality. I thank all of them. Computer and word-processing technologies likewise improved the quality of the work.

This book owes a particular debt to my friend and compatriot, Philip R. Reilly, who has been an important source of encouragement and sound practical advice. Not to mention simply having put up with me and scientific societies for all these years, Jackie McClellan cannot be thanked enough for other considerable contributions, including the study on common members that shows up in chapter 5 following. The enterprise that ends here would have been impossible without her. Others in my circle, upon whom I rely to an unusual degree, will know of my gratitude for their many direct and indirect efforts furthering the appearance of this work.

There are errors of all sorts in what follows. I hope they are few. Unfortunately, I have to take responsibility for them.

INTRODUCTION

Organized Science and
Its Eighteenth-Century Contexts

To say that science is organized is to say several things at once. One notion is that science is an organized intellectual enterprise pursued by a community of practitioners and that it requires the communication of ideas among those practitioners for its existence, pursuit, and advance. In other words, science is part of "public knowledge," the product and responsibility of a community of workers, not the lone investigator or prophet. On the research front of mature sciences, it is evident that the problems attacked by scientists, the ways they are attacked, and the standards for judging success stem from an esoteric and highly defined intellectual, methodological, and philosophical consensus. Even considering science in less mature and stable conditions, the centrality of community and communications in science is apparent.

That being the case for the production of scientific ideas, formal communications among men of science are thus essential to science and an important component of science as an organized enterprise. Examples of real mechanisms for achieving the exchange of ideas immediately come to mind: books, journals, letters, personal travel, the computer, and so on. These are the concrete means for transmitting and helping create scientific ideas in a community context, and in this regard it may mean something unique to say that science is organized, for while other cultural endeavors—religion, the arts, or crafts, for example—may benefit from interchange among practitioners, it is to be questioned whether they require communication among a community in the manner of science.

[xvii]

The practice of science can also be organized as an occupation or a profession, and this notion of the community of scientists as self-organized or self-organizing is another meaning for organized science. The point is again evident in modern science, which is an elaborate professional enterprise entailing education and training, certification, careers, professional societies, refereed journals, and so forth. Whether all intellectual communities of scientists have so organized their human resources and collective existence is to be doubted, but clearly this professional aspect, too, is what one means in speaking of science as an organized enterprise.

A third, distinct way of approaching the topic of organized science is to consider science as a socially organized enterprise, or the place of science in society. Science is obviously highly organized in society today: we spend money on science, we value it, and as a social institution science is well plugged into government, the military, and industry.[1] And clearly, too, the self-organization of science as a profession today would be impossible without larger sources of social support. How much a society values science or incorporates it into its economic or institutional infrastructure seems an important indicator of just how organized science is in a particular context.

Finally, very often the phrase "organized science" is used to mean institutionalized science. Scientific institutions are manifestly a significant component of organized science. Scientific institutions are of interest in their own right, and institutions clearly affect the other ways science can be organized. Scientific institutions promote scientific communications and influence the production of scientific ideas. They are the indispensable institutional base upon which the community of scientists and the modern profession of science are built. And they form the primary link with other institutions and powers in society and a main nexus through which the interaction of science and society is mediated.

The ways in which men of science have organized themselves and communicated have changed considerably over time, as have scientific institutions and the place and importance of science in society. The history of organized science in all its dimensions begins in antiquity, if not the deep past, and this book seeks to trace a small part of the sweep of the history of organized science, in treating the scientific societies of the eighteenth century.[2] But this is not antiquarianism. For the history of organized science, the eighteenth century and scientific societies are both of considerable importance. Indeed, the principal thesis of this book is that organized science in *all* its aspects was entirely recast under the dominion of the scientific societies in the eighteenth century and that the period represents a distinct historical stage, following on the Scientific Revolution and crucial to the further development of organized science as it exists today.

What the scientific societies of the eighteenth century were and what they did will require a lot more explaining, but for the present it can be said simply that they represent science reorganized after the Scientific Revolution and the prime carrying institutions for science, for scientific communications, for the role and profession of science, and for the science-society interface during the one hundred and fifty years between the middle of the seventeenth century and the beginnning of the nineteenth century.

To appreciate the point, one needs to understand that eighteenth century scientific societies straddle two revolutions in organized science: on the one hand that of the seventeenth century, the rejection of university-based scholastic science, and the instauration of scientific societies; on the other hand that of the nineteenth century when the general learned society was eclipsed by other, more modern ways of organizing, producing, and disseminating science, notably in university contexts, into disciplinary specialties, and along professional lines.

In the organizational revolution of the sixteenth and seventeenth centuries, a new identity and a new institutional base were achieved for science as part of the Scientific Revolution.[3] In the seventeenth century organized science turned away from the traditional setting of the medieval universities and established itself in extra-university circles of several sorts. The network of correspondence that arose around Marin Mersenne in Paris in the 1630s and 1640s is one example of new patterns of sociability and communications in science at the time. With the appearance of the *Accademia dei Lincei* (Rome 1603–1630) or the *Accademia del Cimento* (1657–1667) one encounters new kinds of associations devoted in one way or another to scientific inquiry. As we will see, these "Renaissance" academies, as they shall be called here, were not the last of the institutional innovations of the Scientific Revolution.

[xix]

Rather, they were succeeded and ultimately overshadowed by another new kind of scientific institution, the official state science society of types like the Royal Society of London (1662) and the *Académie royale des sciences* of Paris (1666). The emergence of these institutions makes clear that the organizational revolution accompanying the cognitive and methodological redefinitions making up the Scientific Revolution as a whole marks the end of one era in the institutional history of science, that of the medieval university, and the beginning of another, the age of the learned scientific societies.

On the other hand, in the early decades of the nineteenth century organized and institutionalized science underwent another revolution in form.[4] This time one sees the emergence of specialized and professional scientific societies, such as the Geological Society of London (1807) and the British Association for the Advancement of Science (1831), the advent of the specialized scientific journal (e.g., Liebig's *Annalen der Chemie*, 1832), and the return of science to a university context, notably in Germany, of research and teaching.

In the interim between the 1660s and the French Revolution the general scientific society similar to the Royal Society of London and the Paris Academy of Science dominated organized and institutionalized science. Some seventy official learned societies modeled after the Royal Society and the Paris Academy and concerned at least in part with science arose in Europe and America in the eighteenth century, and many private groups of a similar nature augmented these. It would be an oversimplification to suggest that the scientific societies were the only or the only important scientific institutions of the period. In some places, Leyden for example, home of 'sGravesande and Boerhaave, the university continued to be an active and noteworthy center of scientific activity.[5] And other types of scientific institutions, such as observatories and botanical gardens, flourished along with the learned societies.[6] All these institutions must be considered in evaluating the total organizational and institutional development of science in the eighteenth century.

Nonetheless, the scientific societies predominated over other institutions and modes of organization for science in the eighteenth century for a number of reasons. They culminated the seventeenth-century struggle to achieve new institutional forms. Their number grew more rapidly than any other type of scientific institution in the eighteenth century. They provided the primary institutional affiliation for the leading members of the scientific community of the time. They were elite centers in society, and, through a host of institutional activities (including publishing journals, offering prizes, and sponsoring scientific expeditions), they promoted the best science.

The eighteenth century was the hey-day of the general scientific society. By the nineteenth century learned societies ceased to be the premier institutions for the organization and pursuit of science. That role, again, was taken over by specialized and professional scientific societies and by university-based teaching and research. The eighteenth-century type of scientific society survives as a local or provincial social group, as an honorary organization to which one is elected at the end of an active scientific career, or as an overarching bureaucratic entity controlling the research efforts of subordinate units, as in the case of the Academy of Sciences of the USSR.

Related to their special position in the history of scientific institutions, the scientific societies of the eighteenth century also had a major impact on scientific communications. The scientific journal originated with the scientific societies in the seventeenth century (e.g., the *Philosophical Transactions* of the Royal Society, from 1665), and the journals and publications of the scientific societies were the most important avenues for publication of scientific material in the eighteenth century.[7] The ability to control what appeared in their journals and in the works of their members was a significant censorship power in the hands of the scientific societies which greatly affected the flow of information. Scientific societies were centralized clearing houses for communicating science, and perhaps most importantly for scientific communications, the network and system of institutions established de facto by interaction among the scientific societies provided an entirely new institutional means of producing, channeling, and facilitating the exchange of scientific information. In the nineteenth century the central position of the scientific societies in scientific communications receded as alternative channels (notably the specialized journal) developed. Beyond straightforward institutional history, these themes in scientific communications are thus an inextricable aspect of the history of organized science and of scientific societies in the eighteenth century.

[xxi]

But more is involved than the linear, one-dimensional development of scientific institutions or scientific communications over time. Another issue underlying this inquiry concerns changes in the level of social recognition and support for the scientific enterprise in the period between 1700 and 1800 and the role played by the scientific societies as contributors to and as products of those changes. In the eighteenth century the scientific enterprise grew considerably larger and became better integrated in society. The scientific societies represent the vanguard of change in that social integration.

Regarding the seventeenth century, all things considered and despite the profound intellectual reformulations of the Scientific Revolution, science remained a relatively small-scale enterprise in society. Through the 1650s at least, the new natural philosophy was confined to a small core of cognoscenti, few of whom were able to support themselves financially on the basis of their knowledge. Science had yet to fully demonstrate its authority as a means of knowing. It had yet to convince governments of its utility or be assimilated by them. It had virtually no impact on economic activity or the material base. And social orders high and low were as yet impervious to its intellectual or technical charms.

By 1800 the social profile of science had changed considerably. There were many more serious practitioners of science. There were many more niches in society where scientific knowledge was valuable and where the man of science could make a home. Governments at all levels had become convinced that science was useful to the state, and they had incorporated scientific and technical expertise into their service. Intellectually the authority of science had proved itself not only in the domain of nature, but it provided a model for sure knowledge in other areas of human inquiry—witness Comte and positivist philosophy. And socially, while not affecting everyone, movements like the revolutionary "Organization for Victory" in France, mechanics institutes in England, or the romantic reaction against the strictures of science and reason in Germany and elsewhere testify to the degree to which by the nineteenth century science had become tightly woven into the cultural fabric of the West.[8]

This dramatic and continuing transformation in the place and scale of science as an enterprise in society has to be seen as distinct from both the straightforward institutional history of science and from the cognitive development of scientific ideas. The social growth of science since the seventeenth century has been a historical process of such magnitude and importance that it deserves separate denomination as the "socialization" of science.

Scientific societies are by no means the only indicators of this process of the "socialization" of science in the eighteenth century. Analysis of the Enlightenment's fascination with Newton or Bacon, or of the number and kinds of books published, or of public concern for science evidenced in fads for "botanizing," for ballooning, or for Mesmer, for example, show as well the increasing and unprecedented penetration of science into society and culture in the eighteenth century.[9] Yet, eighteenth-century scientific societies need to be seen against the background of the larger "socialization" of science in the period, and they offer especially good evidence for it. For one thing, many and the most important scientific societies were official institutions, legal entities transcending ephemeral groups or movements. For another, they were recognized and supported by governments, and were thus on the cutting edge of relations between science and the state in the period. Finally in this regard, the societies considered here were directly concerned with science. Their formal reason for being was the promotion of science and natural knowledge; unlike other indicators of the social growth of science in the period, the scientific societies were not so much derivative of developments in an autonomous science as responsible for them. It is hoped that focusing on the scientific societies may lead a step closer to understanding the pace and larger course of the socialization of science in the eighteenth century.

The history of eighteenth-century scientific societies impinges on two other related themes in the development of organized science in the period: specialization and professionalization. We will have little to say about specialization in science and the creation of new disciplines beyond noting briefly the ways in which eighteenth-century scientific societies half-heartedly and ultimately unsuccessfully responded to pressures for specialization in science.

The issue of the professional status of the man of science in the eighteenth century is more pointed and directly relevant. Historiographically, it is also more confused. No one disputes the evident development from the sixteenth and seventeenth centuries when men might pursue science or natural philosophy as churchmen, astrologers,

[xxiii]

university professors, lawyers, military men, or whatnot, to the nineteenth century, when a man could gain his livelihood as a professional
scientist. How to characterize that development and do justice to the
eighteenth century has turned out to be more problematic than one
might expect.

For Joseph Ben-David, still the major thinker on this question, science moves closer to the center of society's concerns in the seventeenth
century, and some sort of social role for the natural philosopher/scien-
tist also emerges in the period.[10] But science does not become a recognizably professional activity until the nineteenth century. The problem
is thus how to explain a 150-year "delay." For Ben-David, the "scientific
role" appears in England in the seventeenth century, but "paradoxically" other factors (particularly scientism in eighteenth-century
France) operated to prevent the full flowering of this role until the
advent of modern professional science in nineteenth-century Germany.
The eighteenth century, in other words, was a "scientistic" side-track
to what otherwise should have been a natural progression to the modern
professional scientist.

For another influential author, Everett Mendelsohn, the eighteenth
century is the age of the amateur. He dismisses eighteenth-century men
of science as "amateurs" with merely sufficient wealth or a lucrative
enough medical practice to engage in a "scientific hobby."[11] His brief
mention of the Paris Academy and the Royal Society of London makes
them seem largely dilettantish reunions of these amateurs.

These views are awkward in several respects. There is the unresolved
problem in sociology of what is a professional.[12] One also has to recognize a certain historical and sociological myopia which uses nineteenth-
century standards to define the scientific professional and which judges
the eighteenth century by what should have happened in the seventeenth or by what was to happen in the nineteenth. Then, how does
one account for the final appearance of the nineteenth-century professional if the earlier age had been so devoid of social resources to support
activity of a professional nature?

The way around these problems, explored more fully in chapter 7, is
to recognize that the scientific societies were a separate and key stage in
the professional development and the social definition of the man of
science in the period from the seventeenth to the nineteenth century.

Much argues for this point of view. Eighteenth-century scientific so-cieties were institutionalized centers of power in science. Careers in science *could* be pursued within the orbit of the scientific societies, and they did provide some paid professional positions.[13] The scientific so-cieties were the natural centers in which to work or to which work was sent. The academician or his equivalent got paid or had access through his institutional position to other remunerative sources. By and large he used his knowledge. More often than not a man's work was scrutinized and judged by his peers. Membership in scientific societies constituted entry into a professional cadre with its own standards and values. And, in the positions of academician, F.R.S., or state employed scientific expert society recognized well established social roles. Admittedly, not many men enjoyed all the possible benefits of these positions, but those who did are a far cry from the independent amateur.

Understood this way, the seventeenth century saw the man of science gain an explicit place in society, as exemplified by the foundations of the Royal Society of London and the Paris Academy of Sciences. The eighteenth century did not represent a regression from this initial im-pulse, nor did the scientific societies do nothing to advance science beyond the level of the amateur. In a number of ways to be explored they continued and greatly fostered the social role for the man of sci-ence and standards of scientific professionalism. By late in the century, in and around the scientific societies, a scientific practice developed that was functionally indistinguishable from nineteenth-century pro-fessionalism. The peruked academician, in other words, functioned as his lab-coated successor, and if for some cataclysmic reason the social order had decayed after the eighteenth century, a disinterested observer could not but find that the scientific role and the social definition of the man of science—the scientist—were well advanced by the end of the century. It would seem that changed and more familiar conditions of the nineteenth and twentieth centuries obscure the coherence and im-portance of the eighteenth century as a distinct stage in the profes-sionalization of science.

These preliminary understandings together lead to a historiographi-cal conclusion somewhat at odds with traditional interpretations of science in the eighteenth century. Considering the impressive and un-precedented social, institutional, and professional growth of science in the period, it would seem that the eighteenth century should be considered a coherent and decisive era for the development of modern science. In fact, however, because the eighteenth century was not a period of dramatic cognitive advances in science, it is too often viewed as a barren interregnum marking off the glory days of the Scientific Revolution on the one hand and major reformulations, particularly in physics, in the nineteenth century on the other. Compared to the towering scientific contributions of Kepler, Galileo, Descartes, or Newton in the one century or to the exciting new discoveries and theories of Young and Fresnel, Ampère, Faraday, Joule, Helmholtz, Kelvin, Maxwell, or Darwin in the other, what a pale and dull period the eighteenth century seems. How excited can one get over Aepinus' two-fluid theory of magnetism? Lavoisier and the revolution in chemistry may stand out more, but still the eighteenth century would seem a historiographical backwater isolated from larger bodies of work and interpretation centered on seventeenth- and nineteenth-century science, and when thought of at all, it is too often considered a weak and undramatic bridge crossing the divide between the 1600s and the 1800s.

The sociological views of Ben-David and Mendelsohn are already instances of this lack of consideration. A further, notorious example from the history of ideas is Herbert Butterfield, who considered the eighteenth-century revolution in chemistry really part of the Scientific Revolution of the seventeenth century, only "postponed" because of various intellectual obstacles and hurdles.[14] Marxist historiography likewise plays down eighteenth-century science. For example, J. D. Bernal in his *Science in History* sees the eighteenth century as an epiphenomenal interlude between, on the one hand, the rise of merchant capitalism and the Scientific Revolution in the sixteenth and seventeenth centuries and, on the other, the advent of industrial capitalism and its scientific consequences in the later eighteenth and nineteenth centuries.[15]

An effort is under way to reassess eighteenth-century science in its own terms, and a number of historical studies on which the present work depends are models of historical scholarship which avoid doctrinaire interpretation and are faithful to the nuance and context of the eighteenth century.[16] Still, it needs to be emphasized that to appraise eighteenth-century science and its organization vis-à-vis the centuries preceding or succeeding undermines a better interpretation that sees science as fully and successfully integrated into its various social and intellectual settings at the time.

This account of the scientific societies takes the eighteenth century for its primary frame of historical reference. Scientific societies are not considered here as mere outgrowths of seventeenth-century efforts to reorganize science, nor as leading inexorably to the specialized and professionalized institutions of science of the nineteenth century. Similarly, the man of science is not considered either as a dilettantish amateur or a professional scientist in any nineteenth-century sense. Neither is he thought of as some aberrant social type, awkwardly straddling these alternative roles. Rather, the effort is made to see men and institutions as part of a distinct ancien régime mode for organizing and pursuing science characteristic of the age. Naturally, eighteenth-century scientific societies resolved in some sense organizational themes bequeathed by an earlier age, and likewise they became afflicted with tensions (in scientific communications, for example) which found adequate resolution only in the succeeding era. But just as so much else of the classical eighteenth century seems foreign and removed from more modern ways, so too the scientific societies of that day must be seen as historically delimited and as distinct from later developments as Mozart from Wagner, Franklin from Maxwell, Voltaire from Nietzsche, Montesquieu from Marx.

While outside the primary focus of this book, in this introductory vein it is appropriate to raise the question of the relation of eighteenth-century scientific societies to the development of science considered as a cognitive system of ideas. Sidestepping causative connections, the most elementary consideration of this question reveals strong and significant ties between institutions and ideas in the eighteenth century. The hierarchical and highly professional organization of Continental academies of science, for example, allowed for efficient and informed judgment of extraordinarily technical work in mathematics, rational mechanics, and astronomy—strong suits of eighteenth-century science. Prize contests sponsored by the scientific societies directed and stimulated a great deal of scientific research. The scientific societies were themselves directly responsible for producing considerable new scientific knowledge through expeditions and projects conceived and carried out under their auspices. Through their publications and the control they exercised over scientific communications, societies doubtless further shaped science. Then, as clearing houses for scientific information (in botany and meteorology, among other sciences) and as the most prestigious forums for the debate of scientific ideas, the effect of the scientific societies on the content of eighteenth-century science is indisputably substantial, or so it would seem.

T. S. Kuhn in his article, "Mathematical versus Experimental Traditions in the Development of Physical Science," provides another slant on this issue.[17] Essentially, Kuhn sees two research traditions unfolding in the eighteenth century: one, that of the classical sciences from antiquity (notably astronomy and mechanics); the other, the empirical "Baconian" sciences (electricity, magnetism, heat, etc.) which originated in the seventeenth century. In the eighteenth century the mature classical tradition was secure in its fundamentals, more mathematical and theoretical, and practiced by elite groups of more highly trained men of science. The newer Baconian sciences by contrast were more "democratic," more experimental, and less certain about the theoretical underpinnings of the various fields involved.

In this context Kuhn attributes a special significance to the scientific societies of the eighteenth century. The classical and Baconian sciences were both institutionalized in the scientific societies, and, at the same time, aspects of the organization and character of the scientific societies reflect these different research traditions. For example, while not omitting consideration of university positions, Kuhn makes clear that the scientific societies provided congenial (perhaps even necessary) institutional settings for the classical sciences. The idea, again, is that sophisticated, mathematical problem-solving succeeds best in those contexts where the work of knowledgeable specialists can be actively judged by other experts, and in the eighteenth century that essentially meant the national academies of science on the European Continent. The Baconian sciences, on the other hand, while given second-class treatment in some societies, would seem to have benefitted from the role of these institutions as clearing-houses for science.[18] Kuhn emphasizes that differences in national style in eighteenth-century science—particularly between Britain and France—follow this division of the sciences and the organization of the Royal Society of London and the Paris Academy of Sciences in particular. Furthermore, Kuhn's analysis helps explain why the eighteenth century and scientific societies have been overlooked in the historiography of science, in that, as an extended period of nonrevolutionary, normal science (both cognitively and organizationally!), the eighteenth century offers less intellectual appeal to historians compared to the more dramatic, revolutionary shifts affecting science and its organization in the seventeenth and nineteenth centuries.[19]

A final prefatory topic must be addressed. It is the methodological one of whether the scientific societies of the eighteenth century can legitimately be treated collectively as a whole. Is one entitled to lump them all together? Do they have something in common worth saying anything about? To these and like questions the response is strong. The scientific societies of the eighteenth century all had common historical origins in the seventeenth century and similar fates in the nineteenth. They grew up together as part of a coherent, international institutional movement. They shared common institutional forms. They maintained similar relations with supporting governments and occupied analogous places in their various social contexts. They all engaged in the same kinds of institutional and scientific activities. But more, they formed a collective unity. As a single whole, they shared common members, they undertook common projects, and, through exchange and contact, they established a real and unprecedented network of institutions for organizing science on the international level. In this case the whole is greater than the parts. For any particular society, of course, institutional horizons were more local than not. And, more narrowly circumscribed local histories have a well-merited place. But still, unless the phenomenon of the scientific societies is confronted in its largest dimensions, the big picture of their place, role, and what they represent for the history of science in the eighteenth-century can never emerge, however imperfectly.

The reader should be forewarned that the narrative history of the rise, spread, and interaction of these societies does not pick up until chapter 2. Instead, by way of further introduction, chapter 1 presents a more static taxonomy of the constitution and character of eighteenth-century scientific societies. It will repay to be familiar with the nature of institutions under study here before taking up their historical development.

...nizations. But their differences are ones of degree and not

...al academies and societies of science were, of course, the most ...ive and historically compelling of all eighteenth-century sci-...cieties, but they do not exhaust the set. They were comple-...by a similar group of private academies and societies. Appendix ...s some forty or so additional such societies, for a total of well ...e hundred public and private scientific societies of a similar ...er existing in the eighteenth century. Private societies were es-...y no different from their more established brothers and sisters, ...for the lack of official status and legal recognition. Indeed, most ...academies and societies began as private ones, and some private ...enth-century societies achieved a more public status later in the ...enth century. The private societies, like their official counter-...were to one degree or another concerned with science. Private ...ies likewise had rules, fixed meetings, and elected their own ...ers. Some were equipped with libraries, cabinets, and the like; ...published memoirs, and some even awarded prizes. Private so-...s generally followed the forms established by the Royal Society ...he Paris Academy, although as unofficial associations, their char-...was somewhat less defined. They were not, however, wholly ...mal gatherings. Some private societies (e.g., the *Naturforschende* ...*lschaft* of Danzig and the *Società Italiana* of Verona) were as, if not ...e, important than some official ones. Most private societies, how-...were not that important in the scheme of things, and some (e.g., ...Trenton Society for the Improvement in Useful Knowledge) were ...rely ephemeral.

...ublic and private academies and societies of science like the Royal ...iety and the Paris Academy were the most numerous societies in the ...teenth century. They were the latest and most progressive develop-...ts in organized and institutionalized science, and they constitute ...core of what is most interesting and noteworthy for the history of ...anized science in the period. But, as it turns out, another, related ...d of society needs to be reckoned among the scientific societies of ...eighteenth century. They were carry-overs of a kind of society that ...d its roots earlier in the Renaissance. "Renaissance" academies were ...ds of great uncles and aunts to the younger societies of the eigh-

SCIENCE REORGANIZED
SCIENTIFIC SOCIETIES IN
THE EIGHTEENTH CENTURY

CHAPTER

The Age of Scienti
A Taxono

Aᴘᴘʀᴏxɪᴍᴀᴛᴇʟʏ seventy official acade
modeled after the Royal Society of Londo
Science existed in the period between 1660
ety and the Paris Academy, both founded in
these new types of institution for science. T
a similar sort was the *Societas Scientiarum* of
in 1700, and the rest followed in the eighte
tures characterize these societies, beyond the
another they were all concerned with science
institutions with corporate status; that is, th
by some civil authority: emperors, kings, prir
senates, towns. All levels of the political stru
eighteenth century sponsored scientific societi
organized essentially along one of two lines, eit
by the Royal Society ("societies") or that provi
emy ("academies"). Written rules governed thei
meetings (their primary activity) according to
had officers, and they generally elected a restri
members. They usually possessed their own qu
had such other facilities as libraries, collections of
cal gardens, and occasionally observatories. They
petitions, and, where resources permitted, they
and transactions. There is considerable diversity
tions—notably that some were high-powered and
academies and societies of science, while others w

cial org
kind.

Offi
impre
ence
ment
2 rev
over
char
sent
exc
off
eig
ni
pa
s
n
s

teenth-century type. "Renaissance" academies played a large role in organizational developments in science in the seventeenth century, and they continued to be founded in the eighteenth. A count indicates some additional twenty-five "Renaissance" academies of science spanning the period of interest here. Several features distinguish "Renaissance" academies from typical eighteenth-century academies and societies, notably their short existences and the near fundamental role played by the patron in the lives of these societies. Most "Renaissance" academies were private (though formal) affairs, although a few (like the *Academia Naturae Curiosorum*) did achieve legal recognition from a political authority. Given the high position of most patrons, some "Renaissance" academies existed in a nether world of ducal or marquisate foundation. Later in the eighteenth century the distinction between "Renaissance" academies and other scientific societies (e.g., the *Accademia dei Naturalisti*, Bergamo, 1782) begins to break down. In addition, in the eighteenth century some "Renaissance" academies (e.g., the *Accademia Virgiliana* of Mantua) gave rise to official academies and societies similar to the Royal Society and the Paris Academy. The number of eighteenth-century "Renaissance" academies devoted to science seems small, however, and comparatively speaking, for the overall history of scientific societies in the eighteenth century they were relatively unimportant.

It is to be emphasized that the scientific societies themselves constitute only one subset of eighteenth-century learned societies in general. Indeed, the learned society as a type of institution was *the* characteristic form for the organization of culture throughout Europe and the West in the eighteenth century. Social and cultural enterprises of all sorts were typically incorporated by ancien régime states and social orders into learned societies, several scores of which variously dealt with language and philology, literature and belles lettres, painting and the fine arts, history and archeology, antiquities, architecture, philosophy, medicine, agriculture, economics, the mechanical arts, as well as science. Thus some learned societies (e.g., the Paris *Académie des sciences*) were devoted exclusively to science; some (e.g., the *Académie française*) did not concern themselves with science at all; some (e.g., the many *académies des sciences, belles-lettres, et arts*) incorporated science as one of several areas of institutional concern; and some (e.g., medical or

[3]

agricultural societies) treated science as an ancillary feature of their constitutions. All of these various societies have to be seen as part of a larger institutional phenomenon characteristic of the eighteenth century, although, again, our principal concern is with the science societies and their influence on the scientific enterprise.

The *Académie française* (Paris, 1635) was seemingly the first of this new, characteristically eighteenth-century type of learned institution. Especially after 1700 local and national academies of language, literature, fine arts, and the like arose everywhere in Europe alongside the science academies or sometimes incorporated with them in more general learned societies. Paris was especially well endowed with learned societies. In addition to the *Académie française* (1635) and the *Académie royale des sciences* (1666), there were the *Académie royale de Peinture et de Sculpture* (1648), the *Académie royale des Inscriptions et Belles-Lettres* (1663), the *Académie royale d'Architecture* (1681), the *Académie royale de Chirurgie* (1731/1748), and the *Société royale des Médecine* (1778). Saint Petersburg, Stockholm, and Copenhagen each had their national academies of language and the fine arts, as well as science academies. In Madrid there were learned societies for history, belles lettres, and the Spanish language. In addition to the Royal Society in London, the English capital possessed the Society of Antiquaries (1751), the Royal Society for the Encouragement of Arts, Manufactures, and Commerce (1754), and the Royal Academy (Fine Arts, 1769). Everywhere regional and provincial societies of letters flourished, and in addition there were over one hundred agricultural and economic societies in the old and new worlds. So pervasive was the learned society as the model for the institutionalization of culture in the eighteenth century that the radical thinker, the Abbé St. Pierre, proposed a learned society for every subject![3]

While it is somewhat arbitrary to separate the scientific societies from learned societies in general, the scientific societies were concerned with science, and in terms of numbers, potency, and productivity, they take pride of place as part of this larger institutional and cultural phenomenon. In terms of the history of organized science, these are the institutions to which one must turn. The growth and spread of learned scientific societies from their seventeenth-century roots will be traced in the next chapters. For the present the accompanying map (figure 1) depicts the full geographical extent of scientific societies in 1789.[4]

As can be seen, at that date scientific societies extended from Philadelphia and Kentucky in the west to Saint Petersburg (or arguably Batavia, the East Indies) in the east, and from Trondheim (Norway) in the north to Sicily and Haiti in the south. If we include the *Accademia Scientifica* which existed in Rio de Janeiro in the early 1770s, we can fairly say that eighteenth-century scientific societies extended worldwide.[5] The map reveals in its way the predominance of official academies and societies of sciences over private ones and "Renaissance" academies. By far the largest number of academies and societies was concentrated in provincial centers in France. Italy was possessed of a large number of academies. Several of the states of Germany had science academies (as well as other learned societies). Scientific societies were scattered throughout Central Europe, the Low Countries, Scandinavia, and Great Britain. Two official science societies and four private ones represented North America. Of the major European capitals only those of Spain and Austria were without scientific societies.[6]

The learned society movement in science in the eighteenth century was clearly international in character. That character is of special importance, given the ideological notion peculiar to the eighteenth century of an international Republic of Letters uniting the forces of Enlightenment.[7] Addressing the Royal Society in 1753, Lord Macclesfield, its president, gave one voice to this view.

Learned Men and Philosophers of all Nations ought to entertain more enlarged Notions; they Should Consider themselves and each other as Constituent parts and Fellow Members of one and the same illustrious Republick; and look upon it to be beneath Persons of their character, to betray a fond partiality for this or that particular district, where it happened to be their own lot either to be born or reside. Their benevolence should be universally diffused and as extensive as the knowledge they profess to pursue, and should be sensibly felt by all who in their respective Stations contributed their proposition to the Common Stock of the whole by their endeavours to promote and Advance Science and Useful Knowledge, wherein alone the true interest and welfare of such a Republick consist.[8]

For contemporaries, all of the learned societies but particularly the scientific societies defined the invisible topography of this make-believe republic as institutional outposts of its government and diplomacy. Richard Ruffey of the Dijon Academy put it simply: "Academies [are] the diverse colonies of the Republic of Letters."[9] For the Marquis

□ OFFICIAL SOCIETY   ■ PRIVATE SOCIETY

○ OFFICIAL ACADEMY   ● PRIVATE ACADEMY

▽ OFFICIAL RENAISSANCE   ▼ PRIVATE RENAISSANCE

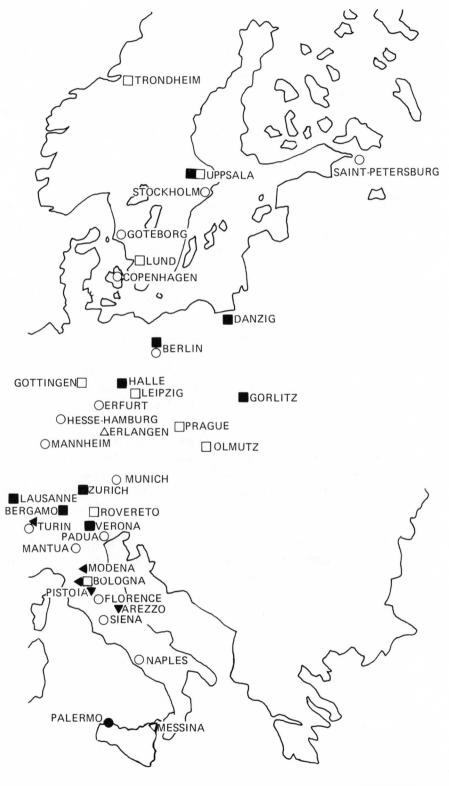

TRONDHEIM

SAINT-PETERSBURG

UPPSALA

STOCKHOLM

GOTEBORG

LUND

COPENHAGEN

DANZIG

BERLIN

GOTTINGEN   HALLE
            LEIPZIG
      ERFURT   GORLITZ
   HESSE-HAMBURG
      ERLANGEN   PRAGUE
   MANNHEIM   OLMUTZ

            MUNICH
         ZURICH
LAUSANNE
BERGAMO   ROVERETO
   TURIN   VERONA
      PADUA
MANTUA
      MODENA
      BOLOGNA
PISTOIA
      FLORENCE
         AREZZO
      SIENA

      NAPLES

PALERMO   MESSINA

BATAVIA

d'Argenson of the Berlin Academy, "Literary and learned Europe makes up, so to speak, only one single society, united by a common goal, which is the progress of the sciences and letters."[10] In this society, he continued, "Each Academy should become a kind of Congress, where the least thing which could touch on the general good of the sciences is decided: a Tribunal, where the least contention can be judged, without any of the interested parties being able to complain of not being heard." For Samuel Formey, likewise of the Berlin Academy:

The Academies are Capitals of the sciences, of which the Capitals of Empires should not or even could not be deprived. Already I seem to see them traversing this sought-after strait, the discovery of which we are on the verge, separating Europe from America and procuring an advantage for our Globe which the sun itself, though active the day, is powerless to provide, that is, to have its two hemispheres illuminated at once.[11]

All scientific societies were located in urban settings. This fact immediately suggests other fundamental parameters. As urban institutions, they fitted in and took their place alongside the whole range of social and cultural institutions inherent in contemporary civilization and city life: royal or princely courts, churches, universities and colleges, municipal administrations, agencies of central government, Estates, Parliaments, legislatures, senates, law courts, theaters, Masonic lodges, and other institutional manifestations of high culture.[12] High culture means literate culture, with all the exclusiveness that that implies for the eighteenth century. It means reading and writing well and being part of the social system that produced and sustained such people: tutors, *collège* and university training, a knowledge of Latin and a familiarity with the traditions of the classics, social connections, leisure time, a place in the world. In practice, members of learned societies were drawn from every quarter of the middle and upper classes in the eighteenth century: kings, princes, titled nobles, lesser aristocrats, local notables, the independently wealthy, high and low churchmen, statesmen, court retainers, office holders, government employees, judges, lawyers, doctors, surgeons, academicians, university professors, and so on, although, significantly, few were industrialists, manufacturers, or business men.[13]

The contrast with urban high culture is raised of eighteenth-century rural economies, surplus production, and the underlying mass of illiterate peasant, artisanal, and underclass life in villages and in cities.

[8]

Given the overwhelmingly agricultural nature of the economic base, the relative unimportance of industrial production, and the ways in which (on the Continent at least) landed rents and income extracted from the countryside fueled the higher levels of society, the learned societies—science and otherwise—were decidedly elite institutions in their respective contexts.[14]

The phenomenon of the societies certainly did not pass unnoticed by contemporaries. By mid-eighteenth-century the proliferation of societies provoked comment and debate on their place and utility. A pamphlet war broke out in France.[15] The societies in part stimulated this debate themselves by proposing a number of essay contests related to the question. For example, in 1753 the academy at Montauban asked whether the great number of academies was beneficial to society.[16] Later, there seems to have been little doubt of an affirmative answer, at least on the part of the academies themselves. As the Society of Sciences of Montpellier expressed it in 1766:

The usefulness of academies is generally recognized today. One is no longer tempted in our century to call that into question, and one can state with sincerity that the rapid progress that human knowledge has made in our days is due in large part to the indefatigable zeal of the considerable number of learned men, who, assembling in diverse societies and uniting their work and talents, have had as their only goal to enlighten their contemporaries and to transmit useful discoveries to posterity.[17]

Another observer of the scientific scene was Grandjean de Fouchy (1707–1788), venerable secretary of the Paris Academy (1743–1776). In a manuscript of 1781, he provides a contemporary witness to the scope of the science societies from his vantage point in Paris.

Do we not see several provinces of the Kingdom decorated with literary companies which animated by the same spirit [as the Paris Academy of Sciences] hasten to emulate its works in their researches? What am I saying? The whole of Europe is almost entirely filled with celebrated academies formed by the care of the most illustrious of learned men under the protection of the greatest princes. The Academy of the Institute of Bologna established by Count Marsigli, our illustrious brother [Marsigli was a corresponding member of the Paris Academy], the Imperial Academy of Saint Petersburg formed by Peter the Great to whom we can give here the same title of fellow member, of which he was proud, and protected by his successors without interruption and at present by this Empress [Catherine], so worthy to succeed him. . . . Like it, the Royal Academy of Berlin, established formerly by the celebrated Leibniz and since reformed and protected by this Alexander of the North [Frederick II] who knows equally how to conquer, govern, and enlighten states. The two Swedish academies are flourishing under the auspices of a young monarch [Gustavus III] whose virtues have already received the admiration of Europe and the love of his people. Those of Naples, of Florence, and a large number of others that the limits imposed on me do not allow me even to name, do they not owe almost all their existence to the noble emulation and the mass of enlightenment that the work of the [Paris] Academy [of Sciences] has spread throughout Europe and, to say it all, also that of this other illustrious company [the Royal Society of London] which England glorifies itself for having produced and which we have always regarded as our sister without rivalry and without the most lively wars which have sprung up between the two nations having ever been able to alter this bond of fraternity that units us? Even America has received the rays of light. In the midsts of the most animated war, Boston and Philadelphia saw flourish in their breast two academies already famous by the cares of this unique man [Franklin] whose multiplicity of talents and the use he knew to make of them have made him at once the Solon, the Brutus, and the Plato of his compatriotes.[18]

The societies which Fouchy so wonderfully evokes functioned as learned assembles, to one degree or another having the advancement of science, natural knowledge, and Enlightenment as their purpose. Reunions of men sharing this common interest in science, their primary activity was to hold meetings. These occurred variously from twice weekly to monthly, and typically involved from one to upward of four dozen men.[19] Officers usually led the business, which consisted of discussion, hearing papers and reports, electing and inaugurating new members, directing correspondence, voting resolutions, and otherwise conducting institutional affairs. Committees often dealt with particular aspects of a society's operation, such as finance, publications, or prizes. The mundane and regular necessities of paying the bills, keeping up the meeting hall, and securing new equipment and resources always demanded attention, and the special project likewise took up time. Some institutions held annual public meetings, where the work of the society would be ceremoniously reported to the larger world. All of these activities took place against the backdrop of a seasonal calendar of ritual: the first formal meeting of the year, periods of election, ordinary meetings, breaks for religious and state holidays, public meetings, vacation.[20]

The publication of a journal was an important activity. The *Philosophical Transactions* of the Royal Society of London, the *Histoire et*

Mémoires of the Paris Academy, various *Acta, Commentarii, Abhandlungen,* and in sum the four dozen or so scientific and technical series sponsored by the scientific societies took up considerable institutional time and resources.[21] There were two types of scientific society proceedings, quarterly or trimestral journals like the *Philosophical Transactions,* and (theoretically) annual volumes like the *Histoire et Mémoires* of the Paris Academy. Publication in the latter type of proceedings was generally restricted to members of the sponsoring institution. Publications of the *mémoires* type far outnumbered those of the *Transactions* type. While not exhausting the scientific press in the eighteenth century, both types of scientific society journal together were the premier places for original (as opposed to secondary or derivative) publication in science in the eighteenth century.

Many societies sponsored prize contests, and these, too, required considerable institutional effort.[22] Some societies (like the Royal Society with its Copley Medal) awarded prizes for work already accomplished. Most (like the Paris Academy) ran elaborate contests which entailed proposing a question for research, advertising and soliciting entries, judging results, announcing and awarding prizes.[23] Hundreds of such contests were sponsored in the eighteenth century by the societies, and considerable sums were paid out as cash prizes: the Paris Academy awarded some 75 scientific prizes based on a capital of over 200,000 livres; the Berlin Academy offered 45; the Danish Academy 125; the Dutch Society 86; the Dijon Academy 53; the Bordeaux Academy 151, the Lyons Academy 163, and so on.[24]

Not only did scientific society prize contests simply stimulate a lot of activity, but there is no doubt they affected the course of scientific research. More study is needed on academic prize contests and their effect. While not actually a "smoking gun," the remarks of Jean (II) Bernoulli give some indication that the research of leading eighteenth-century scientists was channeled into areas decided upon by the scientific societies and that money was a consideration.

Theoria magnetis: This thorny subject had been proposed [by the Paris Academy] for the years 1742 and 1744 without being awarded, so that this time the prize was tripled. M. de Maupertuis, who was in Basel between 1744 and 1746, strongly exhorted us [Jean and Daniel Bernoulli] to enter, saying that that 7,500 livres were really worth the trouble and that, according to the laws of the Academy, the question could not be proposed again. My brother then admitted to me that he had thought about the question on his own accord and that he had even thrown a few ideas on paper, but he was so little satisfied he had completely abandoned his project. At the same time, half in jest, he offered to communicate his ideas to me if I would work them up and share the pie with him in case we succeeded. I took him at his word and the result was the piece in question. . . . Its success surpassed my hopes, as it shared the triple prize with Messrs. L. Euler and du Four, correspondent of the Royal Academy of Sciences. It was published under the names of the two brothers.[25]

It is to be emphasized that the scientific societies were not pedagogical institutions. Apart from the odd subscription course (chemistry, experimental physics, *materia medica*) occasionally and loosely offered under their auspices or the (here) extraneous ties some had to universities, they were not teaching institutions, they did not count among their purposes the transmission of knowledge to a younger generation, and they did not give degrees.[26] This important difference separates the societies of the eighteenth century from the university tradition as a whole and from science and natural philosophy as these were characteristically organized earlier in medieval universities or later in the revitalized universities of the nineteenth century. There is a curious corollary to this nonpedagogical role; that is, there were no formal educational criteria for admission to societies. Some societies did demand accomplishment of one sort or another in a science, and age and "right conduct" were often requirements. But university degrees counted less than demonstrated contributions and commitment to science in the most highly regarded scientific societies or an interested "sociability" in the less regarded ones.

Societies generally stood at the top of the scientific establishments in their various settings. The Saint Petersburg Academy provides the most striking case in this connection. As we will see, the academy presided over a whole scientific bureaucracy, including (at one point) a university and a gymnasium, as well as a botanical garden, observatory, map-making department, and press. Other academies, like those in Berlin and Stockholm, also formally managed observatories and other institutional appendages. In other cases institutional control was somewhat less direct, as in the oversight role of the Royal Society of London as "visitors" to the Royal Observatory at Greenwich. Some learned societies, such as those in Bologna and Göttingen, were officially tied to universities; others, such as those in Montpellier and Edinburgh enjoyed looser but real affiliations with universities in their towns. Universities were independently important centers for science, particularly for experimental physics in Protestant Germany; but where they coexisted with scientific societies, the latter can claim some preeminence as institutions specifically devoted to scientific research and the diffusion of knowledge.[27] In still other instances, patterns of personnel and membership dictated informal ties between societies and other scientific institutions. In this regard one thinks of Buffon's simultaneous tenure as treasurer of the Paris Academy and as director of the *Jardin du roi*, or Sir Joseph Banks' forty-one-year presidency of the Royal Society of London and his even longer hegemony over the Royal Botanical Gardens at Kew. And everywhere the leading men of science, whatever other institutional connections they may have had, belonged to and presented their researches to the societies.[28]

As suggested, learned and scientific societies as a whole can be divided into two major subgroups: *academies* of a kind typified by the Paris Academy of Science, and *societies* typified by the Royal Society of London. Both these institutional types are to be included under the general rubric of scientific society. Further discriminations will be made, but the major differences between academies and societies as types of institutions can be presented initially here.

First, the "academic" form was more common and generally found on the Continent of Europe. The "society" form was less frequent and arose in the social and political circumstances of Britain, the American colonies, and Holland. Academies outnumbered societies on the order of five to one, and differences in their geographical distribution obviously reflect differences in the kinds of social systems supporting particular kinds of institutions. Continental academies were more congruent with centralized absolutist monarchies and principalities; rigid, hierarchical social stratification; limitations to the rise of the middle classes enforced by noble and feudal privilege; predominantly agricultural economies; and the Catholic religion. Institutions of the society type, on the other hand, emerged in more "democratic," decentralized Protestant countries with more fluid social demarcations, less obstructed by traditional feudal restraints, and more oriented toward industrialization, trade, and the sea. In particular instances, the Protestant Berlin Academy, for example, or the ambiguous case of the Swedish Academy of Sciences, these generalizations require further nuance, but by and large this Continental–Anglo-Saxon (or Atlantic) distinction is a fruitful one with further implications for the form and function of academies and societies.

A further difference between academies and societies lies in their ties to the state. It is a myth, though, that academies as a group were created by government fiat (from "on high" as it were), while societies arose ("from below") from private initiative. The Paris Academy and the Saint Petersburg Academy, it is true, were created largely out of thin air by Louis XIV and Peter the Great, while the Royal Society of London could trace its roots loosely through a series of "invisible colleges" back two decades prior to its official foundation.[29] Such famous examples obscure the reality that all but a few institutions—academies and societies alike—had prehistories—some quite lengthy—of private, informal association preceding official incorporation.

Two common patterns of institutional foundation are regularly observed. In one, a private group meets informally, it attracts the attention and support of a notable personage (an archbishop, a court official, a high noble), and formal recognition follows a period of institutional gestation. Roche has demonstrated this pattern for the French provincial academies, and it holds for the Anglo-Saxon societies.[30] The Turinese society provides an extreme example.[31] Originating in 1757 as a private group of friends, it received official recognition from Victor Amadeus of the Two Sicilies in 1759 and was a poorly supported society of science; in 1783 it achieved higher recognition as a full-blown academy of science.

Alternate to the private-group pattern, a particular individual, often with court connections, agitated independently for the creation of an official society. Leibniz, of course, is the outstanding example, as he badgered various governments to establish societies in Berlin, Vienna, Saint Petersburg, and elsewhere during his long career as a propagandist for societies.[32] The case of the Göttingen Society of Sciences is illustrative in this connection also. Between 1734 when Göttingen University was founded and 1752 when the Society of Science finally came into being, no fewer than seven proposals were brought forward by five different men before success was finally achieved.[33]

Nevertheless, even though academies and societies generally share these patterns of institutional formation, it is true that academies more than societies tended to be state institutions of science formally part of larger government bureaucracies. The Paris Academy, for example, was directly linked to royal administration through a secretary of state (*Maison du Roi*), and it had official duties to perform as the scientific and technical bureau for the crown. Similarly, the scientific core of the Saint Petersburg Academy was tied to the czarist administration through a governing academic council and thence to the Imperial Senate and the Court, and it likewise had official functions to fulfill for the state. Similarly, academicians in Berlin and elsewhere were state employees.

For the academies formal status as state institutions brought all-important government funding. The expenses of the Saint Petersburg Academy, again, were entirely and amply provided for out of imperial funds: costs for light, heat, buildings, printing, equipment, salaries, prizes, expeditions, and all extraordinary obligations were completely underwritten.[34] The Saint Petersburg Academy provides almost the "paradigm case" of academies in this regard, but the pattern was taken up elsewhere among other academies of science.[35] Royal and electoral courts in Turin, Munich, Erfurt, Brussels, and elsewhere paid the costs of their respective academies, and although it did not salary every member or fund all prizes, the French government did provide for the costs of the Paris Academy. In yet other instances (e.g., Stockholm, Berlin, or Mannheim), institutions were granted monopoly economic privileges (such as rights over the sale of maps or almanacs) by the state as an independent mechanism for financial support.

This does not mean that every academy was well off; in fact the opposite is probably closer to the truth. The numerous academies of provincial France, for example, rarely received more than token support from the central government and only rarely more than that from provincial or municipal sources. The Berlin Academy was chronically short of funds before Frederick II came to its rescue in the 1740s. The fully or even adequately funded academy was a rarity whose circumstances could change: a half-a-dozen in France, a couple in Italy, a few in Germany, and a scattered one or two elsewhere.

Still, an upshot of academies as state-supported institutions (particularly the major national academies) was that they were better equipped and better able to pursue scientific goals than their "society" counterparts among the learned societies. In particular, they were more likely to possess well-equipped astronomical observatories, botanical gardens, laboratories, and cabinets of natural history or experimental physics. They published the more regular and the more important scientific journals; they dominated the series of scientific prize contests, and they were more likely to undertake large and expensive projects.

But along with state support came state control. Although by far most academies most of the time possessed virtually complete scientific autonomy, they were still not wholly independent institutions. The Saint Petersburg Academy and the early Berlin Academy, for example, were dominated by nonscience administrators and court officials. The Paris Academy had built into its organization a special class of "honoraries" who had no particular scientific competences and who possessed considerable power within the institution. The example of recruiting local luminaries into the body of an academy was regularly followed in the French provinces and elsewhere. As state institutions, furthermore, academies were not fully masters of their own time or activities, being called upon to perform regular and extraordinary duties for secular authorities. Then, although it seems to have happened rarely, they were subject to outright interference by those paying the bill. In two instances, for example, election results in the Paris Academy were overruled by royal command, and the Berlin Academy, except when it was administered by Maupertuis in the 1750s, was little more than the plaything of Prussian kings.[36]

The circumstances of the societies contrasts dramatically. Although recognized by national, state, or municipal governments and even called upon occasionally for expert advice and counsel, societies on the whole had much more tenuous ties to governments and were much more independent than were academies. Not formally part of bureaucratic structures of government, societies enjoyed considerably more scientific and organizational freedom and were better insulated from outside control and interference. But such autonomy and independence brought limitations in their train. For one thing, there was nothing these institutions had to do, no formal institutional mission to fulfill. The full name of the Royal Society of London, for example, committed it only loosely to "the Promotion of Natural Knowledge." The result was that societies overall were less impressive and less productive as scientific institutions. Their activities tended to be more informal and less "professional" than their academy counterparts, and their output, while not insignificant, does not compare with what was going on on the Continent, particularly in astronomy, mechanics, rational mechanics, and mathematics.

Then, because they generally received only token support from the state, societies were almost wholly dependent on their own resources for a financial base. Moneys for their operations came largely from dues paid by members or from private gifts. The result was that, much more so than academies, societies were constantly short of funds, and therefore even less able to support and pursue sustained scientific research. The Royal Society, for example, though by far the richest institution of its type, at one point early in the eighteenth century was forced to sue some of its members in court for nonpayment of dues.[37]

In the abstract, societies of the academy and society types differ profoundly in their internal organization. For present purposes the organizational structures of the Paris Academy and the Royal Society of London can serve as prototypes. Figure 2 depicts the constitution of the Paris Academy as it came to be established after the reform of the Academy in 1699.[38]

The Paris Academy immediately presents itself as a small institution with a heterogeneous and hierarchical membership. Its ties to government are evidenced by the previously mentioned positions of the king, secretary of state, and the class of honorary members, who were usually nobles or high government officials and who thus provided another link between the academy and government circles. The position of honorary required no specific scientific competence, yet honoraries had substantial say in the running of the institution with their votes in both elections and in scientific debates. Royal authority was felt directly through the right of final say by the king in elections to the academy and in appointment of members to positions as officers.

The academy was directed by an executive body of officers, which included a president and vice-president drawn from among the honoraries, a director and subdirector drawn from the *pensionnaire* class, a treasurer and a permanent secretary, both likewise recruited from the *pensionnaires*. This executive thus represented the highest echelons of

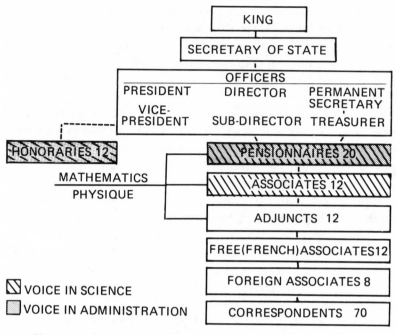

Figure 2: Organization of the Paris Academy of Sciences.

the academy and was responsible for running meetings and directing its internal affairs. The position of treasurer was subservient to an oversight committee, but in the Paris Academy and other academies the permanent secretary was a very important person, tantamount to administrative director, and the record is full of men who brilliantly guided their academies for decades at a time as secretary: Fontenelle in Paris, Formey in Berlin, Wargentin in Stockholm, Zanotti in Bologna, J. A. Euler in Saint Petersburg, and so on.

At the heart of the Paris Academy was an active scientific core composed of three hierarchically ordered classes. On top were twenty paid *pensionnaires*, three in each of the mathematical sections into which the academy was divided (geometry, astronomy, mechanics, and, after 1785, *physique générale*), three in each of the physical science sections (anatomy, chemistry, botany, and, after 1785, minerology and natural history), and the permanent secretary and treasurer.[39] As their title suggests, *pensionnaires* were paid regular pensions for their services to the academy and the government. In addition they received redeemable tokens (*jetons*) for attendance at meetings.[40] In this respect the model of the Paris Academy differs from that of Saint Petersburg (but not of most academies) in that only a few, not all, members received remuneration.[41]

The three classes of *pensionnaire*, associate and adjunct members constitute a professional scientific cadre. Candidates for these positions had to have a demonstrated record of scientific accomplishment. They had to reside in Paris, attend meetings regularly, and report periodically to the academy on the substance of their scientific researches. Honoraries, *pensionnaires*, and associates possessed a voice in issues of science facing the academy—adjuncts did not. Honoraries and *pensionnaires* alone could vote in administrative matters.[42] Furthermore, study has shown that, like modern professorial positions, these three groups formed a career ladder at the heart of the academy, whereby junior members were taken in at the bottom and systematically promoted to top positions.[43]

Peripheral to this central, resident constellation of members were three groups of nonresidents: ultimately twelve regional (i.e., French) free associates, generally eminent men of science from the provinces; eight high-status foreign associates, representing the top levels of science in the international community; and a large, and at first unregulated number of corresponding members. Correspondents of the academy were assigned to resident academicians through whom their occasional reports were delivered, and they alone were not subject to royal appointment. Members of these nonresident groups formally linked the academy to larger national and international scientific communities; all possessed the privileges of communicating with the academy and attending meetings when visiting in Paris, but they held no official voice in academy affairs.

All things considered, then, this kind of organizational structure bespeaks a small and select institution. On the average only 153 men belonged to the Paris Academy at any one time in the eighteenth century, including nonresident members; only 716 men belonged in any capacity in the whole of period 1699–1793.[44] The core triad of *pensionnaire*-associate-adjunct members almost never numbered above forty-five individuals. Scientific qualifications were demanded for scientific positions, and the academy was tightly integrated into the rest of government. The success of the Paris Academy, which was considerable, hinged upon its careful adaptation to the society in which it was created; like the rest of French society in the ancien régime, the Paris Academy was based on privilege, centralized power, and hierarchical stratification.[45] Because of its preeminent position over the rest of French science, the force of its organizational form, and its firm integration into ancien régime France, the Paris Academy was the most impressive scientific institution present on the world scene in the eighteenth century. There is little reason to wonder why other academies of science arose in other Continental societies whose forms of government and social organization were analogous to the French or why the Paris Academy, as the first of its type, was so influential a model in their creation.

In contrast, institutions of the society type were possessed of a much less severe internal organization. They were generally larger, less professional and exclusive, and more homogeneous. The rather more simple organizational form of societies is evidenced in the structural

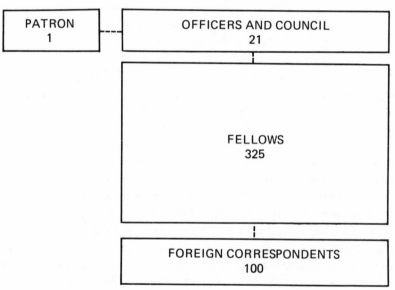

Figure 3: Organization of the Royal Society of London.

makeup of the Royal Society of London (figure 3).[46] The great body of the society was composed of Fellows (F.R.S.) on the "home list." The number of such Fellows was comparatively large, averaging approximately 325 men—seven times the core group in Paris, and there were no other internal divisions into further categories or grades of membership.[47] Thus, in contradistinction to the academy type of organization, societies were structured more "horizontally." Fellows met weekly in the "ordinary meetings" of the society chaired by the president or his deputy; their business was informal and nonadministrative in character. The regular membership did hold one key power, that of election, but it was comparatively easy to get elected to the Royal Society. No demonstrated accomplishment or ability in science was required; all that was needed was support from three members, and a month after the nomination was approved by the council of the society, the candidate was voted up or down, generally up. The Royal Society was thus a larger and socially more open institution than the Paris Academy, and it is hardly surprising that through the eighteenth century "scientific" Fellows of the Royal Society were outnumbered by nonscientist members by better than two to one.[48]

Atop this large and relatively unwieldy mass stood the governing council of the society, composed of officers (president, vice-president, two secretaries, and a treasurer) and elected F.R.S. (twenty-one in total, ten replaced annually). Officers were responsible for the day-to-day running of the society's affairs, and they met monthly or so along with the rest of council, likewise under the chair of the president, to deal with the real business of the society, which consisted of handling its finances and property, contacts with outside institutions and individuals, special projects, and general administration. Unlike the largely honorific position of president in the Paris Academy, the president of the Royal Society came to be its effective leader. The tradition of strong and long-tenured presidents began in earnest with Isaac Newton, who, though nominally reelected annually, held the post from 1703 until his death in 1727. Newton, too, is responsible for shifting power to the council of the society from the more democratic and chaotic fellowship. With Newton the council became the guiding force of the society. The council, though, was even more dominated by nonscientific members than was the regular fellowship, on the order of ten to one.[49] Finally in this regard, the two secretaries of the society (later augmented by a third for foreign correspondence), although they had considerable tasks to perform, were not powerful figures in the London society like their counterparts in academies.

Rounding out the formal structure of the society at the top were the positions of patron (held by the succession of English kings and queens) and at the bottom, nonresident foreign Fellows of the society. The patron was of comparative insignificance to the society. Foreign Fellows averaged roughly one hundred men in the eighteenth century, and their contact with the society was limited to correspondence or the occasional visit.[50] They, too, were of casual importance for the normal operations of the society and the least of its constituent parts.

[A point to be noted at this juncture is that the name of an institution is no guide to its type. The American Academy of Arts and Sciences in Boston and the Royal Irish Academy in Dublin, for example, were "societies" in that their organization paralleled that of the Royal Society of London. The *Société royale des sciences* in Montpellier, on the other hand, was an academy, in that it was modeled directly after the Paris Academy. Then, learned societies are not to be confused with Protestant secondary schools, often called academies (e.g., those at Saumur and Geneva), with military academies (e.g., Woolwich), with riding academies, academies of the dance, and so forth.]

A number of purely scientific societies followed more or less exactly the organizational patterns established by the Royal Society and the Paris Academy. Within limits of scale, the American Academy in Boston and the American Philosophical Society in Philadelphia modeled themselves after the Royal Society of London, while the academies in Montpellier, Turin, Mannheim, and elsewhere were established after the example of the Paris Academy. Among the scientific societies the Haarlem *Maatschappij der Wetenschappen* presents an interesting variation, in that it had two classes of members: nonscientific, dues-paying burgher supporters, and non-dues-paying, unpaid member-workers.[51] Among purely literary academies, a few provincial organizations in France styled themselves after the *Académie française* with every member of equal weight in the institution.[52] But, while many and many of the most important scientific societies of the eighteenth century were exclusively scientific institutions and while some societies had nothing to do with science, the most common variation on both societal and academic organizational themes was the incorporation of science and non-science classes or sections. This kind of "universal" society first appeared in 1700 with the *Societas Scientiarum* of Berlin, which incorporated sections for the sciences, language, literature, and history. And other academies and societies aped the form. Thus, the Royal Society of Edinburgh, while in every respect an institution of the society type, was composed of a literary and a philosophical (science) section; the Royal Irish Academy, likewise a society, had divisions for science, belles lettres, and antiquities. As far as academies are concerned, in the French provinces again, the typical learned society was a tripartite entity committed to science, belles lettres, and arts (meaning the mechanical arts). The academies in Brussels and Munich had sections for science and belles lettres. Academies in the German-speaking world often combined several disciplinary sections under a *wissenschaftliche* aegis; those in Prague and Göttingen, for example, had sections for the physical, mathematical, and historical sciences.[53] The most striking example of a multi-section society was the reformed Berlin Academy after 1744 which included two sections for the mathematical and physical sciences, one for belles lettres and philology, and one, incredibly, for speculative philosophy! The way to see this phenomenon is that learned societies—academies and societies—sometimes were single "discipline" institutions (as in the instances of the *Académie française* and the Academy of Sciences in Paris), and sometimes they incorporated several fields. The existence of purely scientific societies or the incorporation or exclusion of science from institutions were not exceptions to some other rule, but part of an overall pattern characterisic of learned societies in general in the eighteenth century. Generally speaking, when resources were scarce, the multi-area institution prevailed, and when they were great, the single field type arose.

There were some organizational consequences of mixing science and other areas of interest within a single institution. In the French provinces, again, an entire academy met together, probably to the detriment of its scientific and literary achievement.[54] In Berlin and Edinburgh and other multi-area societies, an overarching council emerged, with members drawn from various classes. Such councils, while preserving the fundamental features of academies and societies as types of organizations, doubtless reduced the visibility and voice of their purely scientific components.

The most interesting and important scientific societies share a fundamental characteristic in that they were official institutions given legal recognition by some instrumentality of government. The Royal Society of London was granted a royal charter by Charles II in 1662. The Paris Academy received its regulations from Louis XIV, and other types and levels of government were responsible for chartering other societies. The Swedish, Prussian, and Russian crowns legalized their national science academies. Various German electors incorporated the lesser academies of Germany; Austrian authority legitimated academies in Brussels and Prague, towns over which it had control. Bourbon kings did this for the wealth of academies in provincial France. The General Assembly of Pennsylvania and the Massachusetts state legislature conferred legal status on the American Philosophical Society and the Boston Academy of Arts and Sciences respectively. The Dutch Society of Sciences was recognized by the United Provinces, the combined authority of the Pope and the town senate of Bologna did the same for the Bolognese academy, and so on.

There were several significant consequences of legal status and recognition. First, they distinguished a formal society from merely private groups of interested persons. By far the majority of societies began as private associations, and official recognition after a probationary period sanctioned the effort of private assembly.[55] Legal status legitimated the enterprise espoused by a society, it recognized the promoting group of men, and it elevated a society as an official body into the formal corporate and institutional structure of its particular context. As legal entities, learned societies thus enjoyed an institutional permanence and a status before the law otherwise not available to private persons or groups. It was one thing, in short, for distinguished citizens of Dijon to gather and converse in the *hôtel* of the parlementary president, Pouffier, in the 1720s. It was another to be the endowed, official, and ongoing academy of sciences and letters of Burgundy.

Legal status also brought with it institutional privileges. Some we have mentioned: the right to act as a corporation or possession of monopoly economic privileges. Others were more trivial: annual masses for the institutions, access to the royal presence on special occasions, exemption from military service, tax breaks for members, franking privileges, and so on.[56] Three other types of privilege were important: technological consulting, control of the scientific and technical press, and self-government. All societies were self-governing to the extent that, barring the few exceptions already mentioned, they elected their own members and exercised some control over them and their work. Imprimaturs and censorship control over the press gave academies and societies large powers over the development of science and the work of their members. Control over machines and inventions and service as scientific and technological bureaus made a few scientific societies indispensable to the state and thus even more powerful. Societies of science to a lesser extent than academies possessed rights over the scientific press and what their members published on their own, and they were not state technological bureaus like Continental academies of science. Still, I believe societies are more like academies in these regards than not.

These privileges bespeak a more profound arrangement than has been suggested thus far between institution and state formalized in charters and letters patent. Although contemporaries probably would not have put it this way, with official incorporation institutions and governments struck a deal, a quid pro quo arrangement whereby institutions received recognition, funding, and privileges in exchange for technical service and advice. The base coin of this deal was power. On the one hand, academies and societies possessed the power of expertise and special knowledge of nature which they "sold" to supporting governments. In return they "bought" and the state delivered the operational power to control science and its practitioners.

The idea of state service suggests an important reason for the success of eighteenth-century scientific societies and why governments—particularly the great nation states of Europe—felt it worthwhile to enter into support of scientific institutions and cadres of scientists. It seems that the societies provided a technical base of expertise to help solve problems that fitted into the long-range historical interests and needs of "progressive" elements of nations and enlightened principalities. In England, which had already undergone a bourgeois revolution by the eighteenth century, these forces were in the vanguard of mercantile and industrial development. Others on the Continent were trying to expand economically and were trying to impose rational central rules and standardized administrative procedures, largely in opposition to traditional, privileged, and irrational systems of economic and political governance that had grown up since the Middle Ages. One could point to work done in scientific societies to solve the problem of determining longitude at sea, create reliable chronometers, standardize weights and measures, improve map making, develop geographical and economic surveys of regions and kingdoms, and less dramatic instances where the scientific interests of learned societies harmonized with the political and mercantile interests of government. For Continental academies one suspects that the rational basis of the scientific enterprise itself may have meshed in less technical ways with the goals of central administrations through the eighteenth century.[57] Science and the state were allies in the progressive fight against tradition, cultural inertia, and gross stupidity. Certainly each had something to gain from the other: rational technical expertise on the one hand; recognition, legitimacy, and support on the other. At the least one has to wonder why the Catholic church and various aristocratic and feudal institutions were not in the vanguard of founding learned societies.[58]

In any event, scientific societies did perform many technical and specialized functions for their respective governments. At the extreme, such functions were an established part of the institution's normal business, as, for example, in the control of the Paris Academy of Sciences over French patents.[59] In a like vein the academies in Berlin, Saint Petersburg, and Stockholm were responsible for state calendars and regular astronomical observations. On a less formal basis, many learned societies—as repositories of special knowledge—were called on occasionally to provide a needed expertise. Among dozens of examples that could be cited to this effect, a wonderful case concerns the Montpellier Society of Science which was asked by the local courts to say when the sun set on April 19, 1729, for use as evidence in a murder trial.[60] Whether governments got enough in the way of service from their science societies to merit supporting them is another question. The point is that they bought into the deal.

For their part, representatives and spokesmen for the societies elaborated an explicit ideological role for these institutions based on power and special claims to knowledge and expertise. A sharp ideological

[26]

position for the societies first came into view as part of the "philosophic campaign" associated with the public appearance of the Enlightenment movement in the mid-eighteenth-century.[61] Pierre Louis Maupertuis, newly invested Life President of the Berlin Academy, in 1750 gave voice to the initial formulation of this position in his "On the Duties of the Academician."[62] According to Maupertuis, the principal duty of academies at mid-century was to defend *vrai savoir* against *faux savoir.* The goal and necessity was to protect and extend the true tradition of Descartes, Leibniz, and Newton against forces of ignorance and superstition which threatened it. The point has an Enlightenment flair to it, and it is no accident that the distinction was taken up and broadcast by d'Alembert in the *Encyclopédie.*[63] By the late 1760s, the campaign and rationale of the academies underwent a subtle shift. The initial battle against superstition largely had been won; but another was looming against *demi savoir* or charlatanry.[64] As Samuel Formey, Berlin Academy secretary and champion of the academies, put it: "It is up to Academies to make reign a purified and solid knowledge rich in precious fruits, which gives chase, so to speak, to half knowledge [*demi savoir*] as it was previously given to false knowledge [*faux savoir*]. . . . Academies have hardly a more essential duty to fulfill, a more glorious task to accomplish."[65]

The bases for these claims were that academies and societies, as centers of expertise and special knowledge, were the best institutions to mark sound knowledge from false and the charlatan from the savant; only they could truly judge who should number in their ranks; it was their duty in fact to protect the purity of their knowledge from being sullied by the dilettante, the popularizer, and the fraud. The institution succeeding in establishing these claims was a powerful one indeed. Formey sounded the battle cry for the last third of the century:

The literate public is naturally disposed to consult the learned companies and to regard their responses as decisions of Oracles. That is a great advance. . . . It is now a question of encouraging and directing those in whom natural lights and good intentions are found united, and to dissuade and turn away gently those for whom talents are lacking, to reprimand, to crush if necessary those in whom are associated incapacity with insolence and turpide. A half-century of such a dictature wisely exercised by an academy will produce the most advantageous changes across the breadth of those countries in which its example has an immediate influence and could not be but useful to all the rest of humankind.[66]

[27]

There can be little doubt that societies benefitted from the limited power invested in them by the state to run their own affairs and to control science and the scientific community generally. The scientific societies used this power every day in electing members, writing reports, judging machines, awarding prizes, conducting investigations, and deciding what to publish. As it came to be exercised in most academies, the imprimatur was extended to include control over not just what appeared in learned society journals, but also what members published independently and in some cases the whole of a national scientific and technical press. The Paris Academy again provides the premier example. Publication in its renowned *Histoire et Mémoires* was limited to its members and screened by a publication committee. Then, any member of the academy who wished to publish independently *and use his title of academician* likewise had to have his work reviewed and approved by the academy. Further, members of the academy were regularly appointed as specialized censors and as editors of extramural publications, such as the important *Connoissance des Temps* and the *Journal des Sçavans*.[67]

The administrative power of institutions manifested itself in the election and promotion of members. We have noted already that political intrigue and ministerial meddling were factors on occasion, but far more common was election based on merit and free choice. Extraordinarily, scientific societies were convoked as tribunals to try and judge individual members or other scientists. One recalls the prosecution of Leibniz by the Royal Society of London regarding priority in the invention of the calculus.[68] Another famous instance of this aspect of society power concerned Anton Mesmer, who was reviewed and censured by the Paris Academy of Sciences and the Royal Society of Medicine of Paris.[69] The König affair in Berlin was another notorious incidence; Samuel König was accused of plagiarism by Maupertuis and brought to task by the Berlin Academy.[70] Even in the benign Royal Society of London, where judgmental processes of this type are not supposed to have occurred, one David Riz was ejected "for the benefit of the Society as a proper example to enforce a due observance of their Statutes."[71]

Imagine, then, how a scientific society and the whole network of scientific societies must have appeared to someone on the outside: institutionalized centers of power in science, peopled by the ranking men of science of the age, organizing and controlling science itself. Surely the power and scope of that vision must have been impressive. It could alienate, and increasing numbers rebelled against the "dictature" of the academies.[72] It could awe, and some people begged to join. One asks what the system of scientific societies offered to someone interested and capable of making science a major focus of his life. Serious science was done in the societies, and for a hypothetical young man (or woman) of science (a Clairaut, perhaps, or a real "outsider" like the Russian Lomonosov), he would do well to join one.

While historians might be likely to accept a view of academies of science as powerful government institutions exchanging expertise for autonomy and support, there is a tendency to consider societies differently. Commonly, societies like the Royal Sciety of London are seen as voluntary associations of men arising from private initiative and operating with a noble independence outside of government control or support and furthermore in a tradition directly opposed to that of the Paris Academy and other academies. Charles Weld, the nineteenth-century historian of the Royal Society, provides one expression of this view.

Let it be remembered, too, that from first to last, the Royal Society have received no annual pecuniary support from Government, no assistance of any kind, beyond the grant of Chelsea College, shortly after their incorporation; and more recently, the use of the apartments they now occupy in Somerset House. This is a prominent feature in the history of our great scientific institutions, and one in which the independence of British character shines conspicuously . . . France, it is true, has her justly celebrated Institute [the Academy]; but how wide is the difference between the nature of this establishment and the Royal Society of England! It would be repugnant to the feelings of Englishmen to submit to the regulations of the Institute.[73]

Similarly, a commentator on the history of the Royal Society of Edinburgh remarked in 1862:

Let me interpose a remark on the organization of such societies generally. Even in early times, they differed from one another in respect of being either under the direct influence of the State, or of being merely private associations. This distinction continues to the present day. The French Academies, for example, are national institutions, and the members receive salaries from public funds. The Royal Societies of this country, on the other hand, are free from even the vestige of State control, and pursue their aims without pecuniary objects, and according to their own regulations. . . . The two forms of constitutions—the one creating a power in the State with corresponding advantages to its associates, the other receiving an impulse entirely from within—are really so distinct, that it seems almost invidious to compare them. The latter appears, from the history of our country, to be most congenial to English habits in such matters.[74]

Much commends this view, but, while it is important and necessary to distinguish the model society from the model academy, their differences in structure and organization, and the society's greater degree of freedom in matters of self-government, it is still somewhat false to categorically distinguish societies from academies with regard to their connections to government. A closer look reveals that the Royal Society, again as the prototypical example, was not so unattached or independent as it might seem.

In the first instance, the society's ties to government were not completely *pro forma*. It was the *royal* society, after all, incorporated by the English crown, and it was endowed with certain real institutional privileges.[75] It could act as a corporation, and it held the minor but real privileges of claiming the bodies of executed criminals for dissection and of the free and unhindered use of the mails. It could grant its imprimatur to works, and the imprimatur given to Newton's *Principia* redounds to the society's credit, though it is correct that the society could not compel works to be submitted to it.[76] An early plan for control over patents like that given to the Paris Academy never materialized, but new inventions still had to be exhibited and registered at the Royal Society before the granting of a royal patent.[77] Not the least of the privileges secured by the Royal Society from Charles II was "further the right to hold conventions for the sake of experimentation and investigation anywhere in our domain."[78]

The Royal Society from time to time also secured important financial backing from its patron and the government. Special grants from the crown allowed the Royal Society to undertake several major expeditions and projects that had an immediate bearing on the government's interests in navigation, trade, and colonial expansion. For example, the participation of the Royal Society in the transits of Venus in 1761 and 1769, the establishment of a meridian between London and Paris in the 1780s, or annual publication from 1767 of astronomical observations from Greenwich would have been impossible without government support and funding.[79] Also, in 1780 the government did provide free and

Thomas Sprat, *History of the Royal Society* (1667), frontispiece portraying Charles II, Viscount Brouncker (*left*), Francis Bacon (*right*), and arms of the Royal Society of London.

"for public use" apartments for the Royal Society.[80] It would seem that in those instances where the limited income of the society was insufficient for large tasks, the government did come to its aid, and that Weld's remark, though qualified, that the Royal Society received "no assistance of any kind" is shortsighted.

Then, the Royal Society on occasion undertook special projects "at the request of government," such as those concerning liquor taxation, prison conditions, or the development of the chronometer.[81] A notable instance of the use of the Royal Society for its scientific and technical capabilities to aid and advise other branches of English government occurred when it was called upon to decide the merits of pointed versus blunted lightening rods in the 1770s. The issue involved types of lightening rods and how best to protect buildings and gunpowder stores, and it drew in St. Paul's Cathedral, the government, the Board of Ordnance, and committees of the Royal Society.[82]

The incident illustrates that, as another facet of its real operations in the eighteenth century, the Royal Society maintained a considerable degree of contact with other bodies of English government and society, the Board of Ordnance notably in the case above. In 1714 the Royal Society was appointed "visitor" to the Royal Observatory at Greenwich and after the death of the Royal Astonomer, Bradley, in 1762, it took an active role in the management of that royal establishment.[83] The Royal Society had long-standing ties to the Board of Longitude, exchanging publications with them and furnishing men of expertise to deal with technical aspects of the board's work.[84] Similarly with the Hudson's Bay Company and the East India Company, the Royal Society many times used the resources and facilities of these agencies to further its efforts at scientific investigation, and at one point it formalized a "correspondence" with the Hudson's Bay Company for obtaining botanical specimens from English dominions.[85] Likewise, when it came to any of the several expeditionary missions sent out under its auspices, the Royal Society made contact with the Foreign Office, the Secretary of State for the Colonies, as well as the Admiralty with which the society had continuing relations throughout the eighteenth century.[86]

Another factor tending to modify the traditional view of the "independence" of the Royal Society is that the society was brought into closer and more intimate contact with larger circles of power and officialdom through patterns of recruitment and links forged by membership. The society was obviously not open to everyone. Bearing in mind levels of interest in science and natural philosophy and recalling that

Fellows were required to pay dues amounting to two pounds twelve pence per annum (which also could be bought out for a one-time "composition" fee of twenty-six guineas), it is not necessary to know the exact dimensions of the population of potential members or the exact value of the pound in the eighteenth century to come to the conclusion that Fellows of the Royal Society were drawn from a very narrow and elite base of English society.[87] Not surprisingly, a significant portion of eighteenth-century Fellows of the Royal Society were aristocrats and nobles.[88] Indeed, nobles and peers of the realm were given the small but special privilege of being voted members immediately upon nomination without the usual waiting period.[89] It is likewise not surprising to find that many Fellows also held positions of power and influence in various organs of government: Sir Robert Moray, who played an important role in obtaining the first charter of the society, was Privy Councilor and Lord of the Exchequer; Lord Brouncker, a colleague, was Commissioner of the Admiralty and Chancellor to Queen Catherine; Robert Boyle was on the board of the East India Company; Newton, of course, was knighted by Queen Ann and was president of the Royal Society and Master of the Mint; Sir Charles Morton (P.R.S.) was Commissioner of Longitude; Sir Joseph Banks (P.R.S.) was on the Board of Longitude, a member of the Admiralty, and advisor to George III. These are just a few examples; further study would doubtless reveal that the Fellows of the Royal Society (and even more the leading men among them) formed part of a "floating elite" that in different capacities served as members of the society and at the same time as members of other institutions and government bureaus. There is a curious clause in the charter of the Royal Society, which, although never put into practice, gives substance to this notion. In the event of a dispute arising within the Royal Society, it says, a special committee made up of the Archbishop of Canterbury, the Keeper of the Great Seal, the Keeper of the Privy Seal, the Treasurer of England, the Bishop of London, and the two Secretaries of State should convene and resolve the issue.[90] In this regard it is hard to imagine what was meant by the vaunted independence of the society, since, by and large, it seems part of one institutional system.

All these factors—legal and patronage ties to government, direct and continuing interaction with other governmental and quasi-governmental bodies, and integration into larger circles of government, administration, and officialdom through a floating elite of membership—add up to a vastly different picture of the Royal Society than a more traditional or superficial view would suggest. In essence it is not so much that the relations—the deal—between the Royal Society and English government differed fundamentally from those between academies and Continental monarchies as, simply, that English government itself was different. It depended on a loose, de facto coordination among many bodies in theory separate but in fact linked together by a ruling class. The Royal Society fitted into this situation, and if the reciprocal ties between it and the larger offices of state were not as explicit or as rigidly defined as was the case with academies, it seems that the Royal Society and other societies like it functionally if not officially were perhaps not quite so distinct from their Continental counterparts "that it is invidious to compare them."[91]

After what has been said to this point, is it too obvious to note taxonomically that the scientific academies and societies were not all of equal importance, that they did not share the same reputation and status as institutions? In fact, one observes a pyramidical hierarchy among the scientific societies of the eighteenth century, with a few outstanding institutions at the top and a larger number of less distinguished ones at the bottom. Although dealing with an essentially continuous scale of institutions, a few gross divisions may be made.

At the peak of this hierarchy were a handful of major national institutions of science: the Paris Academy, the Royal Society of London, the Berlin Academy, the Saint Petersburg Academy, and the Royal Swedish Academy of Sciences at Stockholm. These institutions were, in effect, the national science academies of France, England, Prussia, Russia, and Sweden respectively. They were all located in national capitals; they were all chartered by national governments on the authority of a royal head of state (the French and Russian academies again were founded on government initiative), and by and large they were the best supported of the societies, invested with the most powerful and lucrative privileges. With the exception of the Berlin Academy, these national academies were exclusively scientific societies. They were founded relatively early (that is, through the first half of the eighteenth century). And in terms of their membership, their publications, and their general activity, these institutions were the most important centers for science and the production and dissemination of scientific research.

At the next level there followed a larger number of noteworthy regional or provincial scientific societies. In this category one would doubtless include those academies and societies in Bordeaux, Edinburgh, Dijon, Montpellier, Göttingen, Turin, Lyon, Bologna, Mannheim, and Philadelphia. These institutions were located in comparatively large urban centers and major provincial capitals. All of them originated as unregulated and unrecognized private groups before official status was conferred on them. The governmental power recognizing a particular institution of this class was often of a regional or provincial type, as the previously mentioned German electors and the state of Pennsylvania. The point is true in a sense of the French societies which, while chartered by Paris, often became closely associated with provincial towns or Estates.[92] As might be expected, these societies were not well supported financially. None possessed any significant money-making privileges, although given local resources, they were not entirely destitute. This group of scientific academies and societies tended to be founded later in the century than the first group of truly national institutions. Although many were exclusively scientific institutions (Montpellier, Göttingen, Turin, Bologna, Mannheim, and Philadelphia), a larger portion dealt also with nonscientific subjects. The scientific importance of this second-level group was less than their national counterparts, and certainly more variable. Resident membership was less impressive on the whole, and, although all of the institutions in this category did publish scientific memoirs, these appeared more irregularly than those of the major societies and were sought after less. The American Philosophical Society of Philadelphia provides an excellent case in point. The institution achieved international stature in the 1770s with Benjamin Franklin as its president and with the publication of its first volume of *Transactions*. De Fouchy's remarks above reflect something of the visibility achieved by the Philadelphia institution. Later, however, in the 1780s and 1790s after the American Revolution, the institution turned more inward and utilitarian, and its former glory was eclipsed.[93] All other academies in this category went through such ups and downs.

The rest of the scientific societies—the majority—occupy lower rungs on the scale of things. Some of these institutions verged on significant national or regional status, but never quite achieved such a reputation for one reason or another. Although opinions might differ in this judgment, one thinks of the academies and societies in Brussels, Copenhagen, Barcelona, Marseilles, Munich, Rotterdam, or Toulouse as examples. At the very bottom were institutions like the little academy at Pau, a reunion of local elites just above the village level.[94] Societies in these lesser classes were located in lesser towns and received less support. They were founded late in the century as a rule, and they generally had to wait longer—often up to two decades or more—before official recognition was forthcoming. Some of the French societies (e.g., those at Anger, Béziers, Caen, Orléans, and Ville-franche) were unable to muster a continuous existence from their first foundation. Virtually all of the societies at this level concerned themselves with more than simply the natural sciences. The membership roster of these institutions is quite undistinguished, and, although a few sponsored scientific prize contests, none produced regular publications. Still, these institutions, as they all, possessed legal and social status in their communities. And, to a certain degree at least, they represented the wider world of science and provided something of an entrée to it.

Together learned scientific societies of all grades formed a coherent network and system of institutions. This chapter has demonstrated their communality of type, constitution, status, and activity. Subsequent discussion will make evident their common history as parts of an international institutional movement and directly connected through correspondence, exchanges, common endeavors, and overlapping networks of members. Top-quality science was more or less restricted to the top level of this interconnected hierarchy of institutions, it is true, but that level of scientific productivity was supported in some measure at least by the whole interdependent system of institutions. And besides, as one Bitaubé put it in his inaugural speech before the Berlin Academy in 1766:

Is it necessary that the obscurity of some [academies] serve to weaken those that distinguish themselves? On the contrary, do not weak states enhance the splendor of great Empires? One never confuses the Capitals of Kingdoms with less substantial towns and hamlets. If Academies abound, it is the effect of the most universal cultivation of Letters.[95]

Before quitting this initial taxonomy, the scientific societies per se need to be distinguished a bit further from other major categories of learned society and associated movements. Again, learned societies devoted to the traditional humanities (literature, languages, history, the fine arts) form an important subclass. In some instances, again, these disciplines were folded in with science in a single society. The quality of accomplishment and the standards of professionalism were comparable among the nonscience and science societies at comparable levels. They were united in a spirit of inquiry. One senses that the social backgrounds of their members were largely equivalent. The connections between nonscience and science societies (and within them where appropriate) are an important part of the full story of eighteenth-century learned societies as a whole. D'Alembert, for instance, editor with Diderot of the *Encyclopédie* and an important *pensionnaire* of the Paris Academy as one of France's leading mathematicians and mechanists, was also permanent secretary of the *Académie française!* These connections have yet to be really explored, however, and because the history of science is really so different from the history of literature, linguistics, history, or the fine arts in the eighteenth century, societies devoted to these disciplines fall outside the scope of this study.

Academies and societies of medicine are another noteworthy category of eighteenth-century learned society. Formally, medicine does not seem to have been incorporated into learned societies with science. Doctors, of course, were an important component of the overall membership of the scientific societies; in such purely scientific societies as Montpellier, for example, doctors drawn from the noted medical school of Montpellier were key members.[96] Then, the scientific interests of societies in anatomy, physiology, botany, chemistry bordered on and sometimes passed into more strictly medical concerns. The Royal Society of London, for example, was quite concerned with medical topics, and its *Philosophical Transactions* is famous for its unending reports of "monstrous" births.[97] However, within societies devoted to science, both the number of doctor-members and degree of interest in medicine would seem to have declined over the course of the eighteenth century.[98] In the seventeenth and early eighteenth centuries, the profession of medicine provided an entrée into science and a stimulus for scientific societies; later, as science and the man of science became better defined socially and intellectually, the influence of medicine waned. Independent medical societies would seem to have arisen later than the scientific societies and somewhat as their offshoot—witness the *Académie de chiruirgie* (1748) and the *Société royale de médecine* (1778) in Paris.⁹⁹ The model for the organization of independent medical societies derived from the scientific societies. Learned medical societies would seem to be largely separate from the scientific societies, though it may be a shortcoming not to consider them more fully here. As it is, only the *Société royale de médecine* of Paris, an Edinburgh medical circle, the Madrid-based *Real Academia de Medecina* (1732), and one or two other medical associations appear on the periphery of this story.

The subset of learned societies most in need of separate denomination vis-à-vis the scientific societies concerns the economic, patriotic, arts, and agricultural societies.¹⁰⁰ In some ways this important group of societies deserves to be considered together with the scientific societies. In other ways, it does not. The economic, agricultural, and similar societies were applied, utilitarian, and technological in their orientation, while the scientific societies were more natural philosophical in theirs. The historical roots of the economic societies are to be found in the eighteenth century compared to the seventeenth for the scientific societies. Their destinies were fulfilled in the nineteenth century versus the eighteenth-century for scientific societies. Overall, there was comparatively little overlap between and among the scientific societies and the economic societies. By the same token, the economic societies were an important manifestation of the learned society movement in general in the eighteenth century. And the two types of society conjoined in the science-technology interface, such as it existed in the eighteenth century. Certainly a study of the scientific societies needs to bear their economic cousins in mind.

One can distinguish shades of difference among the purposes and goals of arts, economic, and related societies.¹⁰¹ Agricultural societies, of course, were directly concerned with farming and ways to improve agricultural production. Economic societies per se placed a physiocratic emphasis on the land, too, but had goals that tended more toward economic development and increasing state wealth. Patriotic societies were more mercantile in their outlook, with less focus on farming and the land and more on industry and trades, but, like economic societies, they had economic stimulation and increase in state wealth and power as ends. Finally, "arts" societies were more directly committed to technical development in manufacturing, the crafts, and industry. Any one society, though, might share one or several of these institutional orientations.

The origins of the arts, economic, and patriotic movements differ from those of the scientific societies. Only indirectly were they related to the Paris Academy, the Royal Society, and the foundations of the seventeenth century. The grandfather of the economic type of society was the Honourable Society of Improvers of Knowledge of Agriculture in Edinburgh which was active between 1723 and 1745. Other early instances were the *Patriotische Gesellschaft* of Hamburg (1724–1754), the Dublin Society for the Improvement of Husbandry, Agriculture and Other Useful Arts (from 1731), and the *Ökonomische Gesellschaft* of Berne (1759). Based largely on the Berne example, a host of these societies appeared in Germany and Switzerland in the second half of the eighteenth century. Spain became a particularly active locus for this type of institution with a whole subset of societies, the over seventy *Sociedads Económicas de los Amigos del País*, founded and coordinated there after 1770.[102] In France a network of purely agricultural societies developed in the 1760s.[103] Begun initially in Brittany in 1757, twenty agricultural societies were systematically set up by the national government (mostly in northern France). They were divided into forty-five regional "bureaus" and were interconnected through correspondence. The Royal Society for the Encouragement of Arts, Manufactures and Commerce, a noteworthy institution to be sure, began in London in 1754.[104] It was very active and came to be known as the "premium" society for its small honoraria and prizes for new inventions or improvements to existing technology; many of the economic societies adopted this characteristic practice.

Given these different foci of interests and their different historical backgrounds, the scientific societies can rightly be considered separately from the economic societies. They were devoted to science, after all. But the scientific societies and the economic societies were all part of the general learned society movement, and some blurring of the lines and even conflict did emerge. In some few instances, it is hard to tell what kind of an institution a particular society was. The *Akademie gemeinütziger Wissenschaften* of Erfurt, for example, was organized and functioned as an academy of science. But, as its name implies, it was also committed to economic and utilitarian concerns.[105] The *Société patriotique de Hesse-Hambourg pour l'encouragement des connaissances et des moeurs* (1775), on the other hand, was an economic society with extensive contact with regular scientific societies.[106] In other instances, such as the Dijon Academy, for example, regular scientific societies took up agronomy or set up special agricultural or economic sections within their own organizations rather than see a competing agricultural society created in their locales.[107] Then, there was some overlap in membership between regular science societies and the economic societies; Leonard Euler, for instance, was an important associate of the Berne Oeconomic Society. It is noteworthy in this connection, however, that the Saint Petersburg Academy of Sciences twice turned over materials and requests relating to agriculture and rural economy to its economic counterpart, the Free Oeconomical Society of Saint Petersburg, once explicitly as "not the proper business of the Academy of Sciences."[108]

The economic societies constitute an important subgroup among learned societies of all sorts. They arose out of a somewhat separate institutional movement, one which testifies to new and different social and cultural interests developing late in the eighteenth century. One suspects that to a considerable extent their success was made easier by the model and prior institutionalization of the learned scientific societies. Given, too, local elites with local "improving" interests, it is not surprising that some overlap in interest, activity, and personnel between the scientific societies and the economic societies would occur. That notwithstanding, it would be a serious mistake to conflate these two types of institutions, their purposes, or the historical circumstances that gave rise to them separately.

CHAPTER II

Origins: Scientific Societies in the Seventeenth Century

T HE history of learned scientific societies in the eighteenth century is preceded by a separate seventeenth-century stage of institutional development which is of considerable importance for our story. The Royal Society of London and the Paris Academy—the exemplars of eighteenth-century scientific societies—were founded in the seventeenth century in the key decade of the 1660s. But in the seventeenth century they were hardly the institutions they later became, and both underwent formative periods which culminated at about 1700. In addition the scientific society movement—that pattern of modeling new scientific academies and societies after the Royal Society and the Paris Academy—while real in the seventeenth century, began to manifest itself seriously only after 1700. The creation of a third major society, the Berlin *Societas Scientiarum*, in 1700, testifies to the beginning of a new, eighteenth-century phase in the history of scientific societies after 1700. That history will be treated more directly in the next chapters. Here we need to focus on the background to the foundations of the 1660s, the development and effect of the Royal Society and the Paris Academy through 1700, and their transitions at that date. Only then will we be in a position to survey the full development of the scientific societies in the eighteenth century.

The foundations of the Royal Society and the Paris Academy in the 1660s were the major outgrowths of many seventeenth-century efforts to organize science on a new footing. These efforts—the "organizational revolution"—arose in conjunction with the substantive reformulations of the Scientific Revolution of the sixteenth and seventeenth centuries. Organizational alternatives were tried in reaction against late medieval universities and scholastic science. Science had a lesser place in the university, of course, where it was subservient to theology and Church bureaucracy. Its character was Aristotelian, scholastic, deeply institutionalized, and elaborated from an earlier age.[1] It was just this science against which the Scientific Revolution was taking place (for good technical reasons), and, as substantive alternatives were being developed, some men began self-consciously to reconsider the methodological and organizational bases of the new scientific enterprise being created, and new extra-university, secular patterns for organized science began to emerge. The ideological works of Bacon, for example, or the careers of Galileo or Descartes exemplify how little universities were in the forefront of developments in organized science in the seventeenth century. But the foundations of the 1660s were not the immediate products of the organizational component of the Scientific Revolution. They were preceded first by a preliminary stage which included the humanist academies of the Renaissance.

Renaissance humanism is an important movement which itself grew up outside university contexts, and the research of humanist scholars did a lot to foster the Scientific Revolution.[2] The humanist movement made texts available (particularly those of Archimedes), and it opened up new lines of inquiry; Copernicus was a notable humanist scientist. But on the whole, intent on rediscovering the ancient world of learning, humanism initially was less concerned with the substance of science than its sources and those of Western culture in general. But in the fifteenth century already nonscientific Renaissance humanism began to take on organizational and institutional dimensions of real proportion, and literally hundreds of literary and fine arts societies sprang up wherever educated men gathered. One source indicates that some seven hundred new "academies" arose in the sixteenth century alone, mostly in Italy, and Michele Maylender's comprehensive survey of such associations lists approximately 2,500 appearing in the period 1500–1800.[3] These are a distinct type of organization, which in time came to deal with science and natural philosophy and which continued into the period of the societies of the eighteenth century. Our analysis will be forwarded if we label them "Renaissance" academies.

There is some disagreement over which was the first "Renaissance" academy. Commonly Ficino's *Accademia Platonica* (note the switch from Aristotle), founded in Florence in 1442 by Cosimo de Medici, is pointed to as the earliest example, along with the *Accademia Potaniana* which was founded in Naples likewise in 1442 by Alfonso V of Aragon.[4] Maylender, however, wants to distinguish informal associations of humanists from formal academies, and he signals the Venetian *Accademia Aldina* of 1495 as the first of the new type of association.[5] The key features involved in this distinction are formal (although still unofficial) constitutions, a patron, officers, regular meetings, election procedures, and a press.[6]

Skirting the "which came first" question, hundreds of private associations meeting these criteria did spread throughout Italy in the sixteenth and seventeenth centuries. These "academies" took an interest in a broad range of cultural affairs: art, music, literature, language, architecture, history, archeology, religion, the theater, the hunt, and equestrian and military arts. "Renaissance" academies were generally small and for the most part ephemeral. Only the rare organization ever assumed any kind of official, public role; the Florentine *Accademia della Crusca* and *Accademia del Designo* are the notable exceptions.[7] Another key characteristic of the ordinary "Renaissance" academy was the essential role played by the patron. He supplied money, legitimacy, and prestige. When the patron died or suffered a decline in station, odds were that his institution would quickly fold. These academies were the products of Renaissance courts and Renaissance princes, and they delighted in what has been termed "bizarre Renaissance characteristics."[8] Men humorously deprecated themselves and their associations by such names as *della Crusca*, meaning "of the chaff."[9] They often assumed individual code names, such as Jason and the Argonauts, and special rituals (initiation rites, individual emblems, passwords, patron saints, and the like) also typify the "Renaissance" academy and distinguish it as a type from the official learned (scientific) society which succeeded it. "Renaissance" academies appear in many respects more like secret, almost Masonic associations than formal bodies concerned with literature, learning, or science.

The first "Renaissance" academies to have anything to do with science were two shadowy and ephemeral associations in Alessandria and Florence, dated at 1550 and 1560 respectively and both called *Accademia degli Immobili* because of their anti-Copernican persuasion![10] The first "Renaissance" science academy about which much at all is known is the *Accademia Secretorum Naturae* founded in Naples sometime before 1589—possibly in the 1560s—by Giambattista della Porta (1538–1615).[11] It consisted of unknown worthy friends of della Porta who called themselves the *Otiosi*, or idle men. It was a small and private affair; it left no records, and it resembles any number of other "Renaissance" academies except that it concerned itself particularly with natural knowledge and experiment. Della Porta's *Magia naturalis* (1558; 2d. ed. 1589) is a collection of curiosities about light, heat, magnetism, and other phenomena, and the *Accademia's* members doubtless dabbled in these areas. It is symptomatic of the changing and uncertain times in science that della Porta was accused of practicing witchcraft, and the *Accademia* was broken up by Rome. It was a good fifteen or twenty years before another group of note again involved itself in the active investigation of nature.[12]

The next "Renaissance" academy to concern itself with science—this time clearly as an alternative to university organization—was the famous *Accademia dei Lincei*, or Academy of the Lynx-eyed. The *Lincei* was founded in Rome in 1603 under the patronage of Duke Federico Cesi and with his financial support.[13] Begun with a few friends whose purpose was mutual instruction in the sciences, the *Lincei* soon expanded and enjoyed a much larger success than della Porta's academy. The *Lincei* strove for "real knowledge of nature . . . especially mathematics" eloquently expressed. Important to remark: they avoided involvement in politics—a hallmark of all succeeding societies.[14] Meetings and banquets were held, a private record of activities was kept, and scientific works were published under its auspices.[15] Through circles of correspondence, the academy facilitated the quick exchange of scientific news and reports throughout Italy. Ultimately some thirty men of science became members, including the aging della Porta and Galileo, who used the title *Linceo* proudly in print. But the academy encumbered itself with the typical trappings: its name, for example, or its sense of being a scientific "monastery."[16] And after Cesi's sudden death in 1630, it disappeared from the scene.

The famous *Accademia del Cimento*, founded in Florence in 1657, had a distinctly more modern cast.[17] Headed by Grand Duke Ferdinand

Medici and his brother, Prince Leopold, this academy was the first organization founded for the sole purpose of conducting scientific experiments. The *Cimento* was comprised of a small group of about a dozen men, including Borelli and Galileo's pupil Viviani, who assembled under less baroque circumstances than most "Renaissance" academies to perform and repeat experiments. The academy published its famous manual, *Saggi di Naturali Esperienze*, in 1667.[18] This anonymous volume describes instruments, methods, and results for a range of Galilean experimental investigations: the vacuum, heat and cold, optics, electrical and magnetical phenomena, sound, color, and projectile motion. The collective and anonymous enterprise, the explicit dedication to experiment, and the production of the *Saggi* separate the *Cimento* somewhat from the tradition of "Renaissance" academies both literary and scientific and testify to moderate steps by the 1660s to create new institutions for science by exploiting the form of the "Renaissance" academy. Even so, the *Accademia del Cimento* remained completely private and the product of court patronage alone, and whether because of Leopold's appointment as cardinal or because of internal dissention and decline, the Florentine academy ceased operations in 1667.

In terms of the development of science or "Renaissance" academies in Italy in the second half of the seventeenth century, it would be a mistake to see the closing of the *Cimento* as the end of an era. In fact almost the opposite is true, as many more "Renaissance" academies devoted particularly to experimental science were founded in the second half of the seventeenth century than in the first.[19] And "Renaissance" academies continued to be founded into the eighteenth century. By the same token, we can justly leave Italy at this point, as no other institution of science founded there rivaled the impact or the innovative quality of the *Cimento* until the establishment of the Academy of Sciences of the Institute of Bologna in 1714. The model for that institution was not any "Renaissance" academy, however, but the very different Paris Academy.

Turning to England next and the Royal Society of London, one observes that the seventeenth-century search for new institutional roots for science there did not unfold quite so much outside established institutional channels as in Italy or contemporaneously in France.[20]

[45]

Gresham College was a new pedagogical institution, founded in 1598 by private bequest.[21] With paid, resident professors in astronomy, geometry, and "physic," Gresham College provided a noteworthy rallying point for men of science in England in the first half of the seventeenth century. The Royal College of Physicians was another center, and scientific chairs were created in Oxford and Cambridge.[22] On a lesser scale than in France at the time, informal reunions of men were active in the penumbra of these institutions. One in particular was the direct precursor of the Royal Society, a group revolving around Gresham College in the 1640s. During the political unrest of the Commonwealth (1650–1660), several men went up to Oxford (John Wilkins, Seth Ward, Robert Boyle, Christopher Wren, and others), and they constituted themselves as the private Oxford Philosophical Club.[23] With an informal constitution and dues, they assembled weekly to perform and discuss experiments. With the Restoration these men returned to London and, in November 1660, after an astronomy lecture at Gresham, founded another experimental society.[24] Through their court connections (notably through Sir Robert Moray, privy councillor), they obtained royal recognition and patronage, granted by Charles II in letters patent in 1662 and 1663.[25]

A parallel tendency toward new patterns of association in science also occurred in France from the first decades of the seventeenth century.[26] Isolated individuals interested in the new learning—Gassendi, Peiresc, Mersenne, Descartes, Fermat, Pascal—cropped up in Paris and in southern centers.[27] Among this group, the French Minimes monk Marin Mersenne (1588–1648) merits special mention. Early converted to the new science—he and Descartes learned their mathematics together, and later he was the French translator of Galileo—Mersenne, along with his colleague Pierre Gassendi (1592–1655), was, like Bacon, a leading spokesman for the new science generally and mechanical and atomistic explanations in science in particular.[28] Mersenne is also known for his correspondence network which served as a private scientific clearing house for an extraordinary range of seventeenth-century men of science.[29] The elaboration of quasi-formal networks of scientific correspondence on the scale that Mersenne practiced it was unprecedented.[30] In addition Mersenne held reunions of savants in his monk's cell in Paris in the 1630s and 1640s. These developments are further testimony to the continuing organizational needs of the new science at the mid-seventeenth-century mark. But neither correspondence alone nor private meetings were the answers.

An interesting variant was introduced by the Paris-based doctor Théophraste Renaudot. As part of Renaudot's manifold activities on behalf of medicine, the new learning, and public welfare, weekly *public* meetings were held where scientific topics were presented and discussed.[31] Political pressures forced their suspension in 1635, but they failed as well from lack of unity and internal cohesion. Science then turned indoors again in France in a succession of private, "Renaissance" academies meeting in the homes of the Dupuy brothers, Montmor, Thévenot, and others.[32] Some of these were fairly formal affairs with written rules and officers, some were less so.

The Paris scientific community of the mid-seventeenth century was quarrelsome, however, and to arrive at the Paris Academy another historical strand needs to be considered: the French state. A separate tradition of state support for the arts and learning had existed in France since Henri II and the 1580s.[33] Richelieu and the foundation of the *Académie française* in 1635 strengthened it, and the advent of Colbert and the majority of Louis XIV in 1662 set the stage for new levels of cultural mercantilism and absolutist government.[34] Several state academies were founded or protected anew, and, as the idea of a science academy matured in both government and scientific circles, the Parisian *Académie royale des sciences* was naturally forthcoming.[35] Seeking to extend that earlier tradition, to bring science under the umbrella of state control, and to free its development from partisan disputes, Colbert established the academy in 1666 (after consultation) by administrative fiat.

Thus culminated the organizational revolution in Italy, England, and France in the 1660s. The *Accademia del Cimento*, the Paris Academy, and the Royal Society of London were its triumviral products briefly coexisting in 1666–67, and they were distinct from earlier seventeenth-century scientific groups and institutions in several ways. They were distinct from universities. Their concerns were not pedagogical or subservient to larger goals, but rather dealt directly with research, experiment, and the improvement and promotion of science. All three differed from specialized centers such as chairs, observatories, botanical gardens, or medical faculties in that they included all of the sciences within the domain of their interests. The Royal Society and the Paris Academy either had or were to get one further distinctive characteristic that set them apart from *Cimento* and all the rest: corporate status.[36] The person of the patron—fundamental to "Renaissance" academies— was also of less importance to the London and Paris societies. The patronage role of English and French monarchs was always small and less direct, and, as was noted, the *Accademia del Cimento* never received any formal status and proved unable to survive the appointment of its patron as cardinal. Thus the Paris Academy and the Royal Society remained the ultimate and most firmly institutionalized products of the organizational revolution of the seventeenth century.[37] With a shift in the balance of scientific power from the small states of Italy to the larger polities of France and England, the Royal Society and the Paris Academy, as true national societies of science, proved to be the enduring models.[38]

It is a remarkable fact, testifying to the context of their origins, that the Royal Society, the Paris Academy, and the *Accademia del Cimento* all seem to have shared a commitment to the program of scientific activity and the production of new knowledge advocated by Sir Francis Bacon (1561–1626). A major theoretician of the new science, Bacon had a tremendous impact on the early history of the scientific societies.[39] A lawyer, Chancellor to James I, Lord Verulam, and Viscount St. Albans, Bacon was not much of a scientist, but in his *Advancement of Learning* (1605), *Novum Organum* (1620), and the posthumous *New Atlantis* (1646), Bacon argued a devastating critique of the "idols" of university learning and proposed his own empirical and inductive approach to the production of new knowledge. Deeply suspicious of the power of learned authority, Bacon was insistent on the slow elaboration of positive facts built up by careful observations over long periods of time. Experimental testing of hypotheses ("twisting the lion's tail") came after observation and data gathering, and the final formulation of natural law awaited the end of this process, perhaps a matter of generations. A noteworthy feature of Bacon's sense of the new scientific enterprise was the utilitarian ends to which science should be directed and from which practical benefits should be forthcoming. The purpose of science was not refined meditation or the mere transmission of received knowledge, but its application for the benefit of mankind. Bacon pointed to the compass, gun powder, and the printing press as examples of formidable results to be had by society from his new approach. It was arguably a long time before science proved its worth in this regard, but nevertheless the ideological claim had been made for the social utility of the new science, and it was a powerful one that the champions of science and scientific societies would stress in Bacon's name to the end of the eighteenth century, at least.

As part of his general reformulation of the nature, goals, and practice of science, Bacon argued for new institutions to carry out his program. In the *New Atlantis* he described his utopian "House of Salomon."[40] Situated on a fantastic island, Bacon's Salomon's House is a veritable research institution consisting of various departments for investigations into things chemical, astronomical, agricultural, medical, botanical, mechanical, optical, and mathematical. The staff is divided into teams to process and collect data, propose and perform various levels of experimental testing, and, topping the entire edifice, three "Interpreters of Nature" who "raise the former discoveries by experiments into greater observations, axioms, and aphorisms."[41] Bacon's approach to the production of new scientific knowledge may strike the modern reader as visionary and impossible. Certainly he differs from Galileo, where the scientific effort was made to analyze phenomena isolated experimentally apart from their natural contexts, or Descartes, where deductive apriorism was the order of the day. Just the same, Bacon's call for new institutional structures for science reflects, again, the contemporary pressure for the same, and his work was tremendously influential. In particular, members of the early Royal Society and the early Paris Academy tried directly to inaugurate precisely the "Salomon's House" Bacon espoused.[42]

In the case of the *Accademia del Cimento*, however, resemblance to any English-style Baconianism is largely an illusion. Bacon's *Essays* were translated into Italian as early as 1618, but the *Accademia* seems surprisingly unaware of the corpus of Bacon's work.[43] Robert Boyle and the Latin translation of his *New Experiments Physico-Mechanical* (1661) were better known, but, according to Middleton, the *Cimento*'s approach was really more Galilean than Baconian, and the anonymity of the *Saggi* is better explained less as the product of a collectivist approach than as a residual fear of the Inquisition and Church sanctions.[44]

The initial approach of the Royal Society of London was more genuinely Baconian. Thomas Sprat, commissioned to write the apologetic *History of the Royal Society* (1667), emphasized this aspect of the Society's initial interest:

I shall only mention one great Man, who had the true Imagination of the whole extent of this Enterprize, as it is now set on foot; and that is the Lord Bacon . . . The *Society* has reduc'ed its principal observations into one *common-stock;* and laid them up in publique Registers, to be nakedly transmitted to the next Generation of Men; and so from them to their Successors. And as their purpose was, to heap up a mixt Mass of *Experiments*, without digesting them into any perfect model: so to this end, they confin'd themselves to no order of subjects; and whatever they have recorded, they had done it, not as compleat Schemes of opinions, but as bare, unfurnish'd Histories.[45]

But the Royal Society, because it was a society and because it did not therefore receive substantial support from the English government, never effectively became a true Baconian institution. Early on, it did try: Robert Hooke was employed as demonstrator and curator of experiments, and various committees were established to handle—in a Baconian manner—several separate fields of inquiry.[46] In need of operating funds and more dues, though, the small group of original Baconian founders expanded to over two hundred Fellows by 1666.[47] The effect of this expanded membership was that those with a dedicated interest in science and its pursuit through Baconian induction became outnumbered by more casual, dilettantish elements.[48] Thus, instead of becoming a Baconian center itself, what the Royal Society could do was to provide a limited support for the Baconian productions of others and, more generally, to encourage a vaguely defined Baconian empiricism that meshed well with the format of its meetings and the looser interests of its members. Evelyn's *Sylva* (1664), Hooke's *Micrographia* (1665), and Wilkins' *Essay Toward a Philosophical Language* (1668)—reflecting early interests of the Society—are among the books and Baconian natural histories published with support and imprimaturs from the Royal Society.[49] Paradoxically, it was this early failure of its initial commitment to pure Baconianism that forced the society to open up and turn outward and that (in part) explains why the Royal Society was the most active and important learned society in the second half of the seventeenth century.

The Paris Academy of Sciences, on the other hand, *was* a successful

Baconian institution—perhaps the only one in the seventeenth century, and paraxodically again its very success retarded its institutional development.[50] Before its reform in 1699, the Paris Academy was a very different institution than the one sketched in the last chapter. Fully funded by Louis XIV, the early Paris Academy consisted of some two dozen paid professionals who met privately—one might say secretly—twice a week behind the closed doors of the king's library in the Louvre.[51] There were no significant divisions to its membership: no honoraries, no foreign members. But as competent scientists (Huygens and Roberval were early members), supported to do scientific work, and methodologically converted to Baconian causes, the king's men of the Paris Academy together undertook collective experimental investigations into a range of topics: organic analysis, dissections, transfusions, analysis of mineral waters, and so on.[52] In addition, the academy published collectively several large natural histories: *Mémoires pour Servir à l'Histoire Naturelle des Animaux* (1671), *Recueil de Plusiers Traitez de Mathématiques de l'Académie* (1676), and *Mémoires pour Servir à l'Histoire des Plantes* (1676). But because the Paris academicians could afford the luxury of supported Baconian research and having no particular need for outside contact, the academy was inward-looking and self-reliant. It was thus effectively closed off from the larger scientific community and less visible than the Royal Society.[53] It was only with the failure of this initial Baconian impulse and the reform of the Paris Academy in 1699 that it emerged as the leading light in the world of science, reversing roles with the Royal Society.

The early Royal Society was graced with several other characteristics that helped make it the major scientific institution in the second half of the seventeenth century. One was the position of Fellow of the Royal Society (F.R.S.) and the international network established by its large membership. The comparatively few members of the Paris Academy by necessity had to reside in Paris, and there were no foreign-based members in this first period of its history.[54] By contrast the much larger number of F.R.S. spread world-wide.[55] Colonial American Fellows were an important subgroup: John Winthrop (Governor of Connecticut and an original Fellow), William Penn, and Cotton Mather were among a number of American members in regular contact with the mother society. Continental men of science were also early elected fellows, including Auzout, Cassini (both members of the Paris Academy), and Leibniz. And, early on, a separate category of nonresident, nonvoting member was created for those who were unable to be present at the Royal Society for their formal receptions.[56] The international fellowship of F.R.S. would grow to even greater extents in the eighteenth century, but already in the seventeenth men in various locales, licensed by certificates of membership, acted as the extended agents of the Royal Society.

Both the Royal Society and the Paris Academy provided for the office of institutional secretary. The office never amounted to much in the Paris Academy in the seventeenth century.[57] The Royal Society, however, was especially fortunate in obtaining the services of Henry Oldenburg, who turned his position into a formidable platform from which to orchestrate the progress of science in his day.[58] Oldenburg's activities not only redounded to the credit of the Royal Society, but he set the example for the succession of great institutional secretaries of the eighteenth century. Secretary from 1663 until his death in 1677, Oldenburg was widely traveled, at home in several languages, a prolific correspondent, and a figure very reminiscent of Mersenne.[59] Like Mersenne, he was the kingpin of a (larger) international network of correspondents, but, unlike him, Oldenburg held an official institutional position. Organized private correspondence was henceforth dated. Oldenburg's description of his duties for the Royal Society is worth repeating here.

He attends constantly meetings both of the Society and Council, noteth the observables said and done there, digesteth them in private, takes care to have them entered in the Journal and Register book, reads over and corrects all entries, sollicites the performance of talks recommended and undertaken, writes all letters abroad and answers returns made to him, entertaining correspondence with at least thirty persons, employs a great deal of time and takes much pain in satisfying foreign demands about philosophical matters, disperseth far and near store of directions, inquires for the societies purpose and sees them well recommended.[60]

One of Oldenburg's capital accomplishments was his founding in 1665 and editing of the *Philosophical Transactions* of the Royal Society. As it happens, the *Philosophical Transactions* were antedated by two months by the Paris-based *Journal des Sçavans*. These two journals mark the advent of the scientific periodical and an important turning point in the history of the scientific press.[61] That they were founded coincident with the Royal Society and the Paris Academy is not an accident.[62] The *Philosophical Transactions*, a journal containing abstracts from letters, reports, and scientific news was conducted as the private venture of the secretary of the Royal Socety until 1751, but all along it was printed under license and received indirect financial support from the Royal Society.[63] Given Oldenburg's position as secretary, the bounds between the *Philosophical Transactions* and the Royal Society were ever sharply drawn. The *Journal des Sçavans* had a less direct connection to the Paris Academy than was the case in England. The academy itself did not display any interest in publications of this sort, and the *Journal des Sçavans* preceded its foundation. By the same token, the *Journal des Sçavans* was a publishing forum for men of the Paris Academy, and it did publish some reports of their proceedings; later J. B. Gallois, an academician, was its editor.[64] Many other scientific journals of various types appeared after the 1660s and proliferated through the eighteenth century.[65] Most learned journals did not have institutional sponsorship or connections, but the most important ones, loci for the primary publication of original research results, usually did, especially in the eighteenth century.[66]

The effect of early journals on the conduct and structure of organized science was substantial. The *Journal des Sçavans*, the *Philosophical Transactions*, and their later kin introduced a whole new mode for the presentation and dissemination of scientific knowledge. Publication in a scientific journal (either directly or filtered through an editor) offered a speedier means of communications than books.[67] They quickly became the way to establish priority.[68] And they reached a larger, less selective audience than correspondence.[69] The appearance of the scientific periodical in effect created the scientific paper. The paper would develop more fully in the eighteenth and nineteenth centuries, but Newton's 1672 paper on light, for example, published in the *Philosophical Transactions* is a classic, seventeenth-century instance.[70] Not demanding the scope or inclusiveness of a book, yet more formal and directed than private correspondence, the scientific paper and the journal that carried it were important new ways for men of science to communicate among themselves. The distribution of journals, their translation, and republication of articles created new patterns of scientific exchange, and, to the extent that the *Philosophical Transactions* was looked to as the premier scientific journal of the day, this, too, enhanced the reputation of the Royal Society.[71]

In addition to the Royal Society and the Paris Academy, the second half of the seventeenth century saw the continuation of scientific society foundations and the true beginning of what we are calling the scientific society movement. By scientific society movement is meant the succession of scientific society foundations modeled directly or indirectly after the Royal Society and the Paris Academy. Those foundations represent a pattern of historically interrelated institutionalizations with something of a life of its own. The scientific society movement in some sense was a continuation of earlier developments of the "Organizational Revolution" which gave rise to the Royal Society, the Paris Academy, and the *Accademia del Cimento*. But the movement which needs to be highlighted here was the one that derived from the models and examples of the Paris Academy and the Royal Society and gave rise to the eighteenth-century societies. The scientific society movement really did not begin to show its effects until the first half of the eighteenth century, however. Still, as is already evident in the seventeenth century, the movement was a real thing, a genuine historical phenomenon producing new societies after preexisting examples.

The significance and possibilities of the Paris Academy and the Royal Society as new models for the organization of science were not ignored by the men of the time. Oldenburg recognized the ultimate effect the Royal Society, the Paris Academy, and (as he reckoned it) the *Accademia del Cimento* would have on the subsequent historical course of institutionalized science. In a letter of 1667 to Pierre de Carcavy, librarian to Louis XIV, Oldenburg wrote: "I am thoroughly persuaded that the societies newly established here, in France and in Italy will serve as a stimulus, within a few years, to incite all the other nations of Europe to take up the same studies and to oblige them wholly to desert the quodlibetic [scholastic] learning of the schools."[72] In another letter of 1668 to J. B. Gallois, Oldenburg spoke of the Royal Society and the Paris Academy "as being a leaven to raise the rest of the civilized world into following their example."[73]

In these predictions Oldenburg was not wrong. The first indication in organizational terms that there was anything to his persuasions came not long after the foundations of the Royal Society and the Paris Academy with the plan of 1669/1670 of G. W. Leibniz "for the establishment of a Society in Germany."[74] Leibniz was an early convert to the new organizational modes for science, and from this first plan until just before his death in 1716, Leibniz was a tireless promoter of scientific institutions modeled along the lines of the Royal Society and the Paris Academy. This drive can be seen in his next plan of 1675 for a *Societas Eruditorum Germaniae* and its provisions for contact with the French and English societies, his plan of 1676 for a *Societatae Germanicae Consultutio*, or more explicitly in his plan of 1697 for a *Societas Electoralis Brandenburgica examplo Regiarum Londinensis et Parisiensis*.[75] Leibniz' activities are concrete evidence that the scientific society movement was underway in the seventeenth century.

Until the eighteenth century, however, the successes of that movement were few. It was not until 1700 and the creation of the *Societas Scientiarum* of Berlin, Leibniz' first and signal success, that organized science was enhanced by a third major learned society. Even then, it can hardly be said that the movement had "taken off" in any sense. What effect the learned society movement did have in the seventeenth century was curiously uneven. On the Continent, it was very little. Leibniz' efforts proved unrewarding in the states of Germany. The *Collegium Curiosum sive Experimentale* was a science academy of the *Cimento* style established in 1672 in Altdorf; it ceased operations with the death of its founder, the professor Christopher Sturm, in 1695.[76] The *Academia Naturae Curiosorum*, a "Renaissance" medical academy founded in Schweinfurt, Germany, in 1652, did receive recognition and incorporation in 1677 and again in 1687 by the Holy Roman Emperor, Leopold I.[77] The *Academia Caesarae Leopoldina-Carolinae Naturae Curiosorum*, as it came to be known, was one of the few "Renaissance" academies to make the transition to an officially recognized scientific institution with corporate status.[78] It remained a peculiar organization, however, in maintaining special names for its members and in not being spatially located in one place, its headquarters varying with each president. It did not hold meetings as such or operate in any way like a standard, eighteenth-century learned society and it was known almost exclusively through the publication of its *Miscellanea* (from 1670) which largely contained the inaugural dissertations of its medical members. Still, as far as the world was concerned, the *Academia Naturae Curiosorum* was the official science society of the Holy Roman Empire.

In France a goodly number of provincial academies were founded and received royal recognition in the seventeenth century: Arles (1668), Soissons (1674), Nîmes (1682), Angers (1685), Caen (1652), Avignon (1658), Villefranche (1695), and Toulouse (*Jeux floreaux*, 1695).[79] These academies were concerned almost exclusively with literature and poetry and were much more closely (often formally) connected with the *Académie française* than with the Paris Academy of Sciences. The second academy at Caen (1666–75) was the exception in this regard, being devoted to the physical sciences.[80] Many seventeenth-century French provincial academies (e.g., Caen again, or Arles) barely lasted past 1700, and those that did often changed their orientation more toward science and the Paris Academy after 1700. For seventeenth-century science the academies of provincial France were of little note.

The impact of the learned society movement in areas of English influence, on the other hand, while not productive of any major new institutions of science either, was greater and more interesting. As a result of the international character and reputation of the Royal Society of London in the seventeenth century, a handful of new scientific associations were formed in Great Britain and America which saw themselves as outgrowths of the Royal Society in London. Insofar as they were possessed of internal organization, they were modeled along the lines of the Royal Society, but none was incorporated officially. A half-dozen private associations of little long-term consequence, these English societies still say something significant about patterns of organized science in the second half of the seventeenth century.

Actually, the first to fit into this category did not succeed the Royal Society, but rather was its predecessor—the old Oxford experimental club, or Philosophical Society. The Oxford Society continued off and on as a small experimental group until 1690, when it finally petered out.[81] It was in close and regular contact with its London counterpart, apparently at one point even performing a series of experiments on commission from London; Weld calls it "a powerful auxiliary to the Royal Society."[82] In 1683 the comparatively more important Dublin Philosophical Society was established.[83] Begun by William Molyneux (F.R.S.) in the shadows of Trinity College, the Dublin Society was in continuous existence until 1688, and then led an on-again, off-again existence until 1708. Holding meetings and listening to papers were its major scientific endeavors. The Dublin Philosophical Society was

[56]

closely tied to, even affiliated with, the Royal Society. This case may be the first instance of formal inter-institutional association of any real substance among scientific societies.[84] Fourteen of the thirty-seven members of the Dublin Society were also F.R.S., and reportedly it was voted at the Royal Society in London that men who were members of both institutions would have to pay only half dues in London.[85] The Dublin "auxiliary" regularly submitted minutes of its meetings and papers of its members to be read at the Royal Society, and, lacking a publication of its own, the Dublin Society used the *Philosophical Transactions* as a publication outlet for its members with fifty-four Irish articles published there.[86] The Dublin Society was also in contact with the Oxford Society, and Hoppen, the historian of the Dublin Society, sees both Oxford and London as providing an indispensible institutional base for the Dublin group.[87]

Another like organization, the Boston Philosophical Society, was in existence between 1683 and 1688.[88] In this case the Royal Society and Robert Boyle's influence as governor of the Company to Promote the Gospel in New England were spurs to Increase Mather to found the Boston group. Little is known about it, although one may assume that it was in contact with the Royal Society and that it failed for lack of a sustained and congenial cultural context.

Finally in this regard, still in 1685, an effort to establish a philosophical society at Cambridge University was undertaken with the backing of Isaac Newton. This association, too, was to have been directly cooperative with the Royal Society; it failed to materialize because, even in a university context, no one could be found to perform experiments![89] Similarly, in 1670 Bristol, Somerset, a plan for a philosophical society "upon encouragement given from the Royal Society," which was to have used the mails to facilitate communications, never seems to have materialized.[90] Some success was achieved in Edinburgh in 1705, when a small private philosophical society began its short term of meetings. This case takes us into the eighteenth century, and it indicates that the pattern of scientific associations formed in the shadow of the Royal Society by no means came to an end at 1700.[91]

The course of the scientific society movement in the seventeenth century and particularly this blossoming of Anglo-Saxon philosophical and experimental societies reveal several points of note. The Royal

Society is again seen as the most influential institution of the century. Given what else is known about the international scope and open outlook of the Royal Society at the time, there is little surprise here. The circle of British philosophical societies appears as by far the most significant manifestation of the learned society movement in science in the period; there was no comparable development on the Continent. That being the case, it would also appear, especially in light of later developments, that by and large the scientific society movement had made very little progress by the year 1700. This conclusion is reinforced when one bears in mind that, despite the supporting influence of the Royal Society, none of the Anglo-Saxon societies of the seventeenth century could achieve sufficiently deep and secure roots in local cultural and institutional settings to survive on an extended basis.

The most interesting observation to be made about the scientific society movement in the seventeenth century concerns the embryonic institutional system centering on the London-Oxford-Dublin triangle of the 1680s. By dint of their common members, their contact and cooperation, affiliation, exchanges, and patterns of publishing, the Royal Society and the philosophical societies formed a historically real system of scientific institutions composed of a central institution (the Royal Society) and a group of dependent, "satellite" organizations around it. These three societies in London, Oxford, and Dublin—not to mention the Boston group—introduced a potent, although tenuous, new element for organized science, and their degree of collective interaction was without parallel at the time. They set an important but unrewarded precedent, for when the novelty of a cooperating system of institutions reemerged in the eighteenth century as a significant factor for organized science, the old Royal Society system would be outmoded and outclassed by a more elaborate, powerful, and egalitarian network of Continental academies.

Despite their respective standing, the Royal Society and the Paris Academy were by far the two most important scientific societies—if not scientific institutions—in the second half of the seventeenth century. What did they have to do with one another in the years leading up to 1700? The answer is surprisingly little.

At the outset prospects for fruitful contact looked good. In fact, the two communities that went into making the Royal Society and the Paris

Academy had previously made overtures to each other. In 1661, just after the establishment of the Royal Society but still before the Paris Academy, Sir Robert Moray wrote to Montmor, as "president of the Academy at Paris," to formalize relations, and Samuel Sorbière was deputized by the Montmor group to visit the Royal Society.[92] Oldenburg himself was already acquainted with the scientific salons of Paris, and he was excited to learn of the foundation of the Paris Academy in 1666. He was at that time receiving weekly packets from France; his regular source of information was Henri Justel, a Protestant royal secretary (a role later taken over by Francis Vernon, secretary to the English ambassador), and some of the Paris academicians themselves, notably Cassini and Huygens, with whom he kept up a regular correspondence.[93] In his letter of January 1667 to Pierre de Carcavy, whom he mistook to be the presiding officer of the newly-formed Paris Academy, Oldenburg spoke further of his "great joy" and said that he would be "delighted to hear of the consolidation of your institution and to see you make progress worthy of the advantages you enjoy from both the generosity of so great a king as yours and the happy genius of your nation."[94] Nothing came of this letter, although Oldenburg seemed still very interested in the nascent academy.[95] Then in July 1668, J. B. Gallois, acting secretary of the academy, wrote to Oldenburg:

Without having the honor of being known to you, I have hitherto profited from the correspondence you have conducted with Mr. de Sallo [editor of the *Journal des Sçavans*]. But as his great volume of business prevents him from carrying on this correspondence with as much attention as I should wish, I beg you to approve of my taking his place sometimes, and assisting him in developing it. Moreover, since I hold the same office in the French Academy as you do in the English, it is perhaps in the interests of science that we should exchange some letters.[96]

Gallois's offer was lukewarm ("perhaps in the interests of science . . . exchange *some* letters"), but Oldenburg's response was enthusiastic:

To Mr. Gallois, Secretary of the Royal Academy at Paris.
Sir,
    I heartily embrace the offer of your friendship and correspondence which you were pleased to offer me in your letter, and I shall try to cultivate it in the best way possible. I regard these two Academies founded by two great kings for the advancement of useful science as being leaven to raise the rest of the civilized world into following their example; and since you and I have the honor of

wielding the pens of these two illustrious bodies, I agree with you that it would be to the interest of science that we should undertake a cooperative effort, and employ our pens especially towards rousing the intelligent, uniting their forces, and so working that they may mutually assist one another in making experiments and discoveries in nature and the arts. Sir, I am ready to receive your commands so that I may serve you here in any possible way, and after you have broached the business, you will see how prompt I am to follow your example.[97]

It was a novelty for two scientific societies to establish some form of official contact between themselves, and Oldenburg's use of the terms "mutually assist," "cooperative effort," and "uniting forces" is striking. Here is something demonstrably new and potent when scientific societies could speak in such terms. What might it mean and what would be the effect if scientific institutions began an international program of mutual assistance, cooperative effort, and uniting forces?

In the seventeenth century, it meant a little but not a lot. The ill-defined position of academy secretary plus the lack of a strong public face for the Paris Academy did little to help the relationship with the Royal Society.[98] But Oldenburg, Duhamel (the regular secretary), and Gallois were in intermittent contact through the 1670s, and their correspondence concerned securing and exchanging books and periodicals, including the *Philosophical Transactions* and the *Journal des Sçavans*.[99] That much at least was mutual assistance, but then other factors cut the connection short. Oldenburg's death in 1677 was perhaps the major one. Not possessing his language skills or his extensive European contacts, none of his successors was able to duplicate Oldenburg's forceful and respected role as editor, correspondent, and clearinghouse.[100] Indeed, after Oldenburg, who held the post for fifteen years, the average tenure as secretary of the Royal Society seems to be no more than two or three years until Hans Sloane took over in 1693.[101] When Edmond Halley visited Paris in 1680 contact between the two institutions had degenerated to the point where he reported—not visiting the academy—only that Cassini ("who seems my friend") might possibly be able to get "the book on Astronomical voyages."[102] Political tensions were an element of uncertainty throughout the period—they landed Oldenburg a stay in jail for purportedly using his correspondence network as a channel for espionage.[103] And with the outbreak of active conflict between England and France in 1688, conflict that continued almost unabated until 1713, "active scientific and philosophical exchange between France and England end."[104] What began as a promising relationship of contact and exchange between the foremost scientific institutions of the age and an important first for international science had ground to a fruitless halt, and many decades passed before the Royal Society and the Paris Academy once again achieved—then surpassed—the level of contact established in this first period of their common history.

The coming of the eighteenth century marked important turning points in the historical development of the Royal Society, the Paris Academy, and scientific societies in general. With their formative periods over, the relative status of the Royal Society and the Paris Academy as institutions actually reversed as they assumed the definitive organizational forms and characters that would remain virtually unchanged until the nineteenth century. By 1703, when Isaac Newton was elected president, the Royal Society had undergone two stages of development.[105] The first, shorter period witnessed its establishment, an expanded membership, a growing international position, and a high degree of enthusiasm and accomplishment by a distinguished core of men, such as Boyle, Hooke, and Oldenburg. By the later 1670s and 1680s, however, the question arose whether the Royal Society would survive its first, most illustrious generation. Amateurs had diluted its scientific capacity, it became a target for public ridicule (witness Swift's Laputa), attendance at meetings had dropped off, and funding remained a critical problem.[106] Though its standing abroad probably remained high, internally at home in London, the Royal Society was in serious jeopardy. Its salvation came in the stronger role played by the council of the society. Originally conceived to be the servant of the regular membership (the rules of the society prior to 1662/63 called for the monthly rather than the annual election of officers), the council emerged in the eighteenth century as the dominant executive.[107] Balancing the increasingly amorphous quality of ordinary meetings, the council took control, as can be seen, for example, in its screening of candidates for Fellow and in its pursuit of law suits against members defaulting on dues.[108] By 1703, the Royal Society had endured the early threats to its existence; its basic *modus operandi* as a loose scientific association was manifest; it had the beginnings of financial stability; the council and officers had become the major forces in the society's organization, and with Newton's election in 1703 and subsequent re-election until 1727, the tradition of a strong and long-tenured president began in earnest.[109] The Royal Society evolved into a much less spectacular kind of institution than its founders might have anticipated or its early, glorious years might have predicted.

For the Paris Academy, the case is just the reverse. Begun and operated at first as a small, involuted group of experts engaged in a collective, Baconian pursuit of science, in 1699 the Paris Academy suffered a veritable institutional revolution and emerged as a much more powerful, *the* most powerful scientific institution in the early years of the eighteenth century.[110] Noteworthy was the ultimate failure of its early Baconian methodology. For one thing, the academy did not actually function as a collective whole; one or two men would, in fact, do the work on any particular project, and questions of priority naturally arose. A related problem faced by the academy was how to deal with the private publication of academicians outside of the academy. Disrepute could be cast by hasty or ill-conceived publication. The crucial means adopted to meet this situation was that, after 1685, men were required first to submit their proposed work for approval by the academy if they wished to claim the title of academician. This move had the dramatic effect of turning the academy (with ensuing committees and reports) into an active judge of science good and bad, rather than a primary producer of science. This was a large step away from its initial Baconian commitment. New functions also emerged for the academy in its capacity as a government institution. Colbert, while supporting the collective approach to pure science, did not spurn practical benefits to be gained from the academy by the French government and state. Early in the academy's history, its role as advisor to the government on questions of science and technology was carried out in an informal and ad hoc manner.[111] As time went on, this process became streamlined, and with committees regularly handling these matters, the academy metamorphosed into a more explicit kind of government agency. Finally, in this first period of its history, the Paris Academy suffered from the buffeting of political winds.[112] The academy was more vigorously supported under Colbert than it was under his successor, Louvois, and Louis XIV's revocation of the Edict of Nantes in 1685, forcing the retirement of the Protestants Huygens and Roemer, hurt the academy.

In 1699, under the benevolent administration of the Abbé Bignon, the academy received new life and was given its first formal regulations.[113] At the outset of these regulations the failure of its collective Baconian approach was recognized: "Experience having shown too many problems involved in works undertaken collectively by the Academy, each academician will choose rather a particular subject for study, and in his presentation in the assemblies, he will seek by his lights to enrich the members of the Academy and profit from their remarks."[114]

The letters patent of 1699 next legitimated the academy's previously developed practices by giving it explicit and important powers over science and technology:

The Academy will examine works academicians propose to have printed. It will not give its approval except after a complete reading in its assemblies or at least after an examination and report by those commissioned by the Company. No academician can use the title of academician in his published works unless they have been thus approved by the Academy.[115]

If the King order it, the Academy will examine all machines for which a [patent] privilege is solicited from His Majesty. It will certify if they are new and useful, and the inventors of approved machines will be responsible for leaving a model with Academy.[116]

Perhaps to rectify its earlier weakness in this regard, the letters patent of 1699 also added a new international dimension to the academy by creating the categories of nonresident and foreign members and by mandating the duty of the academy to keep up with international developments in science.[117]

The scientific power of the Paris Academy (and academies of science henceforth) was augmented by three other innovations introduced into the ordinary operations of the Paris Academy in or shortly after 1699. Public meetings were one of these. With the semi-annual public meeting legislated by the letters patent, the closed character of the pre-1699 academy was abandoned completely, and from the meetings themselves as well as from reports appearing in various European journals, the work of the academy further entered the public realm. Public assemblies were (and still are) occasions for institutional renewal and display where eulogies of departed members and the reception of new members symbolized the continuity of the institution and where the place of the academy in the larger structure of corporate France was reaffirmed and

Paris Academy of Sciences, *Mémoires* (1699), frontispiece portraying
Louis XIV.

legitimated.[118] Second, after 1699 the Paris Academy began publishing its own proceedings, the *Histoire et Mémoires*. An annual volume appeared throughout the eighteenth century under the direct sponsorship and supervision of the academy. Divided into two parts, the *Histoire* contained the secretary's (Fontenelle's until 1741) review of academic events for the year and his summary of the papers that were included in the second, *Mémoires*, part of the volume. This type of direct institutional publication containing finished scientific papers thus differed from learned journals like the *Philosophical Transactions*, where letters, reviews, short notices, abstracts, and more frequent publication were the order of the day. Academic memoirs provided a major new forum for the original publication and dissemination of finished science.[119]

In the third of its novelties, the Paris Academy began offering prizes for scientific work.[120] Actually, the model comes from the Bordeaux Academy, and the stimulus and funds for this most important of all eighteenth-century prize competitions came not from government sources but from a private bequest from the parlementarian de Meslay.[121] On alternate years from 1720 a prize was offered for a question in mathematics and mechanics (worth two thousand livres) and another in navigation (worth five hundred livres). The mode of competition was noteworthy. In-house members were ineligible, so competitions were necessarily "outward looking." Questions were advertised in advance, and anonymous papers with the author's name contained in a sealed envelope were collected over the course of the year. These were judged by a special, paid committee of the academy; cash awards or honorable mentions were assigned; envelopes of unsuccessful candidates were burned; results and new questions were announced at the next public meeting. The advent of prize competitions sponsored by a scientific society was a significant new feature of the organization and conduct of science in the academy type of scientific society. It was taken up.

By all measures, then, the Paris Academy, as it came to operate after 1699 was in a far stronger position than its antecedent organization. It quickly eclipsed the Royal Society of London, and was the most successful instance of institutionalized science after 1700. Whereas prior to the 1660s, organized science was limited largely to private groups attracting little social or government interest or to reactionary, church-dominated universities, after 1699 a major scientific academy operated close to the heart of a largely secular government with seemingly a great deal of social and political support. The deeper ties between institution and government and between science and society were major advances over the seventeenth-century mode for the organization of science. They presaged a new, more potent character for other scientific academies in the eighteenth century that the Royal Society as a model for imitation was hard pressed to match.

CHAPTER III

The Scientific Society
Movement to 1750

Aﬀter 1700, with the Royal Society of London and the Paris Academy of Sciences having assumed their definitive eighteenth-century forms, the scientific society movement began to pick up steam. Figure 4 depicts in its way the growth of this movement from the middle of the seventeenth century until the end of the eighteenth. The decade of the 1660s provides one terminus with the foundation of the Royal Society and the Paris Academy. The 1790s mark another effective turning point, with the closing of all French learned societies in 1793 during the Revolution and the subsequent disruption of other institutions during the Napoleonic period.

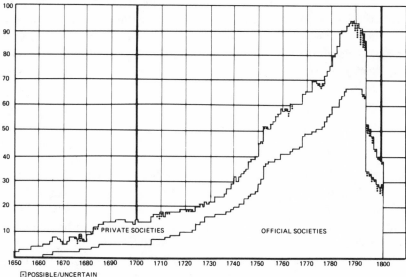

Figure 4: Growth of Academies and Societies of Science: 1650–1800.

Again, this movement was a real thing. As the case of the seventeenth-century "satellite" system of philosophical societies revolving around the Royal Society already makes clear, figure 4 shows not the mere sum of institutions founded but a pattern of interconnected foundations. The institutional model for a planned society was always taken from a successful instance elsewhere, and in many instances the external stimulus was much more direct. The following narrative descriptions will further demonstrate the interplay of outside influences and local circumstances in the creation and unfolding of eighteenth-century learned societies of science.

The growth of scientific societies and the learned society movement is essentially exponential and continuous.[1] There are still stages within this overall pattern, however. Clearly the effects of the scientific society movement in the seventeenth century were not great compared with developments after 1700. As will become clear, there is also a qualitative change in the character of the movement occurring at the midpoint of the eighteenth century. In the first half of the century, with the Paris Academy as the most important example, other major academies of science were founded in cities on the Continent: Berlin, Saint Petersburg, Stockholm, Bologna, and the larger French provincial centers of Montpellier, Bordeaux, Dijon, and Lyon. The net effect of these new foundations was to establish an institutional basis for the system of scientific societies that continued to develop through the rest of the century. In the second half of the century the initial framework of major societies became enlarged by secondary and tertiary institutions, resulting in the full flowering of the scientific society system before the French Revolution.

Berlin

After the Paris Academy, Prussian Berlin became the third major site (after London and Paris) of science institutionalized into learned societies, and the stimulus for developments in Berlin came from Leibniz and from abroad.[2] Leibniz' conception of the institution he hoped to found went beyond anything the foreign models had to offer, however. His great aim was the cultural union of all Germany and the West under the banner of science and a unified Protestant church. His vision

Berlin Society of Sciences, *Miscellanea Berolinensia* (1710), frontispiece with bust of Frederick I.

was universalist, patriotic, and religious, and his idea of an academy or society of science was not something separate from other human endeavors—the sterile pursuit of knowledge—but rather was that of a philosophical church, an evangelical temple, where knowledge and faith worked harmoniously for the betterment of mankind and the greater glory of God.

Through his connections to the Prussian court in Berlin and the ambitious aims of Elector Frederick II, Leibniz at last got a receptive and powerful audience for his plans. In the 1690s Frederick launched a grand program of cultural expansion that included the establishment of a medical college, a beaux-arts college, and the University of Halle. In 1701 his upgrading of Prussia culminated in his elevation as King Frederick I.

Frederick was thus amenable to Leibniz' suggestions for a new scientific institution in Berlin. When the time came, however, Leibniz' entrée was relatively mundane. The calendar was being reformed in Prussia in 1700, and Leibniz proposed a new institution to oversee technical operations in the shift from the Julian to Gregorian calendars. In addition, this proposed learned society was to insure the future production of accurate and scientifically based calendars and almanacs for the kingdom in return for a monopoly on their manufacture and sale. This quid pro quo exchange based on technical needs and available expertise had the benefit of providing Frederick I with good calendars and an institution with high status at no direct cost. The Prussian *Societas Regia Scientiarum* was thus created on May 18, 1700, endowed with an almanac privilege and stiff penalties for those who illegally manufactured or imported calendars into Prussia. The name, Society of Sciences, was chosen because Leibniz feared that the appellation "academy" might cause the institution to be mistaken for a university, so little had the notion of an academy penetrated outward from Paris. Leibniz was appointed the first president of the society.

The *Societas* was not organized directly along the same lines as the Paris Academy; it was more an academy than a society. Its official status in the Prussian kingdom, the privileges it received for the services it rendered, and royal appointment of members makes that much clear. The calendar privilege, in fact, was a major innovation introduced by the *Societas* to the model of learned academies, an innovation that would be taken up as an easy funding arrangement for several other scientific academies. The *Societas* was composed of four different classes: mathematical sciences, physical sciences, German language and history, and literature. This scheme, too, was an important innovation in that nonscientific cultural endeavors for the first time found a home in a learned society alongside science itself. Each class of the *Societas* was governed by a director, and the four directors together with the president and vice-president formed a council to handle the business of the institution. Council government in the *Societas* was reminiscent of the Royal Society and the society model.

The initial plans for the *Societas* were dramatic enough. An observatory was to be built, an astronomer employed to make observations for the almanac, and moneys flowing from sales would fund a laboratory, a press, and other institutional facilities and activities. While fine in theory, practice proved otherwise. It was eight years before a serviceable observatory—the backbone of the *Societas*—was erected, and in the meantime the institution hardly existed apart from its astronomer, Kirsch. Its continuing financial problems—money was drained off by the court despite the agreement—were not greatly relieved by the receipt of an additional privilege/monopoly on the manufacture of silk in 1706. For the moment things looked better in 1710 when the *Societas* was at long last officially inaugurated, provided with quarters for its meetings and work, and began publishing its series of memoirs, the *Miscellanea Berolinensia*. But Leibniz was not renamed president in the 1710 statutes; a court official was. And by then tension was developing between German members of the *Societas* and those who were part of the French Huguenot colony in Berlin. These two factors—court administrators in top positions and conflict between its German and French members—were to characterize and debilitate not only the *Societas* but also the other Prussian scientific societies that followed. After the brief moment of hope in 1710, then, the situation of the *Societas* once again became precarious. Leibniz' leadership of the institution was further reduced between 1711 and 1716 (the year of his death), court appointees came to assume greater control, and the council itself developed an oligarchical attitude toward the rest of the *Societas*. By the time Frederick I died in 1713, little of the early promise had emerged from the weak and court-dominated *Societas*. It was hardly an institution to rival the Royal Society of London or the growing power and prestige of the Paris Academy.

The situation deteriorated further with the ascension to power in 1713 of Frederick-William I, the Sergeant King. Frederick-William I was entirely hostile to the reason and role of his scientific society, and the fate of the *Societas* under his rule can be summed up as continued marginal existence, outright oppression by its source of royal support, and domination by nonscientist administrators. In the end the institution was lucky not to be disbanded altogether. Funds were withdrawn from the *Societas* to support other enterprises deemed more utilitarian, rent was charged for the use of state apartments, and academicians were given the title of "buffoons of the royal court." The only area where the *Societas* made the slightest progress was the creation of an anatomical theater in 1714 and its affiliation to the *Societas*. But the purpose was not to advance scientific medicine, rather to train men for the army medical corps. The *Societas* made a half-hearted turn toward utilitarianism in an effort to preserve itself, but the only real sign of life of the *Societas* in the period prior to the death of Frederick-William I was the appearance of five more volumes of its *Miscellanae* (a total of only seven volumes were forthcoming for the years 1700–1744). And, as Harnack remarks, "The new spirit of science was not too forcefully felt in these volumes."[3] Frederick-William I died in 1740, and from its birth until that date the *Societas* was throughout a hobbled, weak institution with little impact on science or institutional developments outside Berlin. It is best characterized as it was known then: "the anonymous society."

Between 1740 and 1746, however, an institutional revolution occurred in Berlin, which saw the dissolution of the old *Societas*, its replacement by a new *Académie royale des sciences et belles-lettres de Prusse*, and the coming into place of conditions that indeed did make Berlin a major scientific center. Frederick II (the Great) was behind this transformation and the rest of the history of the Berlin Academy in the eighteenth century. Unlike his father, whom he hated, Frederick was enamored of the world of culture and civilization outside barren Prussia, especially things French. Already, as the exiled crown prince, he had voiced his desire to foster a world-renowned, enlightened academy, a cultural jewel to rival Paris and London.[4] The advent of Frederick II augured for change in Prussian science, change based on a new openness to the international community and the Parisian example in particular. On assuming the throne he made his first moves. Originally,

Frederick had wanted Voltaire and Christian Wolff to head his academy, but when these demurred, he settled on importing P. L. Maupertuis from Paris and Leonard Euler from Saint Petersburg. In so doing Frederick brought to Berlin two of the most famous scientists of the day. Maupertuis had achieved a considerable reputation as a Newtonian and as leader of the famous expedition to Lapland sent out by the Paris academy in the 1730s. Euler, in his first years at Saint Petersburg, had gone a long way toward establishing himself as the foremost algebraist and physicist of his time.

Despite Frederick's actions and good will, the disruptions of the Silesian wars prevented more concrete reform, and the *Societas* continued to operate until 1744. In the meantime, Euler and others began assembling as the ex-officio *Nouvelle Société Littéraire*. In the short span of its existence this group met with remarkable success; a large segment of Prussian society participated in its meetings and scientific work. The New Literary Society, supposed Euler modestly, only lacked "an astronomer and a good mathematician to rival the Paris Academy."[5] In a classic instance of institutional reform pitting the insiders of the *Societas* against the upstarts of the New Literary Society—Euler was a member of both—the two groups merged in a fusion of property and privilege, and the new Berlin Academy of Sciences emerged in 1744. With his wars victorious, Frederick officially launched the academy in 1746.

When he arrived definitively in Berlin in 1746, Maupertuis was given free reign as life president to redraft the statutes of the Berlin Academy. Maupertuis used this power to bring over virtually in toto those of his former scientific home, the Paris Academy. Membership in the Berlin Academy was now divided into honoraries, *pensionnaires*, and associates; the privilege of censorship was granted, as in Paris, and the crucial rule of prepublication approbation for works by academicians appearing outside normal academic channels was likewise instituted. The new academy also began sponsoring scientific prize competitions after the manner of Paris, and it published its own esteemed series of *Histoire et Mémoires*. Beyond all that, like its predecessor, the Berlin Academy was organized into science and nonscience classes, restructured this time into the mathematical and physical sciences, belles lettres, philology, and speculative philosophy. The new academy was governed by a council as before; committees handled the separate problems of finance, publication, almanacs, and so on. Following the practice of the New Literary Society, the Berlin Academy used French as its official language.[6] This adoption of French exemplifies further the fashioning of the new academy after foreign models and in part for a foreign audience. Leibniz' ideal of a patriotic and German institution—the *Societas* reflected this in part with its emphasis on German literature, language, and history—was completely lost in the most cosmopolitan, secular academy.

After 1746, then, the Berlin Academy was organized along proven lines. It had the support of a well-intentioned prince, and men of science of world-renown were at its helm. Again like its French sister, the Berlin Academy soon came to pass on inventions and technical processes and to serve as the advisory body to Frederick in scientific and technical matters.[7] An additional privilege on the manufacture, sale, and import of maps was secured for the academy.[8] Along with enlarged markets for almanacs and silk attendant to the acquisition of Silesia, significant new incomes flowed to the academy which, for the first time, put the Prussian institution on a sound financial footing. The academy also enjoyed the largest measure of scientific autonomy yet to be granted in Prussia. Until Maupertuis' death and the Seven Years War Frederick maintained a discrete distance from the academy, allowing Maupertuis, a man of science, to direct the academy's operations with full and unambiguous powers. Maupertuis himself worked hard to raise the academy out of the provincial stupor of the *Societas*, and in large measure his efforts met with quick success.[9] The publication of its memoirs, the election of foreign associates, the activities of its newly elected and dynamic permanent secretary, Samuel Formey, and especially its first prize question (on the cause of the winds, won by d'Alembert of the Paris Academy) at once put the Berlin Academy on the map.[10] By mid-century, Berlin stood just alongside Paris at the peak of learned-society science in Europe.

Saint Petersburg

The Imperial Russian Academy of Sciences founded in Saint Petersburg in 1724 owed an even larger debt to outside, international influences than was the case in Prussia. The Russian Academy was itself only one product of a much larger movement toward the "westernization" of the whole of Russian culture initiated by Czar Peter the Great in the last decades of the seventeenth century and the early part of the eighteenth.[11] Before Peter came to the throne of the Empire in 1682, Russia was essentially isolated from the West; it was dominated by the Byzantine orthodoxy of its independent church, and its scientific attitudes (such as they were) held a "mythical acceptance of the universe as an entity ruled by miracle-working divine caprice."[12] The fruits of the Scientific Revolution had passed Holy Russia. Peter, in an ambitious program that was to last his entire reign, strove to bring Russia into the then modern age. The Church was brought under state control, as were the independent and potentially troublesome noble classes. Emissaries were sent out to assert the claims of empire over the vast territories of Siberia.[13] Peter turned to the West to secure the technical expertise he needed to modernize and industrialize Russia and to facilitate and solidify his enormous undertaking.

Peter himself visited Europe in 1697–98, working incognito in a Dutch shipyard and visiting the Greenwich observatory in England.[14] There he had a discussion with Edmond Halley about encouraging navigation and science in Russia.[15] Between 1697 and 1716 Peter was in contact with Leibniz through correspondence and a series of meetings, and at one point even put him on the payroll.[16] Leibniz, of course, actively sought to persuade Peter to found a learned scientific society in Russia. For Leibniz, Russia was a great untapped resource and a key link between the West and the great cultures of the East, wherein he saw a fertile field for scientific exchange and missionary work.[17] Peter's primary aim was not the creation of a learned society, and for the greater part of his rule he concentrated on importing specialists to run infant industries, such as shipbuilding and mining; he also sought to create technical schools in Russia to train his own experts in navigation, engineering, and artillery.[18] Peter returned to Europe again in 1717, and this time he visited the Paris Academy of Sciences, which, in an extraordinary session convened in his honor, elected him a member "above all ranks."[19] After this visit, Peter thought more seriously about creating an academy, and he consulted Wolff in Halle and Delisle and Fontenelle in Paris about the possibility.[20] After a delay caused by the

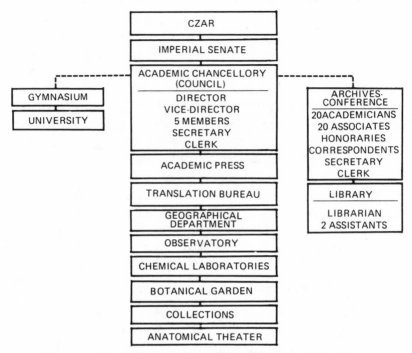

Figure 5: Organization of the Saint Petersburg Academy.

Great Northern War with Sweden (1720–21) and just before his death, Peter capped all of his efforts by approving ministerial plans for the Imperial Academy of Sciences.[21] In this last act, Peter sought to bring over from the West an institution which might continue to guide his work after him. As he said, he built the bridge; it was up to others to get water flowing under it.[22]

The Imperial Academy, as it came to be established and organized after Peter's death, was more than a simple academy in the ways furnished by the examples of the academies in Paris or Berlin (figure 5). It had to be. At a stroke Peter created an academy and a whole educational and support infrastructure to accompany it.[23] In the beginning a gymnasium and university were part of its organization. In the gymnasium and university branches academicians doubled as professors, and students were trained for future service to the crown and for the move to the academy establishment. This additional pedagogical role for members of the academy and for the academy itself was one of the distinguishing characteristics that set the Saint Petersburg Academy apart from other scientific societies.

The Russian Academy was also composed of other departments whose purposes were to aid in the continuing modernization and westernization of the czarist imperium. These, too, collectively distinguish the Saint Petersburg institution. A translation and publishing bureau, complete with its own press and type foundry, was established to bring over scientific classics from the West, to reforge the Russian language into a scientific language, and to publish the only newspaper in Russia, the Saint Petersburg *Gazette* (in Russian and German). An observatory was created to aid navigation and, as in Berlin, to provide an observational base for imperial calendars. A special, geographical department grew up to encourage and coordinate the huge task of mapping the extent of the Russian empire. Likewise, a botanical garden was added to house and catalogue the natural productions brought in from the far-flung Russian territories. In a word, the elaboration of Peter's plan entailed the erection of an entire scientific establishment that in western countries grew up over decades, even centuries.

Included in the organization of the Russian Academy was an academy in the more narrow sense of the term used here. Peter felt more inclined toward the kind of institution exemplified by the Berlin *Societas* over the more narrow scientific model of the Paris Academy, so the strictly academic branch of the academy (called the Archive-Conferences) was initially divided into three classes: mathematical sciences, physical sciences, and the humanities.[24] Ultimately, the Archive-Conferences were composed in that now familiar hierarchical arrangement of twenty academicians, twenty associates, and a larger number of honorary and corresponding members.[25] The academicians met weekly and collectively in the academic library to read, discuss, and deposit papers. The rest of their working time was spent in other departments of the academy. Capping this complex institutional organization was a special administrative bureau (the academic chancellory) composed of a court-appointed director and vice-director and five additional members chosen from among the academicians. The chancellory's duty was to oversee the operations of the entire academy. It was responsible to a committee of the Imperial Senate and thence to the czar. Funds for the new academy, in another of those pay-as-you-go arrangements similar to the financing of the Berlin *Societas*, came from customs duties and license fees assessed in the new Baltic ports acquired by Russia from Sweden.[26]

Conditions of contemporary Russian culture and society clearly had a great influence in shaping this organization. The lack of any kind of science tradition and the virtual nonexistence of an educated middle class meant that the academy had to be fully and completely a government institution; there simply was no other social base in Russia to support the creation of a new scientific institution.[27] This necessity meshed well with Peter's imperialist designs, but it also meant that the fate of the academy was intimately bound up with that of central power and its attitude toward the academy.[28] The deficiency of educational and technical institutions in Russia likewise dictated that the Russian Academy had to assume larger pedagogical and technical functions than was usually the case in western countries. Practically, this same set of realities meant that virtually everything had to be imported from the West, not only institutional models, but also personnel, equipment, scientific values, and procedures.

To staff the new institution, a vast recruiting campaign was launched in Europe.[29] Through newspaper advertisements and the offices of Christian Wolff in Germany attractive offers were made—in the form of five-year contracts—to lure people into academic service in Saint Petersburg. Tax-free incomes, free housing, light and firewood, and reimbursement for travel expenses were sufficient inducements to bring J.-N. Delisle from Paris, Nicholas and Daniel Bernoulli, and later Euler from Switzerland. In all, sixteen Western scholars and men of science along with students (even they had to be imported) initially came to Russia. Because the academy necessarily was imposed from above, and because of the traditional antipathy of the Russian aristocracy toward learning and formally acquired culture, the academy was decidedly a foreign establishment grafted onto backward Russian culture, and it faced many years of slow and often hostile assimilation before it became a completely Russianized institution.[30] By 1725 most of its elements were in place, and the academy opened.

In its first years, the relative success or failure of the Russian Academy depended overwhelmingly on political forces beyond its control.[31] After Peter's death there was a move to quash the budding academy, but it was repudiated by his widow and successor, the Czarina Catherine I (1725–27). The academy's status was not enhanced by the conservative turn taken by her successor, Peter II (1727–30), and his removal of the seat of government back to Moscow. In terms of the political climate, only with the advent of the Czarina Anna (1730–40) did a sufficient degree of political and administrative stability and support allow the academy to develop in relative peace.

Even so, the years through 1740 were difficult for the academy. The various national factions housed within it were often at odds with one another, with the Germans (closely associated with the administration both of the empire and of the academy) generally taking sides against Russian and French members.[32] There were many squabbles with and about the Frenchman Delisle in particular.[33] Language problems also made the work of the academy more difficult.[34] Some exceptions aside, notably Euler, who to his credit mastered Russian, most academicians could not communicate with the Russians in their own language. As a result, Latin was the *lingua franca* of the academy, and Russian students also had to master French or German if they wanted to make any headway in the new intellectual classes the academy hoped to build.[35]

The question of academic administration posed special problems for the scientific academicians of the Archive-Conferences. The academy, although undoubtedly an official state institution, functioned without a charter until 1747. Peter had left a vague set of instructions which gave the academy its status, and Catherine I had amplified these in a set of written rules which informed the academy to a certain extent. But many issues remained undecided and much was resolved in an ad hoc fashion.[36] Peter's initial idea that the academy should be governed by the academicians themselves was passed over entirely, and the academic chancellory came to dominate the academy instead.[37] The director of the academic chancellory after 1730 was without exception a high court noble, and the position proved to be nothing short of an academic czar. Other officers and academicians on the chancellory board were mostly German, and they did not press for greater academic independence. The Archive-Conferences, on the other hand, had no power and could take no step without the permission of the chancellory and its director. Like the *Societas* in Berlin, the Archive-Conferences were without autonomy and were subject to considerable administrative oppression.

Nonetheless, despite the problems facing the early Saint Petersburg

Academy, it achieved a remarkable record of scientific accomplishment in the first two decades of its existence. It published its own set of academic memoirs, the *Commentarii*, which, graced by the works of Euler and Daniel Bernoulli, became a first-rate scientific series.[38] The academy also launched and was responsible for the second Bering expedition (1732–43); this undertaking followed up on the initial work of that explorer and began the continuing tradition of scientific exploration of Russian territories conducted under the aegis of the academy.[39] Concurrently, the academy commenced its long and impressive traditions of astronomical and cartological work. The first general map of Russia was produced in 1734, and in 1745 the geographical department under Delisle published a general atlas of Russia.[40] Other departments of the academy began their operations and met with similar successes. The gymnasium and the university were largely failures, however. The gymnasium had only nineteen students in 1737; the university had none.[41] The academy—meaning the chancellory in this case—soon recognized the utility of western universities as a training ground for its future personnel and began sending its most promising students there for higher education and training.[42]

Despite whatever successes or failures the academy experienced in its first two decades, the lack of a written charter clearly spelling out the various duties and responsibilities of its constituent parts (and hence a document that could be appealed to in instances of administrative arbitrariness) was keenly felt, especially by lesser members of the academic hierarchy who increasingly rebelled against what they construed as infringements on their basic rights as men possessed of scientific expertise. The record is filled with appeals to the crown for settlement of issues of administrative and scientific autonomy in the academy.[43] This was especially true after another period of political instability occasioned by the short-lived reign of Ivan IV and the coup d'état imposed by the Czarina Elizabeth in 1741. Euler quit Saint Petersburg because of it, and Delisle left shortly thereafter.[44] Finally, however, Elizabeth restored a sense of calm to Russia, and the academy finally got its first formal charter in 1747.[45]

In this charter ties between the academy and the university and gymnasium were severed, and the Archive-Conferences were henceforth to deal with strictly scientific subjects, the class of humanities

Saint Petersburg Academy, *Novi Commentarii* (1750), frontispiece with
Peter the Great and Catherine I Ascending.

being eliminated. In the major victory for the reform and scientific elements of the academy, the Archive-Conferences were given the explicit and commanding voice in all matters concerning science. A special "collegio-professorum" was created inside the Archive-Conferences to deal with publishing, foreign correspondence, and scientific questions for all departments of the academy. The chancellory was not to interfere with these activities. Augmenting this new scientific autonomy and following the examples of the Paris and Berlin academies, prize questions were offered for the first time by the Saint Petersburg Academy. If these moves strengthened the position of the scientific academicians, they secured only a small foothold, because the powers and scope of the chancellory and the director over all other academic affairs were likewise confirmed in this charter.

By 1747 several things had occurred. A major academy of science had come to be established in Russia. Despite the internal and external problems facing the Saint Petersburg institution, it had survived and found a place for itself within the larger structure of Russian polity and society. It had achieved a noteworthy record in its efforts to continue Peter's program of westernization, to promote abstract science, and to investigate the Russian realm scientifically.[46] While still dominated by outside, nonscientific forces, the academy had carved out a more autonomous role for itself as a scientific institution. From a Russian point of view, the academy remained a thin veneer of western scientific culture imposed on the comparatively virgin conditions of eighteenth-century Russian society. Conversely, from the point of view of the West, the imposition of this foreign type of institution transformed Russia virtually overnight from a cultural backwater into another major outpost of learned-society science. There was great positive advantage in Peter's plan to build a bridge: the men imported brought with them a host of ready-made European contacts as well as their own expertise. The effect was to make the Saint Petersburg Academy more of an affiliate of the Berlin Academy, or of Paris, than a recognizably Russian institution.[47] One might say that, whereas the academy was physically located in Saint Petersburg, its center of consciousness and real identity lay with its sister academies in the west. The charter of 1747 made the Saint Petersburg Academy a stable beachhead for science in Russia, and the famous first prize question of the academy of 1749 (on whether

[82]

the observed variations in lunar motion agree with Newton's law of gravitation—won by Clairault of the Paris Academy), like the first question of the reformed Berlin Academy, attracted wide attention and did much to enhance the international reputation and standing of Saint Petersburg as a center of science.[48]

Uppsala and Stockholm

In comparison to circumstances in Russia in the late seventeenth and early eighteenth centuries, Sweden had a long tradition of scientific development. The Scientific Revolution came early to Sweden, spurred in large measure by Descartes' settling there in 1649, and soon the new sciences were housed within the medical college in Stockholm and the medical faculty of the University of Uppsala, where a course in experimental physics was taught.[49] Given Sweden's contact with developments in seventeenth-century science, the idea of forming a new scientific society was not far behind. Indeed, there was an early plan for an academy of science originating within the councils of Charles XII of Sweden, but because of the bellicose nature of his reign nothing came of it.[50] Much of the inspiration for the creation of an independent scientific society in Sweden came from England. Several Swedes were in contact with Hooke and Boyle from the earliest days of the Royal Society of London, and the first Swedish contribution to the *Philosophical Transactions* dates from 1666.[51] Erik Benzelius (librarian of the University of Uppsala) and Christopher Polhem (F.R.S., a mining engineer, and "the father of Swedish technology") both visited England, the Royal Society, and Hans Sloane between 1695 and 1700, and they returned to Sweden anxious to create a scientific society in their own country.[52] The English example was obviously an important input to their plans, but they were also stimulated in a negative way by the French. A Swedish astronomical expedition of 1695 to Lapland had come in for criticism from Cassini and de la Hire in Paris; this criticism hurt the national pride of Swedes involved in scientific pursuits, and it motivated them to upgrade their own national science in response.[53]

In 1710–11, Benzelius, Polhem, and a group of mathematicians and doctors from the University of Uppsala collaborated to create the

Collegium Curiosorum. This private association led an on-again, off-again existence until 1728, at which time it was recognized by the Swedish crown as the *Societatis Regiae Scientiarum Upsalansi.*[54] The *Collegium* was the kind of "universal" society envisioned by Leibniz, and was concerned with literature and the humanities as well as the sciences. Its members were part of a particular generation of Swedish thinkers, Cartesians for the most part, and the work of one of them, Emanuel Swedenborg, lent the *Collegium* a slightly mystical and transcendental air. Indicative of the shaping influence of geography on the development of science in Sweden, the Lapps and Lapland were among the major preoccupations of the early *Collegium.* Christopher Polhem drew up an extensive program for scientific investigation of the north that included observation of magnetic, thermometric, air density, acoustical, and optical phenomena unique to that region; in 1711 Benzelius' son undertook an expedition to Lapland under the auspices of the *Collegium* to carry out that program of research.[55]

In 1728 the *Collegium* metamorphosed into the official *Societatis Regiae Scientiarum Upsalansi.*[56] Incorporating the personnel and pattern established by the *Collegium,* the new *Societatis* was a small organization of twenty-four members, vaguely associated with the University of Uppsala. It maintained a broad spectrum of institutional interests and used Latin as its official language. It was recognized but not well funded by the Swedish crown. As undramatic as the appearance of the *Societatis* was perhaps, as Lindroth remarks, "The Royal Society of London and the *Académie des sciences* in Paris, admired foreign prototypes, had now acquired a counterpart on Swedish soil."[57] In testimony to the novel and rich character of Swedish interest in science in the eighteenth century, the *Societatis* was destined to be Sweden's second-ranked scientific society. The major institution, the *Kungl. Vetenskapsakademie,* emerged with another generation of men and a different set of historical circumstances.

The generation to succeed Polhem, Benzelius, Swedenborg, and the others was, on the whole, Newtonian in its outlook, aware of developments in the rest of Europe, and more interested in the utilitarian and practical aspects of science than the staid and strictly academic point of view of the *Societatis.* Martin Triewald, for instance, an engineer and expert on steam engines, lived in England between 1716 and 1726. He

[84]

returned to Sweden to give a series of lectures on experimental physics and Newtonian natural philosophy, and he pressed for the creation of a utilitarian society modeled after the Royal Society of London.[58] Others of this group were Jonas Alströmer, an industrialist; Sten Carl Bilcke, a landed gentleman interested in agriculture; Baron Carl Wilholm Cederheim, known for his interest in forestry and practical agriculture.[59] Among the men more famous for their science were Anders Celsius, Carl Linnaeus, and Samuel Klingenstierna. The latter two had made pilgrimages to England; Celsius had made an even longer journey to Berlin, Bologna, Paris, and London. He became a member of the scientific societies in those cities, and he was able to bring back to Uppsala a rich experience and up-to-date acquaintance with the state of science abroad.[60] More, Celsius returned to Sweden in 1735 as part of the Lapland expedition sent out by the Paris Academy under Maupertuis' leadership. The arrival of this expedition was quite an event in Sweden, and it drew considerable attention from the Court and upper circles of Swedish society to science, to national glory for the support for science, and to Celsius himself.[61] Upon his return, Celsius was especially convinced of the need to invigorate Swedish science.[62]

In 1729 Triewald, as a member of the *Societatis*, proposed the removal of that organization to Stockholm, the adoption of Swedish as its official language, and an increased emphasis on economic and practical matters. In the decade that followed numerous other proposals were put forward for a new science academy.[63] These plans did not occasion any resistance on the part of the *Societatis*; both Celsius and Linnaeus were secretaries of the *Societatis*, and other men pushing for an academy were likewise members of that organization. The idea was not to create a rival to the *Societatis*, but to extend its work into the narrower range of science and economic development and to make the results accessible and beneficial to the larger body of the nation.

The key which led to the resolution of this drive for another scientific society lay in the not quite so extraneous realm of eighteenth-century Swedish politics. After the debacle of the Great Northern War and the fall of the warrior-king, Charles XII, Sweden entered a new period of its history in 1720, the Era of Freedom. The style of government during this time (which lasted until 1772) was that of a titular monarchy backed by a ruling Parliament. In 1739 a factional struggle in the *Rigstad* between the party of the "hats" and that of the "caps" resulted in control of the government shifting to the "hats." This was decisive for the nascent Swedish academy. Marten Triewald's lectures had attracted many prominent "hat" politicians, and he had worked for their support.[64] In particular he had enlisted the enthusiasm and connections of Baron Anders von Höpken, a young peer of Sweden who supported the goals of the loose circle around Triewald, Celsius, and the rest and who eventually became a prime minister under the "hat" party.[65] The "hat" party itself was pro-French, mercantile, and committed to the economic and industrial development of Sweden.[66] It was only natural, then, that when the "hats" came to power in 1739 a new scientific society would result. In June 1739 the Stockholm-based *Vetenskapsakademie* came into being with Linnaeus as its first president and von Höpken as secretary. Its statutes were confirmed by the king in 1741, and from that date it assumed the title of Royal (*Kungl.*) Academy of Sciences.[67]

In surveying the background history of the Swedish Academy of Sciences, it is clear that the Royal Society of London, the Paris Academy, and even the academies of Berlin and Saint Petersburg exercised a certain influence in its creation, but it is likewise evident that the Swedish Academy was to be very different from any of these, if only on account of the interest expressed in utilitarian and economic goals.[68] Indeed, in a title more reminiscent of the economic societies just beginning to spread across Europe, the suggestion was initially offered to call the academy the "Economic Scientific Academy for the Promotion of the Useful Sciences," but this proposal was vetoed by Celsius in favor of the more traditional "Academy of Sciences," to emphasize its debt to Paris.[69] This character of immediate utilitarianism, which grew out of the local conditions surrounding its birth, was not the only feature differentiating the Swedish Academy from the institutional models in other European countries.

The Swedish Academy was organized in a unique fashion, also. Inside the academy features characteristic of both academies and societies can be found.[70] As in the typical institution of the society type, there were no organizational divisions into grades of membership or disciplinary specialties. Yet membership was limited (to one hundred), as in academies. Despite the undifferentiated quality of academic membership, there was a class of students attached to the academy who sat on back benches in academic assemblies like adjuncts in Paris. Like a society, there was no outside control over the election of members; and taking the democratic principles of the model society to extremes, initially officers were chosen by lot after a preliminary election. By contrast, the position of secretary was paid and permanent.[71] Until 1747, too, funds for the operation of the academy were drawn from dues paid by members.

Turning to the ways this peculiar organization fitted into the rest of Swedish society, the situation becomes even more complex. On the one hand, the Swedish Academy was "independent" like the Royal Society in London. It was not a government institution in any formal sense, and it was not tied into other bodies of government administration like other European academies. Yet, from 1739 it was endowed with a special privilege to publish, and in 1741 it was given its royal designation.[72] Furthermore, the academy was called upon to issue reports on various scientific and technical questions and to judge inventions for the *Rigstad* and other government bodies. This it did by forming committees and writing reports which were signed by the secretary in the name of the academy, contrasting with the practice in the Royal Society of London where reports were issued in theory only on the backing of the men who wrote them.[73]

The academy was further schizophrenic with regard to its commitment to science versus economic development. It certainly did see its role as being that of an economic society dedicated to the national interests of Sweden.[74] The Stockholm Academy issued a trimestrial journal, the *Handlingar,* more like the *Philosophical Transactions* of the Royal Society than a series of academy memoirs. To be of greater use, the *Handlingar* were published in Swedish, and over half the material contained in them concerned economic development and agriculture.[75] Similarly, the academy offered prizes, but unlike the prizes offered by other academies, theirs were strictly utilitarian, for which only medals or small cash awards were presented.[76] On the other hand, pure science and pure scientific research did not long remain a subsidiary concern of the academy.[77] Persons such as Linnaeus, Klingenstierna, or Celsius, although they supported the academy's more strictly utilitarian aims, had their proper interests in botany, mathematics, and astronomy, and,

Emblem of the Swedish Academy of Sciences.

being no mean practitioners of these sciences, they raised the level of the academy's scientific work to internationally respected heights. This, of course, was the other face of the Swedish Academy, an internationally admired and sought after academy known for its science.[78]

Between 1747 and 1750, roughly a decade after its first foundation, the *Vetenskapsakademie* underwent a series of reforms and happy turns of events which greatly enhanced its power as an institution and in the course of which its scientific importance was truly established in the minds of other Europeans. In essence the academy shed many of its "society" characteristics—like dues—and assumed more the character of an academy type of learned society. Royal support for the academy jumped markedly in 1747, when it was granted a monopoly over calendars and almanacs (like the Berlin Academy) in preparation for the shift to the Gregorian calendar in Sweden.[79] This decisive privilege freed the academy from dependence on dues. It provided a regular and firm source of funds for the activities of the academy, and it involved the academy even more heavily in its quasi-official role as a state institution. Along with this change, the academy became formally involved in astronomical work. It built an observatory (still to be seen in Stockholm) and began a regular program of astronomical observation. In 1747, too, the first foreign associates joined the academy; they were elected into disciplinary specialties and assigned to individual academicians, as in Paris.[80] Of great luck to the institution was the election in 1748 of Petr-Wilhelm Wargentin as its permanent secretary. One of the great academic secretaries, Wargentin was well liked, an able administrator, and a trained and talented astronomer. He fitted in, and made the academy his life-long home.[81] Among his first acts as secretary Wargentin moved to increase the visibility of his institution in making contact with other learned societies of science, and in this effort he achieved a sparkling success.[82] Wargentin's other early accomplishment was his tripling the price of academic calendars and almanacs; the consequences are obvious. All of these various developments put the Swedish Academy of Sciences in an enviable position by mid-century. Well integrated into the fabric of Swedish society from its beginnings, the Swedish academy was more firmly established than ever by 1750. Considering the size and resources of Sweden, the *Vetenskapsakademie* existed on a remarkably high level internationally, comparable to that of the science academies in Paris, Berlin, and Saint Petersburg.

The French Provinces

In surveying the course of the scientific society movement in the first half of the eighteenth century, one confronts its major manifestation in the academies of provincial France.[83] Between 1700 and 1750 some fifteen academies of science and belles lettres arose in the French provinces.[84] It is beyond the scope of this enterprise to enter into details for each provincial academy founded in this period, but in a way it is not necessary because eighteenth-century French provincial academies formed a relatively coherent group. They were united by language and culture; they were outgrowths of a common institutional movement; they each received recognition from the central monarchy in Paris. Their differences, as, for example, the size or importance of their towns or how many churchmen, nobles, or bourgeois made up their social compositions, are differences of degree more than kind. By and large, they were similar institutions with characteristic features and analogous histories. Rather than discussing each academy separately, it suffices to underscore common traits.

Looking at France and its academies in the first half of the eighteenth century, three major points stand out.[85] One, compared to the seventeenth century, the number of new provincial academies increased dramatically. More than twenty new institutions made their appearance in the period extending to 1760, up from less than half a dozen existing before 1700. And some academies which had languished in the seventeenth century were revitalized.[86] Such an increase indicates that the learned society movement in science was making remarkably strong headway in the obviously fertile fields of the French provinces.

Second, in the course of this development, the major centers of provincial France became learned society cities. While weak and out of the way academies were founded in this period (e.g., in Pau), and, while later the entire spectrum of French cities and towns became dotted with academies of various grades, major provincial academies in the major metropolises of Bordeaux, Lyon, Montpellier, Toulouse, and Rouen came into being comparatively early. By mid-century the academies in Paris were backed up by a network of other academic societies in towns occupying the second level of cultural and economic importance in France after Paris.[87] This was a significant expansion for the learned society movement in France, which upgraded the academic and institutional power of the nation as a whole.

Third in this connection, science became a significant concern of French provincial academies for the first time in this period. (Recall that seventeenth-century French academies were almost entirely literary.) How much provincial academies involved themselves in science varies. Some, like those in Montpellier, Bordeaux, or Brest, were exclusively or to an overwhelming degree institutions of science; others, like those in Caen, Montauban, or Nancy were still devoted more to belles lettres and literature. Some of the seventeenth-century academies which had earlier concerned themselves with literature and belles lettres became decidedly more oriented toward science at this time. The typical academy founded in this period combined both sets of interests, being usually known as an academy of science and belles lettres.[88] Taken together, these developments made France the nation best endowed with learned and scientific societies.

Unlike the national science academies in Paris, Berlin, Saint Petersburg, and Stockholm, where the initiatives behind the formation of institutions stemmed more from the top, French provincial academies emerged from below in ways more like the Royal Society of London. Provincial academies seemingly without exception had their roots in local, private associations where men gathered outside traditional institutional and social settings to engage in new forms of sociability characteristic of the age (concerts, poetry readings, discussion groups).[89] In each instance, private assemblies were felt to be unsatisfactory. They lacked permanance and an official status, and in some cases they left the door open to charges of political dissent or libertinism. As a result, private provincial groups invariably strove to secure letters patent for themselves. Requests from the provinces for letters patent were not immediately honored at Versailles, however, and potential academies all suffered probationary periods of anywhere from two to twenty-five or even thirty years before letters patent were forthcoming. The successful step from private group to official academy was generally taken with the help of someone important and close to central power: the Duc d'Orléans and the Orléans Academy, for example.[90]

The registration and receipt of letters patent held two decisive consequences for any proto-academy. For one, they created a corporate institution and brought the academy into the large corporate structure of the French state. No longer a private association, a newly formed provincial academy moved into a specified niche alongside other government bodies and became a legally recognized institution in the public domain.[91] And then, the official act stemming from the highest sources of power in the nation also elevated the men who initiated action on a local level to a special position within their local social context. Their efforts were recognized and rewarded as being of importance to central authority, and they became entrusted with the running of the institution.

The more prevalent organizational form for French provincial academies was that of the Paris Academy of Sciences. Another, smaller segment of French provincial academies based their organization on the model of the *Académie française* where no internal divisions separated the body of equal members. But this form was more common in the seventeenth century than in the eighteenth. Beginning in the period of concern here, the model adopted from the Paris Academy of Sciences became forcefully felt and was taken up by the majority of provincial academies. The basic organizational structure of these academies was composed of an overseeing body of officers (usually called directors) including a permanent secretary, a category of high-ranking, sometimes paid members (usually called ordinary members), an affiliated group of honoraries, a class of associates or adjuncts, and finally foreign associates (meaning in the case of provincial academies anyone not resident in the town) and correspondents. Provincial academies were not segregated into classes (sciences versus belles lettres), but met in common (usually weekly) in the ritual calendar of meetings that aped the Paris Academy: papers were presented, new members elected, reception speeches heard, public meetings held, vacations taken, and (where funds permitted) memoirs published and prizes offered.

While on the surface provincial academies appear as miniature duplicates of the larger Parisian academies (and the Paris Academy of Sciences in particular), they differed from their illustrious counterparts in several important ways. They did not possess any special institutional privileges, for example. And, although they formed part of and were recognizant of the larger monarchical structure of France, the horizon of their vision and the scope of their activities were limited for the most part to the provincial level. Of great significance was the fact that the provincial academies were not strongly supported financially from the central regime. Some exceptions aside, the mean funding they received from the crown was a token contribution of six hundred livres annually.[92] Compared to the national academies in Paris, there was thus less provincial academies could do to advance their own causes or that of science or literature. Publishing and prize giving were very irregular in the provinces, especially in the period before 1750, and in some instances academies had recourse to dues collecting to maintain an income for their operations.[93] In these circumstances provincial academies could not offer financial incentives to lure people and talent into even a part-time career as an academician, as was the case in Paris and elsewhere abroad. Provincial academies remained overwhelmingly amateur in their membership 'and orientation throughout the eighteenth century. For most men membership and participation in a provincial academy was a social avocation to be taken up in a spirit of continuing sociability after their primary, income-earning responsibilities had been fulfilled. In the Montpellier academy, one of the most prestigious of the provinces, the point was raised in the 1760s that "one knows well enough that the simple title of academician does not always bring with it a station in the world in the provinces; everyone has his particular profession to which are attached a thousand indispensable obligations; academic duties come only afterwards."[94]

Because they were not adequately funded from Paris and because anything more than an amateur class of academicians was an impossibility for them, provincial academies deviated from the national academies in Paris and elsewhere in that they aligned themselves with more local powers and institutions. Given the provincial horizon that circumscribed most academies, the town or province in which they found themselves loomed as an accessible source of funds and a promising field for action. Local nobles, provincial estates, provincial governors, regional intendants, and city governments became important sponsors of academies and academic activities. For the immediate stability gained from better funding, the local role of academies grew correspondingly. No longer private groups or corporate elites recognized from on high, provincial academies became civic institutions.[95] They sponsored and judged prize competitions (ofttimes of only local interest), and undertook projects for local and regional authorities. Slightly later in the century they tended to offer public courses in the sciences for local audiences, and insofar as they were able or called upon, they served as advisory bodies. The elaboration of this local, public service role was facilitated by a pattern of floating local elites and interlocking memberships by which funds were secured and policies advantageous to the particular academy established.[96] More likely than not, for example, local mayors were academic members, and in many cases academies convened in public rooms in city halls. In the extreme case of the Besançon academy, the mayor, provincial governor, archbishop, and first president of the local parlement were all ipso facto academicians (*académiciens-nés*).[97] In the end, while not as strong as the national academies in Paris or elsewhere, academies of the French provinces exercised analogous roles on the provincial levels. They stood in the same relation to local and regional powers as Parisian academies did to the government in Versailles and Paris.

The scientific and technical capabilities and accomplishments of provincial academies, especially in the first half of the century, were not that great. By the same token, they were not nil. For example, in work undertaken for provincial estates the academies in Bordeaux and Montpellier produced maps and natural histories of their respective provinces of Guyenne and the Languedoc.[98] These and lesser undertakings in other academies were of consequence for the development of cartology, geology, and natural history in several regions of France. At least half the time spent in ordinary meetings of provincial academies was devoted to reading papers, and the roughly one thousand scientific papers extant in learned society archives in provincial France demonstrate a widespread interest in medicine, natural history, and other sciences.[99] On the other hand, provincial academies by and large were not strong in mathematics or what we know as physics. Prize competitions did extend the range and power of provincial academies. Although only five institutions offered prizes before 1750, scientific subjects, sometimes taken over from the Paris Academy or other provincial academies were common.[100] The questions of the Bordeaux Academy dating from 1716 were exclusively of a scientific nature and were major competitions in the first half of the century.[101] In posing prize questions, provincial academies overcame their particularism and their closed, elitist social base and opened themselves to fresh scientific input from all over France and, indeed, the international community.[102] Provincial academies also typically had their own *hôtel*, botanical gardens, observatories, collections of various sorts, and laboratories. The establishment of public libraries under learned society supervision or, later, the inauguration of public courses in the sciences added to the scientific potentialities and credit of these institutions.[103]

While little in the history or nature of the academies of provincial France (or, for that matter, provincial academies anywhere) changed the course of science in the eighteenth century, provincial institutions made the academy of science not just an illustrious phenomenon to be found in Paris, but an everyday fact of the social and institutional life of the whole country. The provincial academies introduced science and science education to a larger audience and stimulated a mass of activity that made France by comparison seemingly seethe with scientific productions. They established a de facto institutional network, virtually complete by mid-century, which promoted on a secondary level the scientific work of the more important academies and made the latter that much more prestigious and avant-garde. This network of institutions is important evidence for the emerging system of learned scientific societies. The institutions in place by 1750 formed the backbone of the inter-organization of the French academies and an important element of the whole international system of scientific societies. In addition, for men like Jean Astruc of Montpellier or, later, Guyton de Morveau of Dijon provincial academies provided a stepping stone to scientific careers in Paris.[104] In all, provincial academies formed a formidable institutional base on top of which the Parisian academies and especially the Academy of Sciences stood in resplendent glory.

A few of the academies of provincial France deserve a brief separate mention. These were especially noteworthy as scientific societies. They were comparatively well supported. They were known on the international level, and they approached in stature the major national academies in France and elsewhere.

The most outstanding provincial academy in France was the *Société royale des sciences* of Montpellier.[105] It was an early creation (1706), and it occupied a unique position in the constellation of French academies. Montpellier, of course, had long been an important center for science and learning in France. Its university dated from the Middle Ages, and from the seventeenth century at least the medical school there was especially well known.[106] The medical school was liberal and innovative, and by long tradition it supplied doctors to the royal court at Versailles. This special tradition meant that the medical faculty and Montpellier in general had solid connections to the court and to northern centers. These connections made Montpellier less the provincial town it might have been, and gave Montpellier a reputation as a southern counterpart to Paris itself.

The creation of an academy in Montpellier in the first decade of the eighteenth century reflected and reinforced this longstanding tradition and reputation. In 1702, just after the reform of the Paris Academy of Sciences, Domenic Cassini and a suite of relatives and academicians came to Montpellier as part of the cartological project to map France being carried out under the direction of the Paris Academy. They encountered an informal group of men interested in science, and they discussed with them the events of 1699 affecting the Paris Academy and the possibility of establishing an academy in Montpellier. Cassini returned to Paris with hard plans and pressured the Abbé Bignon, the man responsible for the reform of the Paris Academy, to secure letters patent for the Montpellier group. With such friends operating behind the scenes, letters patent were not long in forthcoming. When they did arrive in 1706, they held some surprises.[107] For one thing, they created an institution virtually identical to the Paris Academy. The letters patent were framed almost clause for clause after those of 1699 for the Paris Academy, the only organizational differences between the two institutions being a slight difference in scale (four foreign associates for Montpellier in lieu of six for Paris, for example). Even more outstanding, the Montpellier academy was not just fashioned after the Paris Academy, they were actually affiliated. The letters patent for Montpellier were presented to the Paris Academy before they became official, and in the register for the meeting for March 13, 1706 one reads:

The King orders that the Society of Montpellier be regarded as an extension and a part of the Academy of Sciences [of Paris]; that they undertake a commerce over the matters that occupy them; that the academicians of one be seated in the other, each according to his rank; that the Society of Montpellier send each year to the Academy the paper read in its assemblies during that year adjudged the most worthy and suitable; and that paper be printed in the memoirs which the Academy will publish for that year.[108]

In not the least of ways, this key quality of being (to use the stock, contemporary phrase) of "one and the same body" with the Paris Academy distinguished the Montpellier Society from all other provincial academies before and after it and made Montpellier that much more special and important. The bond between the academies in Montpellier and Paris formed the basis of the organization of science on a national scale in France in the eighteenth century, and it was a major component of the international structure of scientific societies taking shape in the course of the first half of the century.

But the Montpellier Society did not suddenly transform itself into a high-powered scientific institution as Paris had done six years previously. Much of the prehistory of the Montpellier Society fits into the general pattern described above as typical of provincial academies. Similarly, no large funding was forthcoming from the crown, and the society turned to the estates of the Languedoc and the town of Montpellier for additional support. Over the course of years, it succeeded in this regard better than most other provincial academies, and it became an important local institution well equipped with an observatory, laboratory, botanical garden, and collections. Still, the Montpellier Society was comparatively a modest institution that could not support academicians or undertakings in the way the academy in Paris did. Indeed, the papers submitted by the Montpellier academicians for publication in Paris were sometimes criticized by their more professional northern colleagues and in one instance returned for revision.[109]

There was a town-and-gown air to the membership and character of the Montpellier Society, in that many ordinary members taught in the university, while other posts were occupied by local notables and officials. In its early years the society was especially jealous of its privileged connection to Paris. This explains its active efforts to thwart the establishment of new academies in neighboring Béziers and Toulouse.[110] There is no special break in the history of the Montpellier Society at mid-century. It was not until the 1760s and 1770s, however, that the Montpellier Society reached the full peak of its power as an institution of sciences.

After the academy in Montpellier, the next important French provincial academy was the *Académie royal des sciences et belles-lettres* of Bordeaux.[111] The Bordeaux Academy was an early foundation also, receiving letters patent in 1712, and the institution was heavily committed to science.[112] But, unlike its counterpart in Montpellier, the Bordeaux Academy quickly developed a standoffish and independent attitude toward the central government and the Paris example. It was much more the centrifugal institution. The Duc de la Force was the power behind the foundation of the Bordeaux Academy, agitating successfully for its letters patent and endowing the academy sufficiently for an exceptional financial independence. The academy and its members were in conflict with the king's provincial representatives on several occasions, and its real sources of support, in addition to those from

La Force, were town and provincial authorities.[113] It organized and ran a big public library, published six volumes of its natural history of Guyenne between 1715 and 1739 and *Mémoires* between 1741 and 1749. The Bordeaux Academy was perhaps most influential in its series of scientific prizes, the earliest scientific series in France and one of the most extensive in all of Europe in the eighteenth century.[114] From 1716 the academy offered two prizes annually for questions of *physique* and natural history worth 300 livres each, and it was the Bordeaux example that prompted de Meslay to bequeath funds for the same purpose for the Paris Academy of Sciences.

The academies in Lyon and Dijon, although not quite on a par with those in Montpellier or Bordeaux, were still important learned society centers. The Lyon Academy received letters patent in 1724 with the backing of the local archbishop, but Trudaine quickly took over as court-connected sponsor.[115] The king was the patron of the academy, and the *prévôt de marchands* of Lyon provided funds for *jetons* for the Lyon academicians. The organization of 1724 merged in 1756 with a beaux-arts academy which arose separately, and took the name *Académie royale des sciences, belles-lettres, et arts.* But the academy was very committed to science.[116] It came to have a notable library and natural history collections, and it, too, ran prize competitions. It was active scientifically in developing plans for a natural history of Burgundy, in inquiring into the safety of vaccination, and in investigating Mesmer's claims. The Lyon Academy was one of the first of several academies to subsidize early ballooning trials.

The Dijon Academy, while similar in many ways to the academy in Lyon, was probably of slightly more note as a center for science.[117] Its origins were different from other academies, in that it began in 1725 as a private academy, but with paid academicians and prize competitions for them exclusively. These establishments were funded through the will of the Dijon parlementarian Pouffier. Official letters patent came in 1740, with Montesquieu playing the role of legal engineer. Later in 1757 the nascent academy fused with a rival literary society dominated by other parlementarians, and the Dijon organization became the *Académie des sciences, arts, et belles-lettres.* In the second half of the eighteenth century the academy was dominated by three strong leaders: Richard de Rouffey, Hughes Maret, and Guyton de Morveau, but throughout its history it was torn between its bourgeois members of more scientific bent and more literary parlementarians. Still, science prevailed, and the academy had the standard cabinets and gardens, it published memoirs, it offered prizes almost entirely on scientific subjects, and it taught science-related public courses for the provincial estates. Its leadership seems more active than in Lyon, and it seems to have been more open to other science societies elsewhere in Europe.[118]

The *Académie royale de la marine* in Brest (1752/1763–93) is another provincial academy of note for science. The Brest Academy was the least indigenous of the provincial academies, being largely implanted on that naval center and completely supported by the central government as part of a program of research and training in navigation and hydrography. The marine academy was entirely devoted to science and the "arts," with an interest in astronomy much higher than was typical of provincial academies.[119] Ties to Paris were close, and the Brest Academy has been called a "laboratory" of the Paris Academy.[120]

The academy at Marseille (1726) began as a belles lettres academy, but it turned more to science after 1740, and was reformed as the *Académie des belles-lettres, sciences, et arts* in 1766.[121] It was supported by the town government for ordinary expenses and for publishing its series of *Recueils*. Thirty-one volumes were published between 1726 and 1767 and another fourteen between 1768 and 1786. Science came to predominate in the latter series, and the Marseille Academy also sponsored over one hundred prize competitions; science was the subject of two-thirds of the questions posed. Funded with the usual 600 livres from Paris and supporting fifteen paid *sociétaires*, it is noteworthy that after 1781 the Marseille Academy was charged by the ministry of the marine with running the royal observatory at Marseilles.

In all, these examples and other lesser ones indicate the extent to which the model of the learned society penetrated into the provincial level in France in the first half of the eighteenth century and how much provincial societies became centers of scientific thought and activity.

Bologna

The scientific society movement achieved a notable success in Italy in 1714 with the creation of the Academy of Sciences of the Institute of

Bologna.[122] The origins of this institution in part lie in the tradition of Italian "Renaissance" academies discussed earlier. From the 1650s through the 1690s a number of short-lived "Renaissance" academies concerned with experimental science succeeded one another in Bologna: the *Coro Anatomico*, and the academies *della Traccia, del Davia, del Arcidianono*, and *degli Inquieti*. These all took the defunct *Accademia del Cimento* of Florence as their model. Something of the quality and vitality of "Renaissance" academies in Bologna seems to have slipped as the seventeenth century progressed. Richard Rosen characterizes the early *Accademia degli Inquieti* (founded in 1690 and meeting in the home of its patron, Eustachio Manfredi) as little more than a teenage club interested in experimental science.[123]

Another important element in the prehistory of the Bolognese Academy is the University of Bologna, another venerable university like that in Montpellier. Pressures for University reform were mounting in the 1690s. On the one hand, by long tradition any citizen of Bologna could lecture at the university, and as a result the university was flooded by an overwhelming, impossible number of lecture courses. On the other hand, the new science, particularly experimental science, had made little headway within the university as the seventeenth century drew to its close. Reform was in the air, and it was spearheaded by men who took an active interest in science. As outsiders, Malpighi, G. B. Morgani, and others sought to exert pressure by constituting themselves as an organized group, which they did by taking over the poor *Accademia degli Inquieti*. They revitalized that organization to the extent that "by the middle 1690s, there was hardly a scientist either professional or amateur in Bologna who signed his name without affixing the title 'Inquieti'."[124] Something fundamental happened in 1704 when the *Inquieti* adopted written rules and a constitution from the recently reformed Paris Academy of Sciences. Although it was not recognized by any public power and was still a private association, the *Inquieti* took a giant step away from its "Renaissance" background and tapped the new learned society movement to the north.

These developments within the *Accademia degli Inquieti* attracted the attention of a man who likewise had an international outlook and was interested in university reform for Bologna, Luigi Marsigli. A member of the Vatican diplomatic corps and familiar with institutional develop-

Bolognese Institute, *Commentarii* (1731)
frontispiece.

ments across the Alps, Marsigli became the prime supporter of the embryonic academy and the main instrument of its final success.[125] In 1707 he was in contact with the new Montpellier Society, and he wrote with hope of "the house of Count Marsigli which might well become a true academy in the future; we are constantly working toward that end."[126] From his high social and political position, he proposed to the Bolognese town senate in 1709 a plan for university reform that included the establishment of an academy of science. The plan was debated in 1709, again in 1711, and was finally approved in 1714. An academy of science was created in Bologna.

Marsigli's plan was primarily directed toward the reform of the university. At the heart of this plan stood the Institute of Bologna, a new scientific branch created parallel to existing university divisions. Within the institute five new professorships were established: astronomy, architecture, experimental physics, chemistry, and natural history. In a pedagogical breakthrough, institute professors were expected to teach in new modes, using demonstration, experiment, and practical work as aides. They were also expected to do scientific research, and each was provided with a laboratory or observatory. With the creation of the institute, science became well established in the university, and the problems of outmoded and overloaded instruction were largely alleviated.

The Bolognese Academy of Science was created additionally as an appendage of the institute, to be a learned society connected to it and enlarging its scope. The academy proper was thus integrated into a larger pedagogical institution and university, like the Archive-Conferences of the academy in Saint Petersburg later. The Bolognese Academy did not stand on top of the institutional structure established by Marsigli's reform, however, but very much to the side. It was organized around a core of academic members, the twelve *ordinarii*, by statute made up of the paid institute professors and other men from the university. *Ordinarii* were divided into disciplinary specialties and provided officers for the academy. They were augmented by an equal number of *honorarii*. The *numerarii* formed the equivalent to a class of associates, and *alumni* were the student or adjunct class. The academy was jointly funded and administered by the Bolognese senate and the Vatican, Bologna being part of the papal states in the eighteenth century. With such sources of legal and financial support, the model from Paris and the *Inquieti* was transformed into another official and typical eighteenth-century academy (decidedly not a "Renaissance" academy), and the Pope joined the growing list of absolute monarchs who sponsored such institutions.

Between 1714 and 1745 the academy of Bologna faced several problems which hampered its effectiveness and achievement. It was not immediately accepted by all segments of the university and town communities, and it became embroiled in complicated legal suits over its status. As a result, both material sources available to the academy and the activity of its members remained on a very low level through at least the 1720s. The academy was strengthened by the appointment in 1723 of Francesco Zanotti as its permanent secretary. Zanotti was yet another outstanding member of that illustrious fraternity of academic secretaries, and from the date of his appointment he assumed an almost single-handed leadership of the Bolognese academy. He exhorted work from academicians, took on the task of publishing the memoirs of the academy, and established and expanded contacts with other centers of science to the north. Regarding its memoirs, the *Commentarii*, the academy encountered a problem peculiar to its Italian and Church settings, in that volumes had to be approved by the Inquisition prior to publication. This incursion into academic freedom explains the eight-year delay in publishing the first volume of the *Commentarii* and a similar fourteen-year delay for the second. The fortunes of the academy were enhanced somewhat by the definitive decision of a papal tribunal in 1726, which laid to rest questions concerning its official status, but through the 1740s, Marsigli's grand creation fell far short of the hopes initially invested in it.

The Bolognese Academy underwent an important, revitalizing reform in 1745. This reform was effected by Pope Benedict XIV, a man long associated with the interests of the academy (he served on the papal commission of the 1720s) and particularly attracted to the example of Frederick II of Prussia and his efforts to upgrade Berlin as a center of culture and science.[127] The Pope endowed the Bolognese Institute-Academy with one of the largest libraries of the period, and greatly increased its funding by closing a college for the poor. In the most important move, Benedict created a new class of academicians within the academy, the *benedettini*. The *benedettini* were paid both annually and for regular attendance at meetings, and they were expected in return to produce research papers. The class of ordinary members of the academy essentially merged with the *benedittini*, and after a while the academy became functionally divided into those who were and were not *benedettini*. The long-term effects of this reform proved detrimental to the academy's power as a scientific institution, as Rosen points out, in that the requirement of *benedettini* to produce lowered the quality of the academy's work, but at mid-century the Bolognese academy occupied a stronger position than it had previously, and the prospects no doubt looked bright for its continued success. It was well established in its local setting, it had acquired a new level of funding and support, and it was geared up for the active production of scientific research. By the same token, from an international point of view, the academy of Bologna was not as forceful an institution at mid-century as one might have expected. For reasons that are unclear—perhaps because of is place in the institute and the university, perhaps because the Bolognese scientific and scholarly communities felt complete unto themselves—the Academy of Sciences remained fairly provincial in its orientation (it did not offer prizes, for example). Its members were not very famous or productive, and it was not much more forcefully felt on the international level than, say, the Montpellier Society or other major French provincial academies. Something of this feeling for the Bolognese Academy was captured by a slightly later visitor who wrote, "the Institute [of Bologna] is one of the liveliest ornaments of the town and the memory of the late Pope Benedict XIV by whose efforts this Academy finds itself so admired. It seems that the sciences are well liked here; at least they are protected."[128]

Likewise in Italy this early period of the scientific society movement saw the creation of the *Academia Botanica* of Florence and the beginning of its operations in 1739. The *Academia Botanica* is of particular note as a specialized academy devoted to the single disciplines of botany and natural history. Noteworthy, too, is the fact that the academy was recognized and supported by the Emperor Francis I of Austria, making it an early case of Austrian support for scientific societies. The status of the Florentine botanical academy is unclear, although its foreign members were the elite of eighteenth-century natural history and the life sciences.[129]

Developments Across the Channel

As it turned out, the learned scientific society movement made very little headway in the English-speaking world in the first half of the eighteenth century. The Royal Society of London was, of course, an established scientific society of international repute, and it maintained its existence and reputation as England's foremost learned society while developments on the Continent unfolded. The Royal Society stood alone, however, as no other chartered societies (or academies) were created outside the Continent in the first half of the eighteenth century. What few developments there were in Great Britain and her colonies in the period by and large adhered to the seventeenth-century pattern of institutional growth.

In England itself, after the closing of the Oxford Philosophical Society in 1690, the tradition of private associations, satellite to the Royal Society in London continued with the gentlemen's societies of Spalding (1710) and Peterborough (1730).[130] These Midlands reunions had an even less formal character than the seventeenth-century philosophical societies. Like them, however, the Spalding and Peterborough societies saw themselves as extensions of the Royal Society, and both sent the minutes of their meetings to the London institution for inspection.[131] All things considered, however, these societies are of comparatively little note.

Something of this tradition continued in a new setting and with new force in Scotland. The University of Edinburgh was the important center.[132] Administered by the town of Edinburgh, the university attracted students (particularly dissenting students from northern England) by offering religious toleration, up-to-date instruction, and progressive teaching methods. As the eighteenth century progressed, the University of Edinburgh along with other Scottish universities eclipsed the closed and retrograde universities of Oxford and Cambridge and even Leyden as the foremost centers of higher education attended by English and Scottish students. In the shadow of Edinburgh's vital university, then, scores of clubs and student groups came and went. A somewhat more serious, but short-lived philosophical society did exist in Edinburgh from 1705.[133] In 1731 the less casual Society for the Improvement of Medical Knowledge was established by Alexander Monro I.[134] Centered around the medical faculty, this group of Edinburgh doctors held meetings, heard papers, and published two volumes of *Medical Essays and Observations* through 1739. These *Essays* were transmitted to the Royal Society.[135] In 1739, on the urging of the mathematician Colin Maclaurin, the Edinburgh Medical Society expanded the range of its interests to include science and literature, and became the Philosophical Society of Edinburgh. In the early 1740s under Monro, Maclaurin, and James, Earl of Morton (later P.R.S.), the Edinburgh Philosophical Society was a flourishing enterprise, but it still had no charter or legal standing, it did not publish at that time, and it was shut down by the Jacobite uprising in Scotland in 1745. It did, however, seek the protection and patronage of the Royal Society of London in this period, but a certain change from the seventeenth-century pattern of relations between the Royal Society and groups seeking its support is to be noted.[136] The deference toward the Royal Society remained, but a new center was emerging that did not depend so overwhelmingly on the Royal Society for institutional leadership. The presence of the University of Edinburgh established a cultural context that nurtured an independent tradition of Scottish scientific associations. In the long run, after the Edinburgh Philosophical Society was revived, disbanded, and revived again, this separate tradition bore fruit in the official Royal Society of Edinburgh (1783).

Something similar was happening in Ireland. Just beginning in the 1740s, a small "physical-historical" society grew up in Dublin around Trinity College.[137] Like its seventeenth-century predecessor, the Dublin Philosophical Society, this organization maintained contacts with the Royal Society in London and cultivated a "mutual correspondence" and "friendship" with it. At the same time, there was a nationalistic upsurge in Irish interest in economic and cultural development which spawned the separate tradition of economics and arts societies.[138] Little of note for the organization of science developed in the first half of the eighteenth century in Ireland, but out of this general background the Royal Irish Academy eventually emerged in 1785.

Finally, the situation was not altogether different in the American colonies. After the collapse of the Boston Philosophical Society in 1685, the next plan for a scientific society was put forth by Cadwallader Colden in 1728, again for Boston.[139] It was still too early for such a plan to succeed, and Boston was less and less the leading city on the American east coast. In 1739 John Bartram and Peter Collinson proposed a society for Philadelphia, recognizing that "an academy with paid professors was beyond the capacity of Pennsylvania"[140] The Bartram and Collinson plan was taken up by Benjamin Franklin in his "Proposal for Promoting Useful Knowledge among British Plantations in America" of 1743, which sought to extend the institutional base of the Philadelphia Library Company, the most likely American institution to support a learned society. A private society, called the American Philosophical Society and loosely modeled on the Royal Society of London did result from this proposal, and it spread its roots across Pennsylvania, New York, and New Jersey in 1743 and 1744. Because of the continued thinness of the colonial intellectual community and the lack of support from mercantile interests in Philadelphia, this effort at society building was a dead letter by 1747. While little concrete had materialized out of these developments by mid-century, they nonetheless indicate a certain activity, and they set the stage for the final creation of the true American Philosophical Society in 1769.

In this quick sweep of developments in England, Scotland, Ireland, and the American colonies in the first half of the eighteenth century, several points stand out. Although no new formal scientific societies were created in the period, the various organizational stirrings mentioned here testify that the learned scientific society movement was far from dead in the Anglo-Saxon world. Those private groups and associations which were formed during the period maintained the seventeenth-century pattern of being to greater or lesser extents influenced by and in contact with the Royal Society of London. But at the same time the hegemony of the Royal Society was losing its force. What one sees happening is an increase (albeit slight) in the level of cultural and institutional support available for potential scientific societies outside London and the beginnings of local efforts that ultimately resulted in the spread of learned scientific societies throughout the English-speaking world. These developments notwithstanding, the major point to emerge is the strong contrast with events on the Continent. While major, official academies of science were being created there at a brisk rate, across the Channel things were comparatively quiescent. This situation itself contrasts with the seventeenth century, when societies in the English-speaking world were more vigorous and more vigorous in forming than anywhere on the Continent. Academies and societies of science grew separately and at different rates. For the societal movement (if that term be permitted), the first half of the eighteenth century was a period of incubation and slow development. Its full maturation would come later.

CHAPTER IV

The Scientific Society
Movement: 1750 to 1793

T HE circumstances of the scientific societies and the institutional
movement which gave rise to them were far different in 1750 than they
had been in 1700. In 1750 Europe was graced with several important
scientific academies, each well established in its own context, whereas
fifty years previously hardly a weak handful existed. While the growth
of scientific societies unfolded more or less continuously over the
course of the eighteenth century, still, an important turning point in
the scientific society movement was reached at mid-century.

Evidence for such a point of view derives in the first instance from
the fact that major new foundations of the first half of the eighteenth
century achieved new levels of stability at about the mid-century mark.
The academies discussed in the last chapter were not created at a stroke
at the turn of the eighteenth century, but grew up over its first five
decades. Few of the new Continental academies seem to have been
immediately successful in establishing themselves in their respective
local contexts. In this regard one thinks of the weak and oppressed
Societas Scientiarum of Berlin; political and administrative problems in
the early Russian academy; legal proceedings and the Inquisition in
Bologna; the probationary periods and financial struggles of French
provincial academies. It seems that by and large, while the number of
academies was growing, they early faced a set of obstacles to overcome.
The notion that a turning point in the history of the scientific society
movement was reached at mid-century is given preliminary reinforce-
ment by the series of coincident reforms that affected academies virtually across the board in Europe in the later 1740s. The formal

establishment of the Berlin *Académie* in 1746 put organized science in Prussia on new and stronger footings; the almanac privilege granted to the Swedish academy in 1747 and Wargentin's maneuvers immediately thereafter established that institution in an unprecedented position of strength; the 1745 reforms of Pope Benedict did much the same for the Bolognese academy, and the charter of 1748 likewise strengthened the Saint Petersburg Academy. Taken together, these separate develop-ments—in addition to the mere fact that academies were to be found in these and other locales in the first place—indicate not only that these academies were individually in stronger positions at mid-century, but so also was the growing international network of scientific societies.

Changes affecting the scientific societies and the scientific society movement are representative of larger shifts in European history and culture occurring at mid-century. Historians have long recognized this break in the history of the eighteenth century.[1] Politically, Prussia and Russia had matured as major European powers, and in France changed political and economic conditions set a new historical course that would culminate in the Revolution of 1789. Culturally and in terms of the history of ideas, the mid-eighteenth century was when the En-lightenment made its public appearance: Montesquieu's *Esprit des lois* was published in 1748; the initial volume of Buffon's *Histoire naturelle* came out in 1749; and, most noteworthy perhaps, in 1751 the *Encyclopédie* of Diderot and d'Alembert began to appear. These works signaled the beginning of a new age which publicly espoused the val-ues, thinking, and ideology of the Enlightenment, and for which the term Enlightenment came to characterize its spirit and peculiarity. Mu-sically, the differences between Bach and Mozart capture this differ-ence between the first and second halves of the eighteenth century, and Bach's death in precisely 1750 somehow symbolizes this transition. That notwithstanding, in lesser, but not unrelated ways, some of which have already been touched on, the histories of scientific societies and the scientific society movement must also be divided into these half-century periods.

One sign that a new phase was beginning is what people were saying. Samuel Formey, secretary of the Berlin Academy and ever the spokes-man for it and for academies in general, proclaimed in 1745 that there was no longer a need to justify a *corps savant:* "No one can ignore how the idea of learned companies was born in Europe during the past century and how, vying with one another, they have multiplied in the principal kingdoms in this part of the world."² An even more vivid contrast with the past is provided by Etienne-Hyacinthe de Ratte, secretary of the Montpellier Society. In his discourse of 1751 delivered to a public assembly of the society, he said:

I will admit, however, that if the extreme importance of mathematics and the experimental sciences was felt by but a small number of persons engaged by their natural genius and talents to pursue the same career as we, we could with an abundance of reason complain of the injustice of our century. Happily, our contemporaries are beyond this reproach. Our sciences are no longer regarded as entirely inaccessible to the multitude. The spirit of Geometry and Philosophy is winning on all fronts; and today it seems to want to set the tone for our [whole] century. Let us place ourselves at the time of the birth of this academy [1706]. Our illustrious, departed associate [de Plantade, first secretary of the Montpellier Society] then felt himself obliged to publicly uphold the common cause of mathematics and experimental sciences which the vulgar disdained, and, forearmed, he endeavored to discredit [their opinions]. I know that the Orator, for in him this quality was united with that of astronomer, must have applauded the success of his enterprise and that the blows struck in prevention were not without effect. Yet, would he not have preferred to see the sciences which we cultivate, from that time possessing the general esteem which they merit, than to be reduced to the sorry necessity of making an apology for them. Let us speak no more of a prejudice that the light of reason has caused to disappear! Let us delight in a happy revolution!³

While doubtless not every one would ha:e agreed with de Ratte's enthusiasm, that he could speak in such a way testifies to new social and institutional conditions for science and for the scientific societies.

Another, more forceful indication of changing times was the creation of lesser organizations on the fringes of the major institutions. In the second half of the eighteenth century, it is fair to say, one observes a process of elaborating and fleshing out the structure of initial foundations in London, Paris, Berlin, Saint Petersburg, Stockholm, Bologna, and elsewhere in the first half of the century. That the scientific society movement began to "take off" at about 1750 is evident, for example, in the foundation of a lesser, derivative organization like the Society of Experimental Philosophy of Dangiz (the *Naturforchende Gesellschaft*). In 1743, J. T. Klein wrote an introductory letter to the Royal Society in London, providing details of the new Danzig group. He emphasizes that: "The main of our Design is, that every Experiment occurring in the Transactions of Illustrious Societies or the writings of Learned men shall be resumed in a full Assembly, examined, and all circumstances and Phenomena, whether true and confirmed, whether dubious or in effect contrary, marked down, and the Art of Invention cultivated in the best manner."[4]

Here is an organization that depended on the function of other, established societies for its raison d'être and mode of operations. The idea that it had become even fashionable to found academies was given concrete expression by a M. de Cideville, who said at the opening meeting of the Rouen Academy in 1744, "The capital of Normandy will no longer envy any city the glory of possessing a literary academy."[5] Perhaps the best example of this tendency of second-level societies to feed off of the example and presence of more illustrious scientific institutions occurs with the foundation of the Dutch Society of Sciences (*Hollandsche Maatschappij der Wetenschappen*) in Haarlem in 1752. The historian of the Dutch Society, Bierens de Haan, insists that the original institution was not just another little reunion with an interest in things scientific:

The founder's object was entirely different. They were desirous of giving something to Holland which it had lacked until then, viz. an academy, meaning an institution which was to meet at regular intervals, publish proceedings, offer prizes and medals, etc. France had its *Académie Royale des Sciences*, England its Royal Society and other countries as well had academies. The wish to remedy Holland's lack of a similar institution led to the foundation of the Holland Society of Sciences.[6]

The wish to remedy a lack! So successful were the model of the learned scientific society and the underlying institutional movement that by mid-century the absence of an academy said as much about a country or a city as did the presence of one. The conditions at mid-century were such that the simple desire to keep up with the spreading model was itself a new motivation behind the creation of additional academies and societies. The scientific society movement thus no longer depended (as it had earlier in the century) on a restricted pattern of influences and opportunities, but, like the rolling stone, it gathered momentum and became a self-generating, self-justifying, and self-fulfilling phenomenon.

Notable developments in the scientific press further testify to the new place achieved by the scientific societies by mid-century. In 1755 the remarkable *Collection Académique* appeared.[7] Begun initially in Dijon by a "Société des gens de lettres" and later published in Paris, the *Collection Académique* was a thirteen-volume series which appeared between 1755 and 1779.[8] In some respects the *Collection Académique* is to be linked with the *Encyclopédie*.[9] The *Collection Académique* compiled, translated (into French), and reprinted (wholly and in abridgements) proceedings of scientific academies and societies, journals, and the writings of individual authors. The republication of the institutionalized scientific press, so to speak, loomed large on the horizon of the *Collection Académique*: the *Transactions, Mémoires, Commentarii, Handlingar,* and *Mélanges* of the Royal Society of London, the Berlin academy, the Bolognese academy, the Swedish academy, and a yet to be heard from Turinese academy of great fame were among the institutional sources it actually reprinted. The enthusiasm for the project and the seeming consequence of academies and societies of science were such in the early 1750s that a series of forty volumes was initially envisioned.[10] The appearance of the *Collection Académique* attests to certain limitations to learned societies in handling scientific communications, but the mere existence of this series says something important about the standing of the scientific societies at mid-century.

Perhaps the most significant sign that the middle years of the eighteenth century marked a real and important change in the status and character of the scientific societies lies in the fact that for the first time they began to function as a coherent and interacting system of institutions. While this aspect of the scientific societies will require further treatment, one indication of a new level of inter-institutional interaction is the emergence of a regular pattern of exchanging publications and productions. Academies and societies of science were in only sporadic contact in the earlier decades of the eighteenth century, and it was not until mid-century that systematic exchanges emerged. For example, not until 1749–50 did the Paris Academy and the Royal Society reciprocate the Paris *Mémoires* for the *Philosophical Transactions.* Even more telling in this regard were the first cooperative endeavors undertaken by the scientific societies collectively. The upshot is that with the astronomical voyage of the Abbé LaCaille to the Cape of Good Hope in 1750 the scientific societies in Paris, Stockholm, Berlin, Saint Petersburg, and elsewhere organized themselves to undertake coordinated astronomical observations—the first in a whole series of similar enterprises. With advanced planning, a division of labor, and a sharing of results, no clearer evidence for an active system of scientific institutions could be had. That system more than anything else testifies to the new position achieved by the scientific societies at and after 1750.

The growth of academies and societies of science and the confraternity of such institutions was far from complete by 1750, however. More scientific societies appeared in the second half of the century than the first, although there were no new absolutely first-rank institutions. The initial framework of scientific societies laid down in the first half of the century matured after 1750 and became enlarged and completed with second- and third-ranked societies everywhere possible in the second half of the century. The system flourished until the French Revolution and subsequent events brought its cataclysmic end.

New Foundations in Germany and Northern Europe

An important extension of the developing system of scientific societies in the second half of the eighteenth century was the creation of a number of new societies in the states of Germany and northern Europe. Until the middle of the century, virtually the only academies or societies of science in the German-speaking world were the Prussian *Societas* and *Académie*. The level of cultural, political, and finanical support available for academies in Germany was not sufficiently high or sophisticated to allow for any spread of the academic model. Leibniz certainly tried. In the 1750s and early 1760s, however, the situation changed dramatically, and a number of fairly significant institutions appeared. As this number is not very large, perhaps the best way to approach developments in and around Germany is on a case by case basis.

Göttingen. The history of the Göttingen *Königlich Societät der Wissenschaften* (1752) cannot be conceived apart from that of the University of Göttingen which was founded earlier in 1734.[11] Indeed, at least

Göttingen Society, *Commentarii* (1751),
title page engraving.

two proposals of the 1730s for the new university included the notion of a learned society attached to it in much the same manner as the Bolognese academy was affiliated with the Bolognese institute and university.[12] Many plans followed between the creation of Göttingen university and the founding of the society in 1752, but always the prior needs of the infant university and the question of money stood in the way of a science society.[13] Albrecht von Haller finally provided the solution to the problems involved. A peculiar thing about the legal foundation of this society was that Göttingen formed part of the Hanoverian territories in the eighteenth century, and hence fell under English control. Thus, the Göttingen Society of Sciences, like the university, was formally established by George II, who was also patron of the Royal Society in London. The two societies were not exactly sister institutions, however, although the lack of strong financial backing from London raised the specter of financial necessity for both of them. Haller sought his way around this problem for Göttingen by proposing a smaller and cheaper academy composed of half a dozen paid, core men from the university (including himself as president) and a range of unpaid associates in various categories.[14] The society was also divided into three classes (physical, mathematical, and historical sciences), which made it even more of an academy in the Continental sense than a society in the English sense. Haller also made a compromise with the forces that had and would dominate the Göttingen Society for years—the university itself—in that anatomy, botany, and chemistry were not to be included within the purview of the society, those sciences being guarded within the medical faculty.[15] The society did suffer a long stretch of difficulties and crises.[16] Nonetheless, the society assumed an almanac monopoly, it took over publication of the important journal *Göttingische Anzeigen von gelehrten Sachen*, it issued nearly thirty volumes of memoirs between 1752 and 1808, and it offered over fifty, mostly utilitarian prizes between 1753 and 1782.[17] Haller's international stature and reputation put Göttingen university and its society on the map, but the society itself seems weak and less important scientifically than some of the other German academies created after it.[18]

Erfurt. The foundation in Göttingen just two years earlier was an important example behind the creation of the Erfurt Academy of Useful Sciences (*Akademie gemeinnütziger Wissenschaften*) in 1754.[19] In this case, however, the academy grew up outside the local, Catholic university and was initiated largely by Protestants who had failed in university competitions. Founded under the protection of the Elector and *Statthalter* of Mainz, the Erfurt Academy was more than an academy in the traditional sense. As its name implies, it was also involved in economic and utilitarian concerns. Indeed, Hans Hubrig, who has written on the economic and patriotic societies springing up all over Germany at this time, points to the Erfurt institution as a connecting link between science academies and patriotic/economic societies.[20] The academy was well equipped with a library, observatory, anatomical theater, chemistry laboratory, and botanical garden. It offered four prizes a year on utilitarian subjects, and it published eighteen volumes of *Acta* of a

Munich Academy, *Neue Abhandlungen* (1785),
title page engraving.

more scientific bent between 1757 and 1807.[21] The Seven Years War
saw the near dissolution of the Erfurt *Akademie*, but it was renewed on a
sound basis by a new Elector, Dalberg, in 1776.

Munich. The Bavarian Academy of Sciences (*Churbayerische Akademie der
Wissenschaften*) was the next German science academy to appear, being
incorporated in 1759 by the Landgraff and Elector Maximilian II of
Bavaria.[22] This new official academy arose against a background of
plans for various academies for the region, including humanist associations of Benedictines.[23] J. G. Lori, who for many years attempted a
private animus for a learned society in Munich, noted upon the successful creation of the Munich Academy of Sciences: "I thank God that my
plan which seemed a clear impossibility to most people turned out so

well, that all Europe knows that Bayern has an Academy which the Court from the standpoint of its honor cannot let decline."[24]

The Bavarian Academy was divided into two classes—historical and philosophical (science)—and provided pensions for resident members. It offered prizes and published fifteen volumes of *Abhandlungen* and *Neuen Abhandlungen* between 1763 and 1807. An important development for the academy in this period occurred after the Seven Years War, when Munich fell under the control of the Palatine Elector Carl Theodore. In 1777 this prince renewed the Bavarian Academy (adding a class of belles lettres) and established an extremely close tie (bordering on union) between the academy in Munich and his own in Mannheim. The Munich Academy worked closely with the Meteorological Society of Mannheim, and, through its own permanent secretary, J. P. Kennedy, and Samuel Formey in Berlin, it was in regular and productive contact with the Academy of Sciences in Saint Petersburg.[25] The Munich Academy was the institution with which Mesmer was early associated before he came to Paris, and the loyalist American Benjamin Thomson, Count Rumford, was connected with it from 1786.

Mannheim. As part of the movement to found new German academies, the creation of the Electoral Academy of Mannheim (*Kurpfälzische Akademie der Wissenschaften*) in 1763 owed a debt to the earlier creations in Göttingen, Erfurt, and Munich.[26] But more than was the case with any of these, the Mannheim Academy grew up in response to the largest international dimensions of the scientific society movement, and it was created and supported on such a scale as to rival the largest national academies elsewhere in Europe. In part, the Mannheim Academy was the natural outgrowth of a cultural and artistic movement of the highest order affecting the German Palatinate in the middle of the eighteenth century. The goal of this local renaissance was to transform Mannheim into another "Athens" and a major center of science and culture. The initial plans for an academy came from the Elector of Mannheim himself, Carl Theodore. Carl Theodore had studied in Holland as a youth and had caught the bug of astronomy and experimental physics. As early as 1759 he sought to found his own academy.[27]

When the institution was at last created in 1763, Carl Theodore supported it magnificently with a subvention of six thousand (later nine

Mannheim Academy, *Historia et Commentationes* (1766), seal with Carl Theodore.

thousand) gulden a year, and additionally he provided for prizes, quarters, several sorts of scientific cabinets, a library of 37,000 volumes, and a botanical garden. This was in the grandest tradition of absolutist princes elsewhere in Europe. The academy, called the *Academia Theodora-Palatina*, was modelled after the Paris Academy, but like other German institutions of the same type it was composed of both scientific and historical classes. Crucial to the development of the Mannheim Academy were the personages of one Schöpflin, the Elector's librarian and archivist, and the man brought in as secretary of the academy, Andreas Lamey. Lamey held his post for three decades and exercised it in the classic manner of other permanent secretaries.[28] Of individual importance also was the Jesuit Christian Mayer, who had come to Mannheim in 1752 as professor of mathematics and experimental physics. He was an active and influential member of the academy, specializing in astronomy and cartology. He returned from a visit to Saint Petersburg in 1774 and, with Electoral support, erected a full-fledged observatory equipped with the latest English instruments. With its own press and printing privileges, the Mannheim Academy published a newspaper, the *Mannheimer Zeitung*, seven volumes of scientific memoirs (the *Historia et commentationes*, 1776–94), an index to the botanical garden (1771), and a collection entitled *Flora Palatina* (1775–77).[29]

If all this were not testimony enough to the scientifically rich character of Mannheim during the second half of the eighteenth century, in 1780 yet another scientific academy, the Meteorological Society of Mannheim (*Societas Meteorologicae Palatinae*), was created by Electoral decree alongside the existing Academy of Sciences. This extraordinary institution sponsored and coordinated a major international program of meteorological research involving extensive contact and coordination with other scientific societies, and it published twelve volumes of meteorological *Ephemerides* between 1781 and 1792. More will be said about this episode in a later chapter, but for now the fact that Mannheim so richly supported two scientific societies is indication enough that this mode of scientific organization was spreading into Germany with unexpected force and fecundity.

In this context the *Société patriotique de Hesse-Hombourg* needs to be mentioned. Established in 1775 with the support of the local *Landgraff*, the Hesse-Hamburg society, even more than the Erfurt Academy, grew out of and formed part of the economic society movement affecting Germany in the last third of the century, but it did not spurn science as one of its goals, and it was responsible for a major effort to coordinate the activities of other scientific societies.[30] In its own peculiar way, the Hesse-Hamburg society was one of the more interesting societies to appear.

Apart from these major foundations, a number of lesser groups and organizations arose in the German-speaking world of north and north-central Europe in the second half of the eighteenth century. In Nuremberg the virtually unique Cosmographical Society of Nuremberg was established in 1747. A private group, the Cosmographical Society specialized in cartography and making maps and globes; indeed, it was associated with a globe-making factory.[31] The *Societas Eruditorum Incognitorum*, existing in Olmütz between 1746 and 1751, was another small German society of note; it was recognized and supported by the Austrian Court, and its founder, von Petrasch, was one of those pushing for an academy at Vienna.[32] In Berlin the private but important *Gesellschaft Naturforschender Freunde* came into being in 1773. Licensed by Prussian authorities in the time of Frederick the Great, after his death in 1786 the *Gesellschaft* received significant support from Frederick-William II and became a counterpoise to the by-then declining Berlin Academy. It was an active organization of some renown—though with Masonic overtones, publishing several volumes of memoirs and attracting an international membership.[33] In Halle, a similar and related *Naturforschenden Gesellschaft* began in 1779; it was a rallying point for university and regional scientists, and it continued into the nineteenth century.[34] The private *Oberlausitzischen Gesellschaft der Wissenschaften* (Görlitz) was founded in 1779. Although a private society, it performed many learned society functions: it had a library, it published a provincial newsletter, and it developed a regional topography.[35]

To the east of Germany, the energetic Polish elector J. A. Jablonowski created the official *Fürstlich Jablonowskische Gesellschaft* in Leipzig in 1768.[36] Jablonowski was an *associé étranger* of the Paris Academy, and his noteworthy outpost was a beacon for a number of lesser, private Polish groups and associations interested in science and learning. Included in this number was the *Naturforschende Gesellschaft* of Danzig. The Danzig society dates from 1743, and, although a private organization, it, too, functioned as many official societies, publishing four volumes of transactions (1747–1778); it was equipped with an observatory, a library, cabinets, and laboratories, and was highly experimental in its outlook and approach.[37] While nothing official or lasting came of them, the learned society movement manifested itself in proto-organizations and plans for learned societies in Bratislava (Pressburg, Austro-Hungary), in Warsaw, and in the nether reaches of Transylvania.[38] Even at these extremes of Western scientific and learned culture, the impact of the learned society as a model of organization was felt.

Taking its inspiration from the academies in Göttingen, Munich, and Mannheim, a major foundation on the eastern front was the *Königlich Bömische Gesselschaft der Wissenschaften* incorporated in Prague in 1791.[39] Actually, the history of the Prague academy goes back to 1769 and the creation of the "Privatgesellschaft in Bömen," led by another of those strong personalities, Ignaz von Born.[40] The "Privatgesellschaft" published a number of volumes of *Abhandlungen* before it received legal status, but little direct support, as the Bohemian Society of Sciences in 1784. In 1791 this society was elevated to Royal Society and received a fair measure of privileges and financial support (six thousand gulden annually). As Prague formed part of the Austro-Hungarian Empire of

Bohemian Society, *Abhandlungen* (1785),
title page engraving.

that time, the source of this support was the Austrian Emperor, Joseph
II. Although no scientific society was created in Vienna itself, the
Prague academy is one of several examples of Austrian support for
learned societies outside Austria proper.

The Prague Society was organized into three classes (physical, mathematical, and historical sciences), and it met with moderate success in
the last decade of the eighteenth century, considering the especially
low, essentially feudal level of cultural and intellectual life in Bohemia
in the period. The Prague Society offered prizes on agricultural and
economic subjects; it published over a dozen volumes of *Abhandlungen*
and *Neuen Abhandlungen* between 1775 and 1798, and it sent expeditions
into the hinterland to gather information for a proposed economic map
and survey of Bohemia. It is another interesting example of the extreme
social and cultural niches into which the scientific society movement
had infiltrated by the late stages of the eighteenth century.

Switzerland was more affected by the economic and arts society movements than the scientific society movement strictly speaking. A major and very active institution of the former type was the *Ökonomische Gesellschaft* of Berne dating from 1759. Other Swiss economic societies looked to Berne as the model. But still, the scientific society movement per se was not without some impact on the Helvetian Confederation. *La Société des sciences physiques* of Lausanne published two volumes of memoires in the 1780s, and the *Naturforschende Gesellschaft* of Zurich was a society of some consequence.[41] The latter was established in 1746 after the example of the Royal Society of London, and it mirrored the organization and function of that institution closely. Meeting privately with a limited dues-paying membership of one hundred, the Zurich *Gesellschaft* came to be well endowed with a library, botanical garden, herbarium, and instrument cabinet. It published three volumes of *Abhandlungen* through the 1760s. Like French provincial academies the *Gesellschaft* became closely associated with Zurich city authorities. Under the strong hand of its president, Johannes Gessner, it established an important standing committee, the "Landwirtschaftlichen Kommission," to deal with agricultural and economic questions. Its (minor) prizes were supported by the city, as was the public lottery administered by another society commission. The *Gesellschaft* also was involved in statistical work for the census in Zurich from the late 1750s through 1780. In addition, the *Gesellschaft* had committees to oversee its various departments, and, extraordinarily for a minor society, it sponsored two expeditions, one locally in Switzerland, and another to North and South America.

Still, Switzerland was not a major power among nations with scientific societies. The explanation for this lack must rest on the twin facts that Switzerland had strong universities, such as the University of Basel, which perhaps obviated the need for other institutions for science. Then, too, Switzerland lacked the strong political unity that might have allowed it to create its own national academy or society. The situation is somewhat surprising because, as individuals, the Swiss were among the most outstanding and active members of the community of eighteenth-century scientists in other European academies.[42] The Eulers and the Bernoullis are the prime examples.

Another remarkable efflorescence of scientific societies in the second half of the eighteenth century occurred in Holland. The development of Dutch societies was late, however, and, excepting perhaps the Haarlem *Hollandische Maatschappij der Wetenschappen*, no academy or society of international rank appeared.[43] The reason has to do with the extensive and progressive system of Dutch universities, which, as in Switzerland, early came to incorporate and teach up-to-date science.[44] In this regard one thinks of Boerhaave and his popular teaching at Leyden. Universities also resisted the foundation of science societies, and their eventual creation has to be seen as the lesser outgrowth of non-university provincial culture.[45] The Haarlem *Maatschappij* began in 1752 on the initiative of seven burgher members of the Haarlem town council. It was granted official recognition in 1761 from the Staten of Holland and West Friesland. It was a peculiar organization in that it had two classes of membership: one, dues-paying founders; the other, free active membership of scientific workers. Reflecting the general economic society movement in Europe at the time, the Haarlem *Maatschappij* established its own economic section with corresponding branches throughout Holland and the Dutch colonies. This was after 1777. The Dutch Society published thirty volumes of little note, the *Verhandelingen*, between 1754 and 1793. In 1778 a second Haarlem society, Teyler's *Tweede Gnootschap*, complemented the *Maatschappij;* endowed by private bequest, it was the smallest eighteenth-century scientific society, with only six members.[46]

The Dutch Society in Haarlem became the model for other Dutch provincial societies. The provincial *Zeeuwsch Genootschap der Wetenschappen* was founded in Middleburg (Vlissingen) in 1768. It was recognized by the provincial estates of Zeeland, and it published fifteen volumes of *Verhandelingen* between 1769 and 1792.[47] The *Bataafsch Genootschap der Proefondervindelijke Wijsbegeerte* (experimental philosophy) was established in 1769 in Rotterdam and was recognized by the Holland states.[48] This society was explicitly devoted to investigations in experimental physics, and it published twelve volumes of its *Verhandelingen* between 1774 and 1798. The *Provinciaal Utrechtsch Genootschap van Kunsten en Wetenschappn* was founded in Utrecht in 1777 with the backing of town burghers and William V of Orange as its protector.[49] It was more a literary society, but it was recognized by the Utrecht estates, and its ten volumes of *Verhandelingen* (1781–1821) did contain some sci-

Rotterdam Society, *Verhandelingen* (1775),
title page engraving.

ence. A number of private and ephemeral societies touching on science
seem to have complemented these official Dutch societies.[50] Finally in
this connection, the chartered *Bataviaasch Genootschap van Kunsten en
Wetenschappen*, founded in Batavia, the Dutch East Indies in 1778, is yet
another indication of the flourishing of Dutch societies in the period.[51]
The Batavian society also represents an outer geographical limit penetrated by the model of scientific societies. A bare outpost of learned
culture on the edge of the jungles of Java, this group of mostly Dutch
East India Company employees, promoted scientific investigations of
their region. They sponsored and awarded prizes, and they were in
close contact with Dutch societies back in Holland and with a number
of European scientists.

Contrasting the relatively weak Dutch societies, in neighboring
Belgium, after it had ceased to be part of the Spanish Netherlands and
had passed over to Austrian control in 1769, a more visible and important institution was forthcoming.[52] A private literary society, led by
Joseph Turberville Needham, began in Brussels in 1769, greatly influenced by the Royal Society of London. It offered prizes in history and
experimental physics, but activity fell off almost completely by 1771.
In 1772 Maria-Theresa of Austria authorized the creation of the

Académie Impériale et Royale des sciences et belles-lettres of Brussels. This academy was divided into classes, and it was nicely funded for prizes, pensions, and expenses for publishing its series of *Mémoires*, five volumes of which appeared between 1777 and 1787. Its privileges were renewed with the advent of Joseph II of Austria, and he additionally provided the academy with cabinets for natural history and experimental physics. The Belgian academy was of moderate significance on the international level, although it also shared a localized, provincial outlook.

In Scandinavia in the second half of the eighteenth century a complement of scientific societies rounded out the initial creations in Sweden. The official Danish *Kongelige Danske Videnskabernes Selskab* was an early creation (1742), appearing only a year after the official charter of its Swedish counterpart in Stockholm.[53] The Danish Academy conducted extensive geographical and cartological work within Denmark, it published the results of expeditions to Egypt and Iceland, and it was further known internationally through a vigorous program of prizes.[54] In 1778 the *Kungl. Vetenskaps-och-Vitterhets Samhället* was established on royal authority by the Swedish crown in Göteborg.[55] This institution, a "universal" society incorporating science as only part of its activity, was, overall, a provincial organization, although it gave some prizes and possessed an imprimatur. In Uppsala the private *Cosmographiska sällskapet* was another noteworthy organization existing from 1758; like its counterpart in Nuremberg, it dealt with matters geographical and was asssociated with a globe-making facility.[56] The network of Swedish societies was completed by the provincial *Kungl. Fysiografiska Sällskapet*, which formed in Lund in 1771 and received royal recognition in 1778. Taking up natural history and economy, it possessed a modest membership, published several small volumes of transactions, and had a library.[57] Finally in this regard, the *Kongelige Norske Videnskabers Selskab* was founded in Trondheim in northern Norway in 1767 with letters patent and financial support from the Danish court in Copenhagen.[58] The brainchild of Bishop J. E. Gunnerus, it began as a private association in 1760. It concerned itself with history, philosophy, and the natural sciences. It assembled a respectable library, gave prizes, and published five volumes of *Skrifter*. That an official learned society could arise just three degrees south of the Arctic Circle is eloquent testimony to the power of the learned society movement in the eighteenth century and to how much, again, the learned society was the model for the institutionalization of science and culture in the period.

Italy

As in Germany and northern Europe, many new learned scientific societies were created in Italy in the second half of the century. And, as before, the foundations of new institutions testify to the increasing vigor of the scientific society movement in that period and to its international character.

Augmenting "Renaissance" academies continuing into the eighteenth century and the Academy of Sciences of Bologna—the only major official scientific society in Italy in the first half of the century, the first new Italian academy of science after 1750 became the most important in Italy and a major institution on the international scene: the Turinese Academy of Sciences.

The origins of the Turinese Academy of Sciences are rooted in a larger intellectual shift occurring in Italy after 1740. For a number of reasons, religious and institutional, Italy remained somewhat outside the main currents of scientific Europe during the first decades of the eighteenth century.[59] In the 1740s, however, Italy opened up to a fresh wave of intellectual activity and to a new state of affairs best symbolized perhaps by the first Italian publication of Galileo's *Dialogue* in 1744.[60] The upshot is that science and the scientific progress of the previous decades in northern Europe became established in several of the universities of Italy.[61] Specifically, G. B. Becarria joined the faculty of the University of Turin in 1748, bringing with him a confirmed Newtonianism. In the ensuing battles between Cartesians and Newtonians and debates within the university over the electrical theories of Nollet and Franklin, Beccarria established himself as the leader of the Franklin-Newton faction, and in the process gathered around himself a coterie of progressive, like-minded students, notably J. L. Lagrange, J. F. Cigna (Beccarria's nephew), and the Count de Saluces.[62] In 1757 these three and a few friends constituted themselves as a private scientific society outside the university under the patronage of de Saluces.[63]

This was an interesting development, but there seems little involved in the origins of this private society in Turin to suggest its becoming anything more than a local and minor association formed because of a common interest in science on the part of its members. But the goals of the Lagrange-Cigna-Saluces group were far-reaching, and their character and success made Turin a case apart from all other learned societies.

Quite exceptionally for a private society, in 1759, only two years after their initial organization, the Turinese group published a handsome volume of memoirs, *Miscellanae philosophico-mathematica societatis privatae Taurinesis*.[64] This otherwise obscure volume contained fine articles by Lagrange on maxima and minima and by de Saluces and Cigna on the nature and propagation of sound. Crucially, the authors sent their volume to the major academies and societies to the north where it encountered a very positive reception, particularly from Euler in Berlin and Haller in Göttingen.[65] In part because of the support from these men of science from the international community and in part to model himself after Frederick II of Prussia, Victor-Amadeus, crown prince of the kingdom of the Two Sicilies, extended his patronage to the Turinese group and in 1759 founded the *Société royale des sciences* of Turin.[66] This was a big boost to the private society and its efforts to attract international attention and upgrade its organization.

Indeed, the most peculiar thing about the Turin example was that, from the point of view of certain internationally minded persons in Turin and, more, men of science outside Turin, the situation in Turin was seen and seized upon as an opportunity to expand the community of academies of science in a major way. Already in 1760, only a year after their incorporation as a society, Lagrange, Cigna, and de Saluces in their "Project for the Establishment of a Royal Academy of Sciences at Turin" petitioned Victor-Amadeus III, now king, to upgrade the Turinese Society into an academy.[67] In this curious document one sees clearly that their request was as much for an academy to participate in the larger international community of science and scientific societies as it was for an academy of strictly local benefit. In their "Project" they claim first to have:

observed the good effect that learned societies and academies produce in the interior of those countries where sovereigns have judged it appropriate to establish them, and by their constitutions they tend thus to form, despite the distances between them, a worthy tie between the gifted men of all nations. . . . When one turns to the establishments (of which there are already many examples in foreign countries) noteworthy by their success, one can hardly do otherwise than to imitate them, and it is only a question of choosing in this imitation that which one judges most appropriate. Using this principle we have not thought to be able to do better than to attach ourselves to the statutes that the late king of France, Louis XIV, handed down in the year 1699 for the Royal Academy of Sciences of Paris, adjusting them, however, to [our] particular circumstances. . . . [We seek in letters-patent] the title and prerogatives of a Royal Academy, based roughly on the model of some establishments of this nature which exist in various states of Europe in order that, working on the same basis that has given rise to their institution, the new Academy will tend toward the same goal and contribute in its turn to the attempts and discoveries that are made daily in foreign countries; experience has shown in these last years how much the concourse of nations multiplies the advantages that have begun to be forthcoming from nearly a century of cooperation among learned men.[68]

Included with this proposal was a model set of statutes with every similarity and variation from those of the Paris Academy of Sciences precisely noted. As well, this petition was accompanied by testimonial letters from Condorcet, Euler, and Haller![69] Such a concerted effort and such an intellectual rationale bespeak almost a conspiracy of internationalism.

Something very like that was kept up from the 1760s through the 1780s. Just after the "Project" of 1760, de Saluces wrote to the Royal Society in London of "the second volume of our Society's Memoirs, which will soon appear, perhaps dignified by a very illustrious title."[70] Six months later his report was less optimistic: "We have nothing new here, excepting that, notwithstanding the promises made to our Society, it has received no settlement, the project having alarmed many people."[71] For Lagrange the prospects seemed better; he wrote to d'Alembert in Paris in 1764: "I have been well enough received here by the King and his ministers. I am given a lot of wonderful hopes, but I do not put great stock in them. Your letters have made a big impression at court and around town."[72] Nine months later he added, "Our Society is preparing to have a new volume printed. Would you do it the honor of decorating this Work with your name? That would certainly have a great effect here, and I do not doubt that it would hasten its establishment considerably."[73]

The effort to raise the Turinese Society to the level of a full-fledged academy of sciences did not meet with the success anticipated. For reasons unknown, the king proved unavailing, and after Lagrange left Turin to take over Euler's post at the Berlin Academy in 1766, hopes for an academy were considerably reduced. Lagrange wrote to Condorcet from Berlin in 1775: "I no longer hear the Society of Turin spoken of; I think all the great hopes we had have gone up in smoke."[74] In the same year he wrote to Euler in Saint Petersburg:

The hopes which the Society at Turin conceived upon the advent of the King to the throne seem almost vanished. I do not know what could have given rise to this or if one can flatter oneself to see them reborn. I think, however, that the Society has put the fifth volume [of its *Mélanges* series] to press; at least I hope so because of your memoirs and those of M. Lexel which I hope to find there.[75]

Later in 1777 he tells Condorcet: "I am informed from Turin that there is no more talk about a date for the establishment of the Academy; perhaps because the King has more important things on his mind, or perhaps also because, since the retirement of Count Saluces, the project has been abandoned."[76] Happily, the count must have come out of retirement, because between 1781 and 1783 he again actively petitioned the king for an academy. Success came at long last, and letters patent for the elevated Turinese *Académie royale des sciences* were granted in July of 1783.[77]

The culmination of this long-standing drive to upgrade the Turinese society into an academy did not greatly affect the international stature of the institution. It was already a major scientific center. The Turinese Royal Society, although not well supported, managed to publish four volumes of *Mélanges de philosophie et mathématiques* in the first period of its history, and these became one of the first-class scientific publications of the day. D'Alembert and Macquer joined Condorcet, Haller, and Euler in publishing there. They also joined the society.[78] The *Mélanges*, like the private volume before them, were sent to academies and societies to the north, and they were reprinted as a volume of the *Collection Académique*.[79] The earlier society was an internationally oriented institution of no mean standing. The receipt of new letters patent in 1783 did increase the power of the institution somewhat. As the imitation of the "Project" of 1760 proposed, the new Royal Academy was made up of a core of twenty paid, resident academicians (along with the traditional hierarchy of members and officers).[80] The academy was exclusively a scientific institution, divided into two sections, the mathematical and physical sciences. More importantly it was funded royally by a grant of twelve thousand livres from revenues taken from suppressed abbeys.[81] From 1783 until it was closed in 1792, the Turin Academy maintained a strong international outlook in its operations. Run by de Saluces as president and Cigna as permanent secretary, it published five volumes of *Mémoires* in the period, and the tradition of publication by distinguished foreigners continued.[82] Foreign membership itself expanded to include Laplace, Monge, Franklin, Bergman, Priestley, Boscovich, de Morveau, Lexel, Schelle, and others.[83] The academy continued to send its *Mémoires* to the north, and it maintained active relations with scientific societies there.[84] It does not seem to have offered prize competitions.

On the other hand, the Turinese Academy had a decidedly local flavor, too. The resident and regional members of the academy were all Italians, and their work was mainly of local interest and utilitarian in character, compared to the more purely scientific contributions of outsiders.[85] All in all, the Turinese Academy was an important addition to the confraternity of scientific academies in the second half of the eighteenth century. Its history is exceptional in the degree to which, in turn, the institution served the international community.

But the Turinese academy was not the only Italian scientific society of note in the second half of the eighteenth century, nor the only one to reach the consciousness of other Europeans. A notable development in northern Italy was the creation of societies under the protection of the Austrian imperial government. These foundations, like those in Brussels and Prague, are yet another indication that Austria was an important power after all in founding and supporting scientific societies. The Austrian Emperor, Francis I, as early as 1739 approved and supported the *Accademia Botanica* of Florence. In 1783 this academy was incorporated into the separate and noteworthy *Reale Accademia economica-agararia dei Georgofili*, which had previously been chartered in 1767 and which dates back to 1753 as a private group.[86] The *Georgofili* itself was more an agricultural and economic society, but one with some scientific leanings known beyond Italy. Another Austrian-sponsored institution was the *Imperiale Regia Accademia degli Agiati* founded in

Rovereto in 1752.[87] Based on a preceding "Renaissance" academy, it had an expanded membership of over five hundred members, and after 1765 its interests were focused on agriculture; in fact it became a correspondence and coordinating center for a number of Italian agricultural societies with support from the Austrian government. But the *Agiati* did publish eight volumes of memoirs of some interest for the history of science. Finally in this connection, in 1767 the *Reale Accademia de Scienze, Lettere ed Arti* of Mantua was instituted on the authority of Maria-Theresa and Joseph II through the imperial administration in Lombardy.[88] The Mantua Academy subsumed an antecedent "Renaissance" academy (the *Accademia Virgiliana*), which had received official Austrian recognition in 1752. The Mantua Academy received even more income from Vienna, and it came to have the usual complement of learned society departments. It was divided into four classes, two for the mathematical and experimental sciences. And, like a full-fledged academy anywhere, it possessed powers of censorship and control over copyrights.

Other authorities, too, were involved in establishing new scientific societies in Italy in the second half of the eighteenth century. In Siena the *Reale Accademia delle scienze di Siena* began in 1759, upgraded from an earlier *Cimento*-like academy.[89] It published six volumes of *Atti* between 1761 and 1794, and a number of distinguished foreigners were counted among its more than 125 members, including Lagrange, Linnaeus, and Beccaria. It had the peculiarity that professors from the local university could join automatically. In 1778 Ferdinand IV de Bourbon established the *Reale Accademia delle Scienze e Belle-Lettere* in Naples.[90] Modeled after the Berlin Academy, the Naples Academy had ten pensioned academicians, paid officers, and sections for science and literature. It required work from its members in typical academy fashion, too, and it published a single volume of *Atti* in 1787. In 1779 the more important *Accademia de Scienze, Lettere ed Arti* of Padua was founded with letters patent from the Venetian senate.[91] It received a large donation and a regular income from the Venetian state, for whom it served as technical consultant. Like the Berlin Academy, it was divided into four classes: experimental and mathematical sciences, belles lettres, and yet another class of speculative philosophy. Like the Bologna Academy, the Padua Academy was tied to the University of

Padua, with pensions going to university professors. Twenty-six of its thirty-six resident members received annual pensions, and it published three volumes of *Saggi scientifici e litterari* between 1786 and 1794.

Among a number of lesser societies, the private organization *Società Italiana delle Scienze* is of note.[92] It appeared in Verona in 1782, largely the creation of one man, the chevalier Anton Maria Lorgna. The *Società Italiana* did subsume a predecessor institution, the *Accademia degli Aletofilli*, previously recognized as a ducal academy by the Venetian Senate, but Lorgna was decidedly the moving force behind the *Società*. He saw to the publication and international distribution of its seven volumes of *Memorie de mathematica e di fisica* (1782–96), and with his death in 1796 the *Società* merged into an agricultural society in Modena.

Elsewhere in Italy, scientific societies arose in Arezzo, Bergamo, Modena, Palermo, Trapani, and Turin.[93] The *Accademia dei Dissonanti* of Modena is interesting because, like a number of Italian academies already mentioned, it evidences the tendency of "Renaissance" academies to metamorphose into more typical, official academies. The Modena academy began as a private "Renaissance" academy in 1680; in 1729 it merged with the *Accademia Peloritana dei Pericolanti* of Messina and received Spanish support; in 1752 it became the ducal academy of Francesco d'Este, and later in 1817 it became the *Reale Accademia Modenese di Scienze, Lettere ed Arti*.[94] But, as evidenced by the *Accademia Aretina* of Arezzo (1787), "Renaissance" academies continued to be founded well into the eighteenth century. The *Accademia dei Naturalisti* of Bergamo (1782) and the *Accademia Filopatria* of Turin (1782 also) would seem to evidence a new note among Italian academies, more like the various *Naturforschender Gesellschaften* appearing in Germany and elsewhere at the same time.

The French Provinces Revisted

In the second half of the eighteenth century, as in the first half, the French provinces continued to be a vital region for new learned societies. In all, a dozen academies were newly created or reorganized between 1752 and 1784. These additional provincial academies resembled their predecessors and colleague-institutions in most crucial respects: the patterns of their prehistories and securing of letters patent, their modes of organization and operations, their integration into local society and politics as civic institutions, and so on.[95] Learned societies of science et alia created in this period did tend to be academies of a lesser order which extended the model academy down from Paris and the major provincial cities of the first half of the century to smaller and less important centers. Probationary periods for these institutions were longer: witness the twenty-five year wait for letters patent for the academy at Châlons-sur-Marne, for example. A minor point of note, the model and example of the learned academy had so far penetrated into social practice in France by the second half of the century that the new academies more often took their inspiration from other provincial examples than from Paris, as was the case in the first half of the century.

In several other ways, however, the period until the French Revolution saw a new stage of development affecting all the provincial academies of France, as indeed elsewhere. One especially noteworthy change in this period is a shift away from an interest in the pure sciences (so characteristic of provincial societies in the middle years of the century) to a greater emphasis on utilitarianism and the practical applications of knowledge. Considering prize questions posed by the academies of Bordeaux and Lyon, for example, one sees this "clear evolution" toward more practical work and concerns: wine-making, canal-building, bleaching, grains, and so on, and in general, industrial and agricultural subjects take over from prior inquiries into medicine, chemistry, electricity, magnetism, meteorology, botany, natural history, and astronomy.[96] Provincial academies likewise became involved in the topical subjects of the day, such as lightning rods and especially balloons and ballooning.[97] Pierre Barrière summarizes and adds another note:

The role played by the French provincial academies became less great as the interest in novelty disappeared, as the societies multiplied, as science developed, and as its methods demanded qualities less easily accessible to the autodidact and the amateur. . . . Toward the end of the century, literature and the arts tended to regain the allure that science had stolen from them at the beginning.[98]

For all these reasons the provincial academies have to be reckoned less significant scientifically in the second half of the century than in the first.

But their lack of commitment was made up somewhat by their activity. There was a great increase in the number of prize competitions on the provincial level. Roche informs us that, whereas before 1750 only five academies held prize contests, fifteen institutions did so between 1751 and 1770, and on the eve of the French Revolution twenty-four academies were offering prizes for a range of subjects.[99] In all, he notes a twelve-fold increase in the number of competitions in the course of the eighteenth century with the number doubling in the last twenty years. Similarly, the level of funding from various local, provincial, and national sources was higher in the second half of the century than before.[100] This increase allowed provincial academies to better carry out all their activities, and along with prize offerings, publication of provincial *Mémoires* picked up in the period. The Bordeaux Academy published regularly between 1741 and 1779.[101] The Montpellier Society was able to secure funds for its two volumes of *Mémoires* of 1766 and 1778.[102] The Dijon Academy produced its first volume in 1769, nearly thirty years after its beginning.[103] The Toulouse Academy published two volumes in the period.

An important but somewhat overlooked aspect of provincial academies in the second half of the century is that, in their roles as civic and service institutions, they became much more involved in science teaching and pedagogy. In Montpellier, for example, in 1764 the Society of Sciences took control of the royal chair of mathematics and hydrography (hitherto in the hands of the Jesuits) in the University of Montpellier.[104] One suspects that similar developments occurred elsewhere where learned academies existed in close contact with regular pedagogical establishments.[105] The more significant trend was for provincial academies to offer public courses of their own. In the 1740s the Abbé Nollet offered his course in experimental physics at Bordeaux, subsidized in part by the academy there, and after 1750 the Bordeaux Academy itself sponsored courses in experimental physics and natural history.[106] After mid-century the practice of offering public courses in the sciences and applied science seems to have become widespread. In 1767 the Dijon Academy proposed an ambitious program of seven courses: experimental physics, architecture, anatomy, geometry and mechanics, design and painting, veterinary medicine, and music.[107] In practice only three courses (botany, chemistry, and *materia medica*) were paid for by the provincial estates. Roger Tisserand mentions various other courses in chemistry, botany, astronomy, mathematics, *physique*, and medicine offered by academies in Angers, Orléans, Rennes, Reims, Caen, Grenoble, Metz, Châlons, Amiens, and Marseille.[108] In Montpellier, again, there were strenuous efforts to get public courses going in 1768 and in 1776. In 1780 the estates of the Languedoc finally provided the funds for courses in experimental physics and chemistry to be administered by the Montpellier Society.[109] Discussion of the proposal of 1768 gives an indication of the thinking and motivation behind these courses: "This day the Company [met] in extraordinary session. Mr. de Causan expressed that the aid necessary for instruction in experimental physics, which today forms an essential part of public education, is lacking in the province and in this town in particular, although this science is one of the principle objects of research of the royal society and that it is of the greatest utility to students of medicine and surgery."[110]

Not very exciting, but it is apparent that there was a lack of institutionalized, up-to-date instruction in the sciences to which provincial academies were responding in offering their own courses. Although not primarily pedagogical institutions, provincial academies in this way opened science instruction to a wider audience than previously. In the long run the future of science instruction and initiation did not lie in this direction. One can only wonder as to the short term impact of academy-sponsored courses.

Significantly in these regards there was a new sense of identity and a greater degree of contact and interaction between and among the academies of provincial France in the second half of the century. After 1750 several plans and actual trials for union of provincial academies happened. These developments reflect the greater maturity of the system of scientific societies in France and Europe at the time.

In the last decades of the eighteenth century, however, provincial academies lost their hegemony as the almost exclusive forms of social and scientific organization present on the provincial scene in France. Essentially, three new types of institution grew to rival and compete with the older, academic form of sociability and institutionalized culture. They were the agricultural society, the Masonic lodge, and the *Musée* (free school).

The system of agricultural societies was mentioned earlier. These organizations were the French equivalent to the various economic and patriotic societies arising elsewhere in Europe in the period. They were not supposed to be in competition with normal academies, yet they had that effect. Academic resistance was strong, and in some instances academies took up agronomy or set up special agricultural sections within their own organizations rather than see a competing agriculture society created in their locales.[111]

Masonic lodges were another form of organization with implied and avowed cultural purposes very different from regular provincial academies.[112] Lodges began in Paris in the 1750s and rapidly spread across France. They numbered some six hundred groupings on the eve of the French Revolution. They offered a new and more attractive form of sociability. They were more "democratic" and egalitarian in their overtones, and were less positively committed to science as an especially important end to pursue. The movement behind Masonic lodges was really incommensurate with the scientific society movement in provincial France, as Daniel Roche's comparative figures on membership indicate, but the phenomenon does point up that the learned society movement and learned society affairs were no longer receiving the kind of larger social support in the second half of the century that they had earlier.[113]

Finally in this connection, *Musées* and various free schools (design, baking, etc.) were established in several cities in France in the later eighteenth century.[114] Their objects were not especially scientific either. Rather, they sought to unite the sciences and the arts and to make technical and practical knowledge more accessible to the general, working-class public by open programs of instruction and education.[115]

What these latter developments indicate is that the old, almost unchallenged position of French provincial academies was breaking down in the latter part of the eighteenth century. They were proving themselves to be too rigidly ingrained in their local contexts of power and privilege, too narrowly grounded in a social base of elites, and in the end not sufficiently flexible to meet the changing needs of the society in which they operated. The academies of provincial France, while innovative and exciting institutions for science and society through three-quarters of a century, were rapidly becoming outdated as France approached its Revolution.

[137]

Lisbon Academy, *Memorias* (1780–1788),
title page engraving.

Rounding out our survey of the spread of scientific societies on the Continent of Europe, mention must be made of the Portuguese *Academia real das ciências* in Lisbon.[116] This academy began in 1779 and received royal designation from Maria I and Peter III of Portugal in 1783. It complemented the much earlier *Acadmia real de Historia Portuguesa* founded in 1720. The Portuguese science academy was organized, supported, and active, like any state academy we have seen. It was divided into three classes, two for the usual mathematical and physical sciences, and one for the moral sciences (!) and belles lettres. Academicians were required to produce papers, and they received pensions and *jetons* for attendance at meetings. These were funded at first from lottery proceeds. The *Academia* had its own press, but most of its publications date from the 1790s, as it developed strong economic, agricultural, and industrial outlooks. All things considered, the Portuguese Academy of Sciences did not command a large international reputation.

Considering continental Europe at the peak of the scientific society movement there in the 1780s, almost every nation supported a learned scientific society to one degree or another. Only Spain and Austria were without major national academies of science in their capitals. But in both countries the scientific society movement was felt. In Spain plans were well advanced in the 1780s for a science academy in Madrid.[117] The motivation for this effort stemmed from the decline of humanistic academies in Spain and the successful example of science academies outside the country. The effort of the 1780s was led by Count Florida-Blanca, who contacted the Royal Society in London, the Saint Petersburg Academy, and (probably) the Paris Academy for guidance and support for the proposed Spanish Academy of Sciences.[118] With the death of King Carlos III in 1788 and the succession of Carlos IV, not a man interested in protecting the sciences, this plan fell through.[119]

While the Madrid effort failed and it was not until 1847 that the Spanish Academy of Sciences finally came into being, Spain did possess a significant provincial scientific society in the *Real Academia de Ciencias Naturales y Artes* of Barcelona.[120] Begun in 1764 as the *Real Conferencia Física,*it was upgraded into a full-fledged academy in 1770 by Carlos III under the presidency of the captain general of Catalan. Like its French neighbors, the Barcelona Academy was in these regards a provincial institution. While of scientific bent—it followed closely developments in Lavoisierian chemistry—the horizon of its interests was more limited to Catalan, and after 1790 it took up questions of industrial growth and development.

Regarding the Austrian case, we have already seen how the Austrian crown came to recognize and support a number of scientific societies in areas under its imperial control: several Italian societies and ones in Brussels and Prague.[121] One might recall that the *Academia Naturae Curiosorum* was supported by the Holy Roman Empire, and thus the Austrian Hapsburgs indirectly sponsored yet another scientific academy. Vienna, however, had no science academy of its own, but not for lack of efforts to create one. Leibniz' memorandum of 1704 was merely the first, and it was succeeded by several others of his own making, two prospoals in 1749 by Gottsched and von Petrasch, and another in 1774 by I. M. von Hess and Maximilian Hell.[122] These plans failed because they could not deal with the financial, religious, and staffing problems involved.[123] While Austria must not be ruled out as supporting academies, personally Maria Theresa seems not enthusiastic about science academies (perhaps because Frederick II of Prussia was so attached to his), and she was loath to relinquish the privileges and monopolies that would have allowed any of the proposed academies a measure of independence and stability.[124]

In sum, then, by late in the eighteenth century every nation and principality on the Continent of Europe either had an academy of science or felt the effects of the scientific society movement. Compared to the relatively pathetic organizational and institutional developments of the seventeenth century or even those of the first half of the eighteenth century, the elaboration of Continental academies of science through the second half of the eighteenth century was remarkable. It would seem an unprecedented development in the organizational and institutional history of science, equaled only perhaps by the emergence of medieval universities in the twelfth and thirteenth centuries.

The Maturation of Societies of Science

The growth of scientific societies in the second half of the eighteenth century was hardly limited to developments on the Continent, however. If, as was seen earlier, efforts to create specifically *societies* of science in the Anglo-Saxon world in the first half of the eighteenth century were weak and out of phase with the corresponding movement to create academies on the Continent, the discrepancy was rectified after 1750, and the maturation of societies of science in the English-speaking world constitutes a major thematic development in the history of the learned society movement in the second half of the century.

On the American continent, after the collapse of the first "American Philosophical Society" in the later 1740s, organizational stirrings continued to affect the colonies. The medical community began to get organized in America in the early 1760s, and "arts" societies sprang up in Boston and New York.[125] A private club, called the American Society and interested in scientific questions, met in Philadelphia in the late 1750s and early 1760s; there was also a plan developed in 1766–67 for an American Academy of Sciences in New Haven.[126] None of these continued efforts had yet produced a permanent scientific society in America by the late 1760s. The impulse that finally gave rise to one came from the unexpected quarter of political dissent. The victory of the English in the Seven Years (French and Indian) War and their further political and economic exploitation of the colonies led to a new sense of nationalism and cultural purpose on the part of the colonists. This included a fresh desire to create their own learned society in America.[127] The American Society of Philadelphia reemerged in 1766 in response to this new set of conditions, and from a closed club it metamorphosed into a record-keeping and more public type of society called the American Society for the Promotion of Useful Knowledge. Another, similar group, the American Philosophical Society, appeared in 1767, and thus Philadelphia was graced with two organizations that took the advancement of science as one of their primary concerns.[128] The differences between the two Philadelphia societies were mostly social: the Society for Useful Knowledge was predominantly Quaker and mercantile in composition and outlook, while the American Philosophical Society was largely Anglican and proprietary. Their competition for members and status greatly spurred the vitality of the two groups. In the long run, however, such competition was seen to be detrimental, and the two societies merged in 1768 to form the American Philosophical Society for the Promotion of Useful Knowledge. Discounting the ephemeral Boston Philosophical Society of the 1680s, this was the first successful extension of the learned society model to the American continent.

The American Philosophical Society was a typical institution of the society type in terms of its organization and semi-public status. It was composed of a large and undifferentiated body of membership and an administrative committee of officers.[129] Admission fees and annual dues represented the main source of funds for the society.[130] The society was incorporated by the Pennsylvania Assembly, and through the patronage of the English and later American governor of Pennsylvania, it received funds to cover extraordinary expenses.[131] In addition, the American Philosophical Society appealed to private sources (such as the merchants of Philadelphia, army officers, and sea captains) for help in securing funds or special equipment.[132] In 1785 the society was endowed by the Englishman and Fellow of the Royal Society of London, Magellan, with a prize for the improvement of navigation worth two hundred guineas.[133]

It would be a mistake to see the American Philosophical Society as more than a weak and provincial institution. Attendance at meetings was poor, and more importantly, political and military events disrupted the society in 1774 and closed it completely between 1776 and 1779.[134] In the long run, the effect of these disruptions was to dampen the enthusiasm of the Americans for their own national science that had characterized the prerevolutionary period.[135] After the Revolution American science and the American Philosophical Society in particular became less vital than before and more intent on utilitarian ends. Projects concerning surveying, map-making, engineering, and the improvement of agriculture and manufacturing, for example, were among those undertaken after the war.[136] Even at a late date in the eighteenth century, the poverty of the American intellectual and institutional environment was strongly felt.[137]

Were it not for two fortuitous occurrences, the American Philosophical Society would have been considered by the more powerful and illustrious scientific societies of Europe for what, in fact, it was: a small and unimportant outpost of science stuck on the very edges of civilized culture. First, the American society bathed in the reflected glory of Benjamin Franklin. By the 1770s, of course, Franklin had achieved a wide and popular reputation in Europe, partly on account of his scientific work proper, partly on account of his politics, and partly on account of the Rousseauistic background out of which he came.[138] Franklin was elected in absentia as president of the American Philosophical Society in 1769, and he held that position until his death in 1790. To Europeans, the Philadelphia Society was Franklin's society.

Second, the American Philosophical Society was founded in 1769, just in time to be a major institutional participant in the second transit of Venus of that year. The 1769 transit of Venus was an event similar to the LaCaille observations and transit of Mercury of the early 1750s and a "most fortunate opportunity" for the nascent society to achieve an international reputation.[139] The society served a major role and was a sponsor and clearing house for some twenty-two sets of American observations.[140] These were published in 1771 as a major part of the first volume of the society's *Transactions*. In a calculated effort, copies of this publication were sent to "the most considerable Philosophical Societies in Europe," including those in London, Uppsala, Stockholm, Berlin, Saint Petersburg, Göttingen, Paris, Bologna, and Turin.[141] Franklin, in Europe for most of the 1770s and 1780s, was instrumental in distributing the *Transactions* to the appropriate institutional and individual recipients.[142] This volume received elaborate praise from all over Europe as evidence for a new level of scientific and institutional maturity in the Americas.[143] Also, by dint of Franklin's connections and by its own lights, the society established formal contacts with institutions overseas and elected some of the most noteworthy European scientists and academicians to its membership.[144] Says Brooke Hindle of the society at this point in its history, "Although no comparable attainment was registered in the balance of the colonial period, the character of the American Philosophical Society had been adequately established to satisfy most of its European well-wishers."[145]

Merited or not, the American Philosophical Society held an honorable position in the world community of scientific societies. But it was not the only representative of the societal model to make its appearance in America in the second half of the eighteenth century. In 1780 the American Academy of Arts and Sciences was established in Boston.[146] The Boston Academy was similar in organization to the American Philosophical Society and the Royal Society of London, but because of anti-British and pro-French sentiments following the Revolutionary War, the name "Academy" was chosen.[147] The stimulus for the academy came from John Adams and his desire to do for Boston what Franklin and the American Philosophical Society had done for Philadelphia. Its main sources of support above and beyond its dues-paying membership were Harvard College and the Massachusetts legislature, which incorporated the academy. Unlike its Philadelphia counterpart, it was run by a council and shortly also had a distinguished list of foreign members. The Boston Academy did some work in astronomy, but it, too, was mostly concerned with utilitarian questions. The academy published its own series of *Memoirs*, and with the publication and dissemination of the first volume in 1785, writes Hindle, "The American Academy was accepted in America and Europe as one of the two learned societies in the United States.[148]

The appearance of the Boston Academy represented a challenge to the unique status of the American Philosophical Society. The following inaugural communication from the vice-president of the Boston Academy to the Philadelphia Society reveals something of their sense of themselves and their common situation:

We are pleased with the distinguished figure your Society has already made; and we are persuaded, it will increase in reputation and grow continually more useful to the world. I doubt not, these two public institutions of America will always harmonize, and that all the strife which will ever subsist between them will be, which shall be [sic] most beneficial to these United States and most useful to mankind. This will be a noble emulation, from which the most happy consequences will result: there seems to be a greater call than ever, for men of science in America to exert themselves to promote every kind of useful knowledge, as we are become an independent people, and are reckoned among the distinct nations of the earth.[149]

In addition to these two official societies, the American scene was enriched in the last decades of the eighteenth century by the appearance of several short-lived private societies which looked especially to the American Philosophical Society and thence to the Royal Society of London for their inspiration. The Connecticut Society of Arts and Sciences and the New York Society for Promoting Useful Knowledge are the most noteworthy of these, but one can mention the Trenton Society for Improvement in Useful Knowledge, the Kentucky Society for Promoting Useful Knowledge, and the Virginia Society for the Promotion of Useful Knowledge.[150] The efflorescence of these small American societies, like those elsewhere, defines and testifies to the scope of the scientific society movement in the eighteenth century.

Likewise in America, attention should be drawn to the curious *Académie des Sciences et Beaux-Arts des États-Unis de l'Amerique* (Richmond, Virginia, 1786–1789).[151] In the pro-French atmosphere of the 1780s in America, one A. Quesnay de Beaurepaire, grandson of the physiocrat François Quesnay, proposed this roughly Academy-like institution for the United States. The American *Académie*, though, was to be more pedagogical than the usual academy, in hiring a teaching staff. The project was undertaken with the encouragement of the Parisian Academy of Science and Academy of Painting and Sculpture![152] It was never fully operational, although one professor of chemistry, Jean Roulle, did come from Paris. The Virginia experiment folded with the French Revolution, and in the end it proved "beyond the capacities of the United States to support."[153] The instance is noteworthy as evidence of the range of social and cultural environments into which it was thought possible to implant a learned society and as the extreme, westward extension of the Continental academic movement in the eighteenth century.

But the more successful and remarkable instance of French influence on learned society development in the Americas is the straightforwardly colonial *Cercle des Philadelphes* which was established in Haiti in 1784. Originated by a French elite of rich local planters and doctors tending a morass of black slaves, the *Cercle* was in fact typical of provincial academies back in metropolitan France.[154] It elaborated a complex private constitution in 1785, received approval and funding from local authorities, and through friends in high places in 1789 was elevated by Louis XVI to the status of *Société royale des sciences et des arts du Cap Français*, although given the subsequent turmoil in France and in Haiti, the history of the *Cercle* is essentially complete by 1789.[155] The *Cercle* was composed of twenty resident members of a social composition not unlike other French provincial elites, with corresponding regional and overseas members; Franklin was a foreign member. It held public meetings and sponsored prize competitions. The *Cercle* published a volume of memoirs in 1788, and it possessed a library, botanical gardens, and cabinets. It even struck *jetons*. In all of this the *Cercle* was wholly imitative: a tropical *académie*. Yet, in an exceptional move for a group of its stature, in 1789 the *Cercle des philadelphes* made contact with the Paris Academy of Sciences with which it not only instituted an exchange, but sought and received an affiliated institutional status, just short of the union of the Paris Academy with the Montpellier Society of Sciences.[156] Nothing productive followed from this exceptional institutional relationship. But the incident illustrates, again, the unusual niches in which societies arose and the extent of their spread by late in the eighteenth century.

Institutional developments in the British Isles during the second half of the eighteenth century displayed a pattern similar to the maturation of the North American societies. The prehistory of scientific societies in Scotland and Ireland has already been touched on. Noted were several efforts through the middle of the century, the most successful of which was the emergence of the private Philosophical Society of Edinburgh in 1738.[157] After disruptions between 1745 and 1752, the Scottish society, still unofficial and centered around the university of

Edinburgh, was revived under the leadership of a new circle of men, including David Hume, Alexander Monro II, and Joseph Black. The society published three volumes of *Essays and Observations, Physical and Literary*, in 1754, 1756, and 1771, but it lapsed again in the later 1770s, and Scotland was once again without even a private scientific society of note. In the early 1780s, internal disputes within the university over control of the Regius chair of natural history, the formation of a competing Society of Antiquaries in Edinburgh, and its solicitation of a royal charter brought forth a new drive to establish a learned philosophical society in the Scottish capital.[158] Led by William Robertson, principal of the university, this latter drive was successful, and in 1783 the Royal Society of Edinburgh received a royal charter from London. The new Royal Society assimilated as Fellows members of the previous Philosophical Society.[159]

The Royal Society of Edinburgh very much resembled its sister society to the south, the Royal Society of London. It was run by a council like the London Society, it elected its members in a similar way, it published *Transactions* from 1788, and so on. The major differences between the two societies were that the Edinburgh Society was composed of two classes—literary and philosophical—and that it met monthly rather than weekly.[160]

The fact that the Royal Society of Edinburgh was established so late in the eighteenth century already puts it a bit beyond the scope of this history; it simply does not figure as that vital or important an institution. In part, this judgment derives from the fact that in its early stages the literary department of the society was the more productive of the two divisions.[161] Also, with comparatively infrequent meetings the internal work of the society was not very sustained; its main business consisted in conversation and hearing papers. In two respects, however, the Royal Society of Edinburgh is of note. On the one hand, it was another learned society on the international scene, and it attracted its share of notable members. Then, although meetings cannot be hailed for their scientific content, the *Transactions* of the Edinburgh Society fairly quickly became an important scientific forum.[162] However one evaluates it, the final creation of the Edinburgh Society in 1783 capped the efforts of many decades standing and established in the British Isles

Royal Society of Edinburgh, *Transactions* (1788),
title page engraving.

a second official scientific society to rival the heretofore undisputed position of the Royal Society of London.

This same process of an official institution emerging out of a long prehistory also came to fruition in Ireland at about the same time. Here the story is somewhat more vague. A "physico-historical" society was in existence in the 1740s, and it lasted as an informal reunion through the 1770s.[163] The Royal Dublin Society—the first of the economic societies—was founded in 1731 and incorporated in 1750. In 1772 it evidenced a special interest in Irish antiquities and set up a separate committee on the topic. In 1782, yet another group was established on the fringes of Trinity College, Dublin. Out of its broad interests (and subsuming the antiquarian branch of the Royal Dublin Society), the Royal Irish Academy for Science, Polite Literature, and Antiquities was formally created in 1785.[164] The Royal Irish Academy—so named because of its three constituent classes—was also essentially a society like its English or Scottish counterparts, but in no way can it be said to have rivaled them. Richard Kirwan did do significant work in chemistry within the academy, but the thrust of the institution seems to have been more toward regional, economic, and cultural development than toward an institution devoted to the expansion of the sciences.[165] The Dublin academy is important to this survey as further evidence that efforts to establish societies of science in the English-speaking world were reaching a new stage in the later eighteenth century.

The late decades of the eighteenth century saw a singular manifestation of the learned society movement in England. One has in mind the so-called Literary and Philosophical societies which arose in the Midlands and north from the 1780s: the Manchester Lit. and Phil. (1781), Derby (1783), Newcastle-upon-Tyne (1791), etc.[166] To a degree, these societies continued in the tradition of the Spalding and Peterborough societies. To a degree also, they represent the English equivalent of the French provincial academy movement, only a little later. On the other hand, the Lit. and Phil. societies do seem a bit different. Mostly that difference stems from the fact that these were dissenting associations, and they are early associated with dissenting schools.[167] Then, significantly, the Midlands societies were definitely part of the industrializing tradition of the English countryside.[168] None of the Lit. and Phil. societies received official recognition, and the full flowering of such institutions did not occur until well into the nineteenth century.[169] The so-called Lunar Society of Birmingham, as a private social group, does not really qualify as a Lit. and Phil. society or any kind of formal organization, but the people and the spirit involved are remarkable in this context and evidence the trend.[170] The major example, fully the equal of a solid provincial academy in France, was the Literary and Philosophical Society of Manchester. Though not officially incorporated, the Manchester Society was organized with rules and a limited membership. It elected honorary members from among the leading members of other academies and societies. It maintained contact with these other institutions, and it published a respectable series of *Memoirs* (5 vols, 1786–1807). Its archives were destroyed in WWII.

In what seems to be a response to the new level of institutional and organizational sophistication in science in the second half of the eighteenth century in the English-speaking world and on the Continent, the Royal Society of London itself made moves to bolster its position in the learned world after mid-century.

Note, first of all, that the vitality and prestige of the Royal Society had fallen off considerably from the early years of the century when Newton and Sloane led the institution, and even more so from its seventeenth-century height.[171] In 1743, for example, the Royal Society Club was established. This "club within a club" was essentially a dining club, although it had statutes and kept minutes.[172] The existence of this social appendage to the Royal Society is testimony to its unmanageably large membership and to more interest in sociability and gastronomy than science. Within the Royal Society itself there were complaints about members not paying attention, holding private conversations, and bringing unannounced guests into meetings.[173] The international reputation of the society declined correspondingly. In 1750, for example, one of Formey's correspondents wrote to congratulate him in Berlin upon his election to the Royal Society, saying, "The Society is a little numerous, but associates of your merit will always have standing over those who are merely honorary."[174] Ten years later a report directly from London was more severe: "This famous body has greatly declined from its former glory; today true men of learning disdain having themselves made members."[175] In 1765, Dr. Matthew Maty, secretary of the Royal Society, boasted about his position to the Duc d'Orléans, saying, "For the rest, the duties are not too demanding at all and take nothing away from my [medical] practice."[176] Comparing this secretary of the Royal Society who had time to take off for his medical practice with the full-time permanent secretary of any Continental academy, one gets a sense of the casualness that had come to characterize the London institution.

In the late 1770s and early 1780s, however, the Royal Society seems to have changed its attitude and reasserted its place as England's foremost scientific society. The major personnage involved in this about-face was Sir Joseph Banks, of course. A haughty aristocrat, Banks had been a Fellow of the Royal Society since 1766 and a veteran of some of its more notable activities.[177] In 1778 he was elected president of the Royal Society, a position he held until his death forty-two years later in 1820. In the long run, Banks' tenure in that position served the Royal Society poorly, but initially, in striving to make the society an instrument of his will and thereby to rid it of its sloppy habits, he was a major force in upgrading the organization.[178] There had been attempts to reform the society earlier in the century, either by introducing disciplinary classes or by demanding that Fellows produce one paper a year.[179] But it was only by taking the tradition of a strong presidency to the extreme that Banks and his supporters believed they could put the Royal Society on a par with Continental academies.

Banks first moved to make election to the society more difficult and more an honor. The number of foreign Fellows had been limited (to one hundred) just prior to Banks' presidency, but he went further and insisted on a virtual personal veto over all prospective candidates. These efforts produced a rebellion within the society—closed meetings, resignations, and pamphlets—but Banks' authority prevailed.[180] At this time, strengthening the international visibility of the society, the assistant secretary for foreign correspondence was elected to full status as foreign secretary.[181] Finally, through Banks' court and aristocratic connections, the Royal Society moved out of its cramped quarters in Crane Court and into sumptuous rooms at Somerset House provided by their royal patron at public expense and "for public use."[182] William Pringle, outgoing president of the Royal Society, made clear the society's motives in making such a move: "We would wish to have it considered, that this alotment of publick apartments to the royal Society, will be understood by all Europe, as meant to confer on them an external splendor, in some measure proportioned to the consideration in which they have been held for more than a Century."[183]

Once the Royal Society was installed in Somerset House, Banks himself continued on this theme in expressing "how much above all foreign academies" the Royal Society now stood on account of this new indication of royal support.[184] Later, in 1785, in full control of the society, Banks again expressed his feelings about the status of his institution:

Praise from Men of Science to a Monarch who protects with such unbounded liberality is but a feeble testimony of Thankfulness; we should remember with gratitude the various benefits Learning has received through our Hands, when he has made our Society the Dispensers of his Royal Bounty. [A survey of Royal Bounty follows.]. . . . Let us then seated as we are in an Apartment Convenient and Elegant beyond what we ever possessed, likewise the Gift of his Royal Bounty, Look around us and consider whether in Europe Another Monarch can be found who patronizes with equal Bounty a Society equally free by its constitution from all servile dependence.[185]

It seems from his actions and these remarks that Banks was indeed anxious for the Royal Society to appear as fully equal (if not superior) to other academies and societies of science that had grown up so powerfully from the earlier decades of the eighteenth century. In a backhanded way, these developments within the Royal Society are testimony to the fact that the learned society movement had made considerable progress throughout the world and that the Royal Society had some catching up to do.

By the later 1780s, then, Europe and the West were populated by a virtual host of learned societies of all sorts. The scientific societies were among the most numerous and impressive. Organized and institutionalized science had achieved an impressive new institutional base for themselves with a whole constellation of first, second, and third order institutions of characteristic types. The eighteenth-century personification of science—depicted on many a frontispiece—must have felt proud. The future looked bright. Did she know that the French Revolution would bring the whole edifice down? When it was time to rebuild, the age of scientific societies would be over.

CHAPTER V

The Communications Network of the Scientific Societies

T HE efflorescence of scientific societies traced in the last chapters is obviously of major importance for the history of organized and institutionalized science. A whole new group of scientific institutions came into being over the course of the eighteenth century. Several were powerful state institutions for science, and many lesser societies extended the model and presence of the learned society deep into diverse social and institutional settings. They crossed national divides, and in general they provided a remarkable new institutional basis for the scientific enterprise. Science had never been so well or powerfully organized.

The collective history of eighteenth-century scientific societies must take into account a further key feature of these institutions: the ways in which they were interconnected and functioned as an interacting system of institutions. The scientific societies were not wholly independent entities. Rather, through complex interaction, individual institutions transcended their purely local roles and separate existences and collectively forged a larger institutional network and system. This system of eighteenth-century societies was new and completely without precedent in the history of organized science, and its effects on the international organization of science were unparalleled and unmatched until well into the nineteenth century.

Something of the character of eighteenth-century scientific societies as a collective system of institutions should already be apparent from the ways various societies emerged as part of a common institutional movement. Further, we have seen something of the promise of inter-institutional cooperation in the seventeenth-century circle of societies satellite to the Royal Society of London. In this chapter we survey the ways in which the network and system of eighteenth-century societies arose more formally through patterns of institutional recognition and exchange, common members, and plans for union of institutions. In addition, certain limitations to the growth and success of this collectivity of institutions—the problem of communications—will be treated here. The chapter following continues this investigation into the systematic interactions among the scientific societies by detailing cooperative endeavors undertaken by them collectively.

Contacts and Exchanges

Just as the scientific societies themselves began to spread outward from London and Paris after 1700 and became established in Berlin, Stockholm, Saint Petersburg, and elsewhere in the first half of the eighteenth century, so too were the first contacts made among institutions and the first steps taken to create a budding system of scientific societies. Similarly, just as the history of many individual scientific societies through the first half of the century shows them to have been relatively weak, their systematic interaction in the same period was also tentative and not very productive. Around the now familiar mid-eighteenth-century mark, the elements of the mature system of scientific societies fell into place, and the stage was set for more sustained and productive relations among scientific societies in the second half of the century.

Actually, the history of contact and interaction among scientific societies extends back into the seventeenth century. We have already seen how the Royal Society of London and the Paris Academy of Sciences failed to maintain ongoing relations after the 1670s and how, on the positive side, the Royal Society and its attendant "satellite" societies formed the first example of a cooperating system of institutions. We also saw that the Royal Society kept up these domestic connections, as it were, with its sister societies in Britain through the first decades of the eighteenth century. Until 1727 and Newton's death as its president, however, the Royal Society was by and large insulated from the Continent and what little learned society activity there was there.[1]

The first significant break in this essentially quiescent pattern of inter-institutional relations occurred in 1726/27 and was occasioned by the foundation of the Imperial Academy in Saint Petersburg. Given the ways in which the Russian academy grew out of and depended upon foreign contacts, one would expect a certain openness toward the international scientific community and other scientific societies. But it is still noteworthy that one of the first official acts of the Saint Petersburg Academy was to contact other scientific societies in Europe. On the same day, September 6, 1726, it addressed introductory Latin letters to the Royal Society in London, the Paris Academy, the Berlin *Societas*, and the University of Uppsala (and hence the *Societatis* there).[2] The Saint Petersburg Academy clearly felt the desire and saw the opportunity for fruitful contact with other scientific societies in the west.

There is a poignancy to the letter the Saint Petersburg Academy sent to the Royal Society in London, in that it arrived and was read at the last meeting of the Royal Society chaired by Isaac Newton.[3] Newton's reign was not strong on international contacts, and the Saint Petersburg letter marks a new beginning, not only for the Royal Society but for all eighteenth-century academies and societies. It deserves to be quoted at length.

To the renowned President of the famous English society and its worthy and most honorable fellows, the St. Petersburg Academy sends its greetings. Inasmuch as the foundation of your Academy and the celebrated discoveries it has achieved have been admired by the French, no less than by the Germans (so much so that, roused by the English example, they too have founded similar societies and academies), we doubt not that all men of learning will deem the decision of the Emperor Peter the Great (confirmed by the generosity of the Empress) to found the St. Petersburg Academy worthy of the highest praise: that it might not only disseminate learning throughout the Russe Empire (where it was but thinly spread before) but also enrich medicine no less than mathematics with new discoveries and methods. We hope that our observations will be useful above all for the development of astronomy, which should be all the more interesting to you in that up till now these have rarely been prosecuted in the North. While other transactions of the Academy are contained within the volume that is being sent presently, we thought it appropriate to enclose two contributions by our academicians with this letter. For us nothing can be more gratifying than the growing approbation of Your Learned Society, first by virtue both of its foundation and the authority it enjoys, and yielding to no other Society in both the number of its fellows and their services and merit.4

This communication was followed shortly (April 27, 1727) by a letter from G. B. Bullfinger, a member of the Petersburg Academy, to James Jurin, secretary to the Royal Society: "containing an Invitation to Establish a free Correspondence between the two Societys on Philosophical Subjects, Assurances of their Readiness to perform any Observations or Experiments in that Countrey which shall be proposed to them and the reciprocal Obligation for the favour of any timely Notices which shall be made to them of any thing new or curious here in England."5

This more explicit proposal for cooperation between the two institutions actually resulted in an experimental program being drawn up by the Royal Society for the Saint Petersburg Academy: the precise position of Saint Petersburg, the length of a pendulum beating a standard second there, ice in winter, atmospheric refraction, meteorological data.6 These instructions were forwarded to the Saint Petersburg Academy. Two years later, in 1729, an official exchange of publications between the Royal Society of London and the Saint Petersburg Academy (the *Commentarii* for the *Philosophical Transactions*) was begun, with Delisle as intermediary.7 Discounting the seventeenth century, this was seemingly the first official exchange of publications and establishment of "diplomatic relations," as it were, between two major scientific societies in the eighteenth century. Subsequently other books and materials were exchanged between the English and Russian institutions; the Royal Society received visitors from the Saint Petersburg Academy (Martini in 1729, Müller in 1730), and news of the two institutions was regularly conveyed by mutual correspondents Johan Amman and Prince Cantemir.8 An effort was made within the Saint Petersburg Academy to translate into French the index and tables of contents of the *Philosophical Transactions* in order to make the work of the English more accessible to the Russian academicians.9 Finally in this connection, the Royal Society and the instrument maker George Graham furnished astronomical instruments for the Imperial Observatory in 1737.10 From this evidence, it is clear that the Royal Society and the

Saint Petersburg Academy fairly quickly established a durable institutional connection and an important link in the overall structure of scientific societies and international science emerging in this period.

The 1726 letter of the Saint Petersburg Academy to the Paris Academy was preceded by contact between the Paris Academy and the court at Saint Petersburg. The election of Peter the Great as a special member ("above all ranks") of the Paris Academy in 1717 has been mentioned. Upon that occasion the academy in Paris shipped a complete set of its *Mémoires* from 1699 to Saint Petersburg, and Peter named his first doctor, Blumentrost, to handle a "commerce of letters."[11] The introductory letter from the Saint Petersburg Academy itself spoke of Peter's attachment to the Paris Academy, the policy of scientific exchange between the two countries, the formal foundation of the Russian Academy under Catherine's leadership, and its interest in science and natural history.[12] The letter further proposed in the name of truth and friendship coordinated plans and common endeavors to be undertaken by the two scientific societies. Unlike the case in England, however, this letter and its proposals met with no immediate response from Paris. Its receipt is not recorded in the records of the Paris Academy, and between 1717 and 1742 contact between the two institutions was limited to that between individuals.

The 1726 letter sent by the Saint Petersburg Academy to Sweden was addressed to the rector and the "learned professors" of the University of Uppsala, no doubt bringing it into the hands of the *Collegium Curiosorum* and the nascent *Societatis* there.[13] The letter spoke of the recent peace between Sweden and Russia and proposed not only an exchange of publications and information but actual common meetings (*commune conferendi*). The reception of this overture was not as dramatic as in England either, but a steady relationship gradually built up between the academy in Saint Petersburg and the neighboring scientific community in Sweden. In 1735, for example, Benzelius and Benzelsternia, librarians of the University of Uppsala, were in contact with the academy in Russia; in 1737 Celsius entered the picture, and from then the *Acta* of the Uppsala *Societatis* were regularly received in Saint Petersburg.[14] The record speaks of a regular *commerci epistolici* between the two institutions.[15] Surprisingly, there was no early contact with the

[157]

Swedish Academy of Sciences. Even though it was founded in 1739 and Celsius was one of its leading members, the Swedish Academy had no official contact with its cousin in Saint Petersburg through the 1740s.[16]

The 1726 Saint Petersburg letter to the *Societas Scientiarum* in Berlin was the most interesting.[17] It spoke of the number of new academies founded of late, how no one could ignore the *Societas*, how there were actually more Germans in Russia than in Germany, and how the two institutions ought to cooperate in joint work (*iunctis operis sociatisque ingeniis concurramus*). The result was disappointing and predictable, given the lamentable state of the Berlin institution in those early years: one communication in 1731 from J. T. Jablonski, court appointed president of the *Societas*, and three sets of astronomical observations from the astronomer Kirsch between 1735 and 1737.[18] The two institutions really had very little to do with each other until the time of Frederick II and until Euler and Maupertuis immigrated to Berlin.

In addition to these contacts with foreign academies and societies, the Saint Petersburg Academy was also in close touch with Jesuit missionaries in Peking. A regular "commercium litterarium" began in 1734 and from there a vigorous program of exchange (both ways) continued at least through the 1740s.[19] Indeed, in 1739, one of the Jesuits in China, Father Gaubil, was declared by consensus ("in recognition of his services to the Academy through his correspondence and his aid in new discoveries") a member of the Saint Petersburg Academy.[20] The Saint Petersburg Academy was in surprisingly close contact with the Portuguese *Academia Real da Historia*.[21] Large packets of books were sent to Lisbon in 1735 and again in 1740.[22] In general, a regular correspondence was maintained over the expanse between the two capitals.[23]

In England, Newton's passing marked a watershed in the external relations of the Royal Society. Hans Sloane succeeded Newton as president of the society, and in general Sloane pursued a more vigorous administration of the society than Newton in his declining years. An assistant secretary (later a full secretary) for foreign correspondence was created by private bequest in 1727.[24] One P. H. Zollman held this office until 1748 and actively performed his job as translator/secretary.[25] Sloane reformed election procedures for foreign members, and he strove especially to establish better contacts abroad.[26] Considering seriously the nature of foreign membership for the first time really, and establishing firm regulations for election to that category, the Royal Society voted that foreigners did not have to be present for official reception into the society and that, while no fees were due either, letters of recommendation were required.[27] In a letter of 1740 to the academician Duhamel du Monceau in Paris Sloane spoke of this last requirement.

As far as M. Pitot is concerned, I am so thoroughly persuaded of his talents that I do wish to see him a member of our body. But there is this prior condition required for the election of all foreigners, that is, unless he has communicated observations to the Society, he must be recommended by two or three members from his area. Thus, Sir, as soon as you will find it convenient to send me a certificate for this savant, signed by a few of our Parisian members, I will be happy to assist him, as much for myself as for my friends.[28]

This regulation insured at least minimal institutional control over the new Continental members the Royal Society received with a seriousness unequalled since the days of Henry Oldenburg.[29]

We have just seen how the Royal Society was able to capitalize on its post-Newtonian position to make ties with the academy in Saint Petersburg. Much the same pattern repeated itself with regard to the Royal Society and the learned men and institutions in Sweden. From what has been seen of the international background to institutional developments in Sweden and traditions of contact with England and the Royal Society, this comes as little surprise. The Royal Society in London was in regular contact with Benzelius and the professors of the University of Uppsala from 1724, and in 1730 the charter of the *Societatis Scientiarum* was forwarded to the Royal Society for inspection.[30] The *Acts* of the *Societatis (Acta Literaria Suecia)* were regularly communicated to the Royal Society after 1730 by Triewald and Celsius; from 1742 the *Philosophical Transactions* were sent in return.[31] This exchange between the Royal Society and the Swedish *Societatis* was another major and regular inter-institutional exchange established between scientific societies, though it postdates by over a decade that between London and Saint Petersburg of 1729. After 1739, the *Handlingar* of the Swedish Academy of Sciences in Stockholm were forwarded to the Royal Society in London in a seemingly unofficial exchange through the same private channels as the *Acta* of the Swedish

Societatis, but somewhat surprisingly the Royal Society had no official contact with the Stockholm Academy until 1750.[32]

Italy was another place where the Royal Society of London established early contact. In 1722 it actively sought to promote "a philosophical correspondence with the learned of Italy" through the offices of one of its Fellows, Thomas Dereham.[33] For many years Dereham was an important intermediary in providing the Royal Society with news literary and scientific from that part of the world, including correspondence with the secretary of the Bolognese Academy.[34] In 1729, after the legal problems of the Bolognese Academy were resolved, Marsigli sent its charter directly to the Royal Society, and in 1732 Zanotti instituted an early official exchange of publications (the Bolognese *Commentarii* for the English *Philosophical Transactions*) between the two institutions.[35] And in 1737 the Royal Society had Graham prepare astronomical instruments for the Bolognese observatory.[36] In discussing the sending of this present, Dereham spoke of the Bolognese Academy as "in Alliance with our Society."[37] Here is another major case and a third for the Royal Society in the first half of the eighteenth century (Saint Petersburg, Bologna, and Uppsala) where an official and recognized tie between societies (and the Royal Society in particular) crossed national boundaries and added to the emerging network of European scientific societies.

The case of the Royal Society and the Berlin *Societas* is more straightforward. The London and Berlin societies were in official contact from 1724 over a matter to be discussed momentarily. Through 1731 Jablonski communicated regularly with the Royal Society, and in one instance, say the records of the Royal Society, "expresse[d] throughout the warmest disposition of that Society to Contribute whatever lyes in their part towards Cultivating a good Correspondence with this."[38] In 1731, after Jablonski's death, the new secretary of the Berlin *Societas*, Ludovicus Casper, wrote "of his readiness to keep up a correspondence with th[e] Society," but relations were broken off.[39] The two institutions never had much to do with each other in the first half of the eighteenth century, and they had no official exchange of publications. The instance testifies to the weak condition of the Berlin *Societas* and is a good reminder that the scientific societies did not immediately establish close and formal relations.

In addition to its comparatively active role in establishing contact with other societies and communities of scientists in the later 1720s and 30s, the Royal Society of London was further responsible for a significant innovation in the pattern of inter-institutional relations and a major step forward in the elaboration of the system of scientific societies in its proposal for a cooperative project involving systematic meteorological research. In 1723 Jurin suggested a plan for the "joint observations of the weather" in various parts of the world.[40] One needs to consider how such a program meshed with the institutional capabilities not only of the Royal Society, but of scientific societies in general. As institutions, not many projects were open to them. All institutions—academies and societies alike—depended primarily on their members to make scientific contributions. But a systematic investigation of the weather was one area where institutions could take charge and actively work to advance the cause of science. This is especially true of the Royal Society with its traditional Baconian interests in natural phenomena and its international network of F.R.S.[41] Indeed, Jurin's initial proposal did not mention cooperation with other institutions per se, but stressed meteorological observation by "gentlemen" associated with the Royal Society.

As it turned out, however, this initial program, implicitly limited to the seventeenth-century mode and a dependence on foreign F.R.S., soon became mixed with and augmented by institutional cooperation with the scientific societies emerging on the Continent. The correspondence with Jablonski of the Berlin *Societas* largely concerned joint observations of the weather.[42] Weather reports were also received from Sweden (the Uppsala-*Societatis* group) and the Saint Petersburg Academy.[43] The Dereham-Italian connection served to forward observations from the Bolognese academy and Naples.[44] (Additional reports were received from individuals as far away as Bengal.[45])

One problem facing such a project was the fact that observations had to be made with comparable instruments (barometers and thermometers largely) if meaningful conclusions were to be extrapolated from the data, and, to this end, the Royal Society sent eighteen sets of instruments to its correspondents and institutions abroad.[46] After some years of this project, one reads in the records of the society for 1731 that a volunteer, Dr. Derham, "had Undertaken of Examining and Comparing the Journals of the Meteorological Observations made in different parts of the World in Order to Collect and form some general Remarks."[47] As impressive and as path-breaking as it was, this first attempt at large-scale meteorological data-collecting by institutions would seem largely a failure. There were no formally announced procedures, standards, or reporting channels; the balance between individual and institutional contributors was not clearly worked out, and the project does not seem to have attracted great or sustained interest.[48] It was, over all, a tentative undertaking, one that prefigures a much more important weather project later in the century and the general pattern of cooperative endeavors pursued by the scientific societies.

Connections between the Royal Society and the Paris Academy of Sciences in the first half of the eighteenth century have yet to be discussed. It is not the case that the two institutions were entirely isolated from each other. There was a fair amount of indirect contact between them. In London, for example, extracts from the Paris *Mémoires* were presented regularly by individuals, and the prize questions of the Paris Academy were received regularly there also.[49] Similarly, excerpts from the *Philosophical Transactions* were occasionally read in the meetings of the Paris Academy.[50] A certain Woolhouse, living in Paris, was a key intermediary (like Dereham in Italy) between the two institutions.[51] But not much can be made of this contact. From the Paris point of view, membership in the Royal Society seems to have been little more than a superficial honor with little or no scientific follow-through.[52] Harcourt Brown has investigated the relations of Buffon and Voltaire with the Royal Society, and Douglas McKie has done the same in the case of Fontenelle.[53] Generally speaking, their studies reinforce the notion that the Paris and London societies were loosely connected at best. Indeed, only five letters from Fontenelle, permanent secretary of the Paris Academy, to the Royal Society survive from his long tenure in that position.[54] The report of one of these gives the impression that the two institutions were not operating at all on the same plane: "A Letter of thanks from Mons[r] Fontenelle, Secretary of the Royal Academy of Sciences at Paris, upon the Occasion of his being elected a member of this Society was read, whose compliments thereupon, tho very considerable were observed to proceed upon a Mistake, that the Constitution of this Society is like that at Paris, where the favour of Election is bestow'd on singular merit, without asking."[55]

English participation in and communications to the Paris Academy were somewhat more regular. Hans Sloane (P.R.S.), Cromwell Mortimer (S.R.S.), and Martin Folkes (P.R.S.), were faithful correspondents of the Paris Academy.[56] The communications from the institutional representatives of the Royal Society were not official communications, however. The records for 1709 of the Paris Academy show only, for example, that "a letter was read from Mr. Sloane to the Abbé Bignon in which he thanks [Bignon] for having gotten him into the Academy and promises [to send] different observations or curiosities."[57] There was no formal institutional recognition or commitment of exchange included in this letter, and indeed the subsequent letters of Sloane, Mortimer, and Folkes were limited to casual observations like rattlesnake-bite cures, effects of castration on fish, a new fire pump, and so on. Of note, too, in this connection, is the fact that in 1727 Edmond Halley lost out on his bid to be elected to the *associé étranger* position created by Newton's death, "On the only account," says Woolhouse, "that the English might not pretend to an hereditary right to succeed one of their Compatriotes."[58]

In only one major instance in the first half of the eighteenth century was this informal and low-level pattern of relations between the Royal Society and the Paris Academy augmented by official contact and exchange. Coincident with the expeditions of the 1730s sent out by the Paris Academy to Lapland and to Peru to measure the shape of the earth, Godin, Paris academician, French royal astronomer, and editor of the *Connaissance des Temps*, proposed to the Royal Society in February of 1735 that the two societies exchange a set of standard weights and measures.[59] The utility of such an exchange can be seen in the following report on the project made to the Royal Society.

Some curious Gentlemen both of the Royal Society of London, and of the Royal Academy of Sciences at Paris, thinking it might be of good use, for the better comparing together the Success of Experiments made in England and in France; proposed some time since that accurate Standards of the Measures and Weights of both Nations, carefully examined and made to agree with each other, might be laid up and preserved in the Archives both of the Royal Society here and the

Royal Academy of Sciences at Paris: which proposed having been received with general approbation of both those Bodies, they were thereupon pleased to give the necessary directions for the bringing the same into Effect.[60]

In October of 1735 a present of the same from the Paris Academy was received in London, and in February of 1736 in London, it was "ordered that Mr. Graham be desired to procure, at the Expense of the Society, the English Standards of Weights and Measures, for a Present to the Royal Academy at Paris, in the name of the Society."[61] It was not until 1742, however, that the English set was completed and sent off to Paris.[62]

That the English and French science societies would attempt to harmonize scientific experiments is certainly noteworthy—a fine example of institutional and scientific cooperation. But it is hard to understand why the two societies were not exchanging publications in this period. It is a curious question not just on its face, but also because a French translation of the *Philosophical Transactions* was being made in Paris at just this time, in part under the auspices of the Paris Academy.

François de Brémond (1713–1742) seems to have begun his translation as a private endeavor in 1734.[63] By 1737 Brémond had issued perhaps two of the three volumes of his *Transactions philosophiques* (this series ending in 1740), covering the English numbers for the years 1733–36. Brémond's effort attracted the attention of the authorities, notably the Chancellor of France, d'Aguesseau, and the Paris Academy of Sciences.[64] Brémond's accomplishment apparently matched the felt needs of the Paris Academy, for in 1737 Brémond was taken on officially to translate and work with the English *Philosophical Transactions*.[65] Perhaps it was in this context that Brémond put together a complete, annotated index of the *Phil. Trans.* from 1665 to 1735, which appeared in 1737. But Brémond had a tendency to reorganize the contents of the *Phil. Trans.*, and his translation was felt to be unsatisfactory. Nevertheless, in 1739 Brémond was elected to the Paris Academy as an adjunct in botany. In all likelihood it was at that point that a new, official plan of action was sought to handle this important translation effort. Brémond's éloge by Mairan, read before the Paris Academy in 1742, picks up the story.

The importance of the subject having attracted the attention of the learned, the Chancellor (who has as much at heart the advancement of the sciences as he does mastery of them), having been informed of the work and of the capacity of M. de

Brémond, assembled in his quarters several members of the two Academies, Sciences and Belles Lettres[!], to deliberate on the manner in which to render this translation more complete, more useful, and more agreeable to the public and the Company to which it especially pertains.[66]

As a result of these deliberations, Brémond retranslated and reorganized his volumes for the English years 1733–36, and then he produced an additional three volumes for 1731–33. These appeared in 1740 and 1741.[67] Unfortunately Brémond died suddenly in 1741/42, but the Chancellory still considered the project "a subject worthy of the attention of government," and the Abbé Gua and the doctor Demours were appointed to succeed in the work.[68] A volume of one sort or another may have appeared before 1743, when Gua and Demours broke down in their collaboration.[69] In 1745, Morand, librarian to Louis XV, wrote to Mortimer of the Royal Society that the translation had stopped entirely "on account of some difference of opinion in those concerned."[70]

The Royal Society heard news of this activity in 1737 and again in 1739.[71] The Royal Society elected Brémond a Fellow in 1740, being "well known for his accurate Translation of the Philosophical Transactions into French."[72] The Brémond translation was making its way to the Society of Sciences of Montpellier, a second edition of his Index to the *Phil. Trans.* appeared in 1739, and there seems even to have been a (French) edition published in Bologna.[73]

By all indications, then, before its collapse after his death, the translation by Brémond would seem to have been a thriving enterprise with several interesting twists. The high-level government interest in the productions of the Royal Society and the initiative displayed by d'Aguesseux and the Chancellory seem unusual. One wonders about the nature and extent of the interactions between the French academies of sciences and belles lettres in their dealings over the Brémond translation. The story shows, further, that the academicians of the Paris Academy were not ignorant of the contents of the *Philosophical Transactions* at the time, and, indeed, they seem to have expressed a keen concern for getting an adequate translation made. But for present purposes, certainly, it is extraordinary that the Brémond episode occurred with no other direct contact between the Royal Society and the Paris Academy over the matter. No official notice of the activity reached London; no offer of the *Phil. Trans.* reached Paris. It was only after this episode that the Royal Society and the Paris Academy did, in fact, begin sustained contact and a direct exchange of publications.[74]

Compared to the Royal Society of London or the Saint Petersburg Academy, the foreign relations of the Paris Academy itself in the first half of the eighteenth century were meager. Of course, as part of their "union," the Paris Academy was in regular and close contact with the Montpellier Society of Sciences.[75] The Paris Academy also had brief contacts with the academies in Bordeaux, Béziers, and Toulouse; only with the latter institution, however, did this reach the level of a correspondence between their two permanent secretaries, and this was only in 1746.[76] Generally speaking, although provincial academies throughout the eighteenth century were indebted to the Paris academies and the Academy of Sciences in particular for their institutional models and for creating an overall atmosphere facilitating their development, they were hesitant to subvert their provincial status and commitment to the larger national design of Paris institutions.[77]

The range and depth of international contacts of the Paris Academy outside of France were hardly more extensive. The Paris Academy had some direct and indirect contact with the Berlin *Societas* while Leibniz was its president, but after 1713 this little died out completely.[78] There was no formal contact with Sweden until 1740, and even then contact was limited to communications from individuals: Linnaeus, Celsius, and Wargentin.[79] There was no direct contact with the *Societatis* of Uppsala. Similarly, there is no evidence that the Paris Academy received communications from the Bolognese Academy during the period under consideration here.

The Paris Academy did maintain strong ex-officio contact with the Saint Petersburg Academy through individual members. Dortous de Mairan, future secretary of the Paris Academy, was one prominent foreign correspondent of the Saint Petersburg Academy over the period.[80] Of signal importance to the relations between the two academies and the international face of the Saint Petersburg Academy as a whole was the astronomer Joseph Nicolas Delisle. An associate member of the Paris Academy, Delisle went to Saint Petersburg in 1727, seeking the career awaiting him in the academy there.[81] His ties to Paris and the Paris Academy were more than just pro forma after he left, however. One clause of the contract that brought him, his brother, Delisle de La

Croyère, and the instrument maker Vignon to Saint Petersburg notes that they can make "astronomical observations at the times and in the parts of Russia it will please them to choose, with the approbation of the Court, and can send their astronomical observations to the Academy of Paris without their being retained, retarded, or otherwise impaired."[82] The 1726 Russian letter of introduction to the Paris Academy also made note of Delisle's special position in both institutions. Delisle himself corresponded directly with the Paris Academy, the Royal Society in London, and even the Montpellier Society, and he was a major intermediary between the Saint Petersburg Academy and such other foreign correspondents as de Mairan and Réaumur of the Paris Academy and Sloane, Graham, and Mortimer of the Royal Society.[83] Delisle returned to the Paris Academy in 1741 and remained an important figure on the international scene until his death in 1768.

Through roughly 1740, then, while a number of overtures had been made and some initiatives taken, positive contacts between and among scientific societies were limited. The Royal Society had the most extensive formal contacts and institutional exchanges through its ties to the Saint Petersburg Academy, the academy at Bologna, and the Swedish *Societatis*. The Saint Petersburg Academy was open to contact in all directions, but little of substance emerged from its willingness to entertain relations with other societies. The Paris Academy did not actively pursue relations with other scientific societies, and was comparatively closed off despite the fact that it was well known and influential internationally. The Uppsala scientific community was active in many sectors (but not in Paris); the Swedish Academy of Sciences in Stockholm, however, does not figure in the chronicle of these events; neither does the Berlin *Societas* to any signficant extent. Overall and despite some instances of substantive cooperation and actual common endeavors, the scope and depth of the international relations of the scientific societies does not seem very strong in the first half of the century.[84]

Then, in the 1740s and more especially in the 1750s, the situation turned around completely. We have already observed that the major institutions of the first half of the century faced and overcame difficulties and that the character of the scientific society movement changed at mid-century. To find that, as a collective group, the various parts rather suddenly began to interact formally and officially around mid-century

only reinforces the point about a transition at that time. But it is true—
a whole constellation of official contacts precipitated through the 1740s
and became commonplace in the enlarging system of scientific societies
after that.

The transformation of the Berlin *Societas* into the *Académie royale* in
1744–46 brought great promise. Contact between the new Berlin Academy and the Saint Petersburg Academy picked up considerably when
Maupertuis and especially Euler went to Berlin. Then began what
Messrs. Youschkevitch and Winter have called the "union" (*Verbindung*)
between the Prussian and Russian academies of science.[85] Euler, of
course, brought with him a host of contacts and connections from the
Saint Petersburg Academy. And Maupertuis was an international figure in science and already a favorite of the Russian Academy—witness
his special pension of two hundred rubles a year as a foreign honorary of the Russian Academy.[86] Maupertuis also made overtures to the
Royal Society of London in 1745, as reported in the minutes of
the Royal Society.

The President [Martin Folkes] said that he had lately received a Letter from Mr.
Maupertuis who informs him that he is now going to leave France, upon an
Invitation made by the King of Prussia; And that he is to be made President of
the Royal Academy at Berlin; And expresses how sensible he would be of the
pleasure and satisfaction of maintaining and cultivating a good correspondence
with this Society in Philosophical matters.[87]

Such communication argues for a new beginning to the relationship
between the Berlin Academy and other scientific societies, but, as it
turned out, the Berlin Academy remained surprisingly diffident toward other societies.[88]

The year 1742 brought renewed contact between the academies in
Paris and Saint Petersburg. By dint of their indirect connections
through individuals, news of events in the respective academies was
reciprocally and regularly communicated, and prize questions of the
Paris Academy in particular were seen in Saint Petersburg every year.[89]
In 1739 individual copies of the Russian *Commentarii* were sent to Cantemir, Fontenelle, Dortous de Mairan, Maupertuis, and Réaumur, but
there was still no official contact between the two academies.[90] Finally,
in 1742, fourteen years after their last official contact, the following

mundane but significant reports appear in the registers of the Paris Academy:

18 April: The Prince Cantemir sends to the Company the seventh and eighth volumes of the Memoirs of the Academy of Petersburg, on the part of the Academy. . . .
21 April: In consequence of what was said in the preceding Assembly about the Memoirs sent by the Academy of Petersburg, the Company judged it appropriate to send to this Academy the last four volumes of our Memoirs.[91]

From that point on, the two institutions regularly and directly exchanged publications and other productions, and added another important element to the emerging system of scientific societies.

The decisive missing link in the overall scheme of things was between the Paris Academy and the Royal Society of London. It is unclear which side finally initiated the exchange. The secretary of the Paris Academy records in its registers for May 14, 1749: "I read a letter from Mr. Mortimer, Secretary of the Royal Society in London, by which he informs me that this Company proposes to send to the Academy the Philosophical Transactions and to accept in exchange its *Mémoires*. This proposition was accepted and the Academy as a result charged me with making him a response."[92] Yet two letters from Mortimer to Grandjean de Fouchy (then secretary) written at just this time make no specific mention of this proposed exchange.[93] Buffon, who actually sent the Paris *Mémoires* to London, intimates that the Paris Academy was making the overture. He wrote to Folkes in April of 1750:

I have sent a few days since directed to you a case, wherein you will find an entire set of the memoires of the Academy of Sciences, this case will be sent to London free of all charges of carriage or duty, and I pray you to get it delivered to you: It comes from the Academy who have commanded me to send it to your Royal Society and to pray their acceptance of it. . . . The Academy will constantly take care to have all volumes they shall hereafter publish, sent to your Society in the same manner as soon as they come out.[94]

In July of 1750 sixty-nine unbound volumes of the Paris *Mémoires* did arrive in London.[95] The volumes were ordered to be bound and thanks returned to the Paris Academy for its "noble present."[96] There was some confusion in the council of the Royal Society as to how to procure a complete set of the *Philosophical Transactions*, whether to send them bound or unbound, and so on, but in 1753 a set was at last sent and received in Paris, and henceforth the two institutions reciprocally began to exchange their productions on a regular basis.[97] Regardless of which party first initiated it, one wonders what beyond the new mid-century climate prompted the Royal Society and the Paris Academy to relinquish decades worth of standoffishness and begin a new partnership.

For its part, the Royal Society generally seems to have increased its contacts with foreign institutions at this time. Its unofficial and indirect exchange of publications with the Swedish Academy in Stockholm was transformed, likewise in 1750, when: "The President informed the Council that Dr. Abraham Back, Physician to the King of Sweden, had presented to the Society the Treatises published by the Swedish Academy from 1741 to 1748 inclusive. [It was] order'd that the Society in return, do make the Academy in Sweden a present of the Philosophical Transactions from 1746 to this time."[98]

In 1752 the editing and publishing of the *Philosophical Transactions* were officially taken out of the personal control of the secretary of the Royal Society and made official functions (run by committee) of the Royal Society itself.[99] In 1753, now in full control of the distribution of the *Philosophical Transactions*, the council of the society reassessed its policy of exchanges and presents abroad and: "Ordered, That copies of the new Volume of the Transactions be presented as follows: To the Royal Academies of Paris, Berlin, Gottinghen, Sweden, Madrid [*Real Academia Medicina?*]; the Universities of Oxford & Cambridge; the Societies at Spalding and Peterborough; to Signor Zanotti at Bologna, Dr. Trew at Nuremberg, and Dr. Weidler at Wittemberg."[100]

At this time, too, it was decided that the *Philosophical Transactions* would make "a proper present for the Missionaries in China Corresponding with the Society."[101] And in 1755 the list was amended so "that the 48th volume of the Philosophical Transactions, and such volumes thereof as shall be hereafter published, be sent as a present from the Society [to] Count Rosamonski [sic; Razumovskii, the director of the Saint Petersburg Academy] in Russia."[102] While the Royal Society had entertained correspondence and exchange with all of these groups and institutions prior to the 1750s, it is clear that the society was at that time taking new stock of its international relations with other scientific societies and pursuing them in a more organized fashion.

In terms of new exchanges and general activity on an international level, the most remarkable development of the mid-eighteenth century was the emergence of the Royal Academy of Sciences in Stockholm as a new institutional power. The key man here was the new secretary of the academy, Wargentin. Wargentin was a pupil of Celsius and had studied with Delisle in Saint Petersburg. Once he returned to Stockholm and assumed his new position as academy secretary in 1749, he immediately launched a new policy for the academy of direct contact with other European scientific societies.[103] The new official exchange with the Royal Society in London was handled first through Dr. Back, but Wargentin himself wrote to Mortimer very shortly thereafter and "invite[d] the Doctor to a mutual communication of the most important discoveries and observations made by the two Societies in England and in Sweden." Unlike other such proposals and as he did with all his correspondents, Wargentin sent substantive communications, in this case Swedish astronomical observations and records of magnetic variations at high latitudes.[104] Regarding the Paris Academy, Wargentin had been elected its and Delisle's correspondent in 1748, but in 1750 the relationship between the two institutions was formalized and yet another official program of exchange instituted.[105] Similarly, between 1750 and 1752, the Swedish Academy made contact and began regular exchanges with the academies of Saint Petersburg and Bologna.[106] Wargentin himself later spoke of his efforts to extend the foreign contacts of the Swedish Academy during this period.

Before my time, the Academy had (so to speak) no correspondence with foreign countries, but, as I had already earlier exchanged letters with a few astronomers of other countries and as I had always found a great pleasure therein, I sought, in my name and that of the Academy, to give even more of an extension to this correspondence; all of which permitted me to establish some ties (like a confraternity) between our Academy and a few of the most famous foreign ones. Now, every year they send us their acts and other works, and our Academy sends them its own to which, in the desire to increase our prestige and our glory, I generally join other works appearing here which seem to me to honor our nation and the tastes of foreigners, such as the works of Linnaeus and others of the same type. . . . A few literate foreigners have, on account of this, been obliged to learn Swedish; others have had translated certain articles of our Acts.[107]

The advent of Wargentin as secretary of the Swedish Academy spelled the end of its isolation from the rest of the international community of scientific societies and marked its forceful entry onto the international scene.

With all of these fresh, new contacts and exchanges between and among the major international scientific societies and with its own institutional position that much more secure after 1746, one would expect the Berlin Academy to have likewise joined in the quickened pace of inter-institutional activity. But such was not the case. Maupertuis *was* broadcasting his new position at Berlin, and prize circulars and news *were* sent out to other institutions.[108] Conversely, the Saint Petersburg Academy sent its *Commentarii* and *Novi Commentarii* to the Berlin Academy regularly, beginning with ten volumes in 1748; and, the Royal Society of London also sent its *Philosophical Transactions*.[109] The Berlin Academy itself, however, was not sending its *Mémoires* to foreign institutions at this time, and it had no official programs of exchange.[110] It is a surprising case of institutional isolation that is hard to account for given the strong position of the Berlin Academy in the later 1740s and early 1750s.

The Berlin Academy was an exception to what was obviously becoming a general rule at mid-century. As a regular feature of their interactions, scientific societies entered into formal and reciprocal exchanges of their publications and other productions. The exchange of publications may not seem like much, but certainly, as a normal part of inter-institutional relations, it bespeaks a more sophisticated quality to international science and connections between scientific societies than had been the case previously. Indeed, in entering into reciprocal exchanges, academies and societies necessarily made direct and official contact with one another and hence instituted ongoing relations for which the actual exchange of books was merely the most concrete symbol. Thus, for example, the Dijon Academy sent its first volume of *Mémoires* unsolicited to the Montpellier Society of Sciences in April of 1770; in June the Montpellier Society reciprocated by sending its *Mémoires* in return.[111] The registers of the Dijon Academy describe the thinking of the two academies.

Mr. D'Aigrefeuille to whom the [Dijon] Academy also sent a copy of its *Mémoires*, expressed his gratitude in the strongest terms, and speaking of the volume which the Royal Society of Montpellier presented to the Academy, he said, "The reciprocity of the present that it [Montpellier] offers you [Dijon] should make you understand how much it desires to maintain a forthcoming correspondence that will be cultivated with care." This correspondence is equally the desire of the Academy, which senses all the advantages that it can get out of it, and it was decided that the secretary in a letter he will write to the members of the Royal Society of Montpellier will convey to them that we will spare nothing to keep up this useful correspondence.[112]

In the case of the exchange and ensuing "correspondence" between the American Philosophical Society and the Turinese Society of Sciences, begun in 1773, the terminology is more striking.

The Royal Society of Sciences in Turin have received with the utmost Gratitude, the Scientific Repast of the American Philosophical Society, and have (with the approbation of their President) accepted of the Correspondence of a Society compos'd of such Illustrious & celebrated members, [and] as a token of their gratitude & Esteem, they have sent them such of their Transactions, as were already printed, assuring them that they will send from time to time all their future publications in order to maintain (with greater security) their Desire of an honble Union for ye Advancement of ye Arts and Sciences.[113]

These examples bear out the view that exchanges between institutions were the occasion for establishing more formal institution-to-institution contact, in the guise of "correspondences," and that these might be likened to the establishment of diplomatic relations between nations.[114] In the long run, the fact of such contact between institutions has to be reckoned of great import for the international organization of science in the eighteenth century and the developing system of scientific societies.

Beginning in the 1750s, then, and continuing throughout the rest of the eighteenth century, with the exception of the Berlin Academy, exchanges and correspondences increased steadily and constituted the most prevalent form of inter-institutional relations among academies and societies. They were the groundwork on which more impressive cooperative endeavors were built. Indeed, the sending of memoirs and the solicitation of exchanges became largely routine after 1750 for any new academy or society. We have already seen in the case of the American Philosophical Society and the first volume of its *Transactions* how important such an act could be for the reputation and standing of any new scientific academy or society. The Dutch Society of Sciences provides another example. Founded in 1752, it contacted the Royal Society of London and the Paris Academy of Surgery about exchanges in 1756, and, after a wait appropriate to its station, in 1766 it similarly contacted the Paris Academy of Sciences.[115] Two more representative examples: At the Paris Academy, "A letter was read from M. Desrocher, Secretary of the Brussels Academy, who announces that the Academy of Brussels is sending to the [Paris] Academy the collection of its works. It was decided that the [Paris] Academy would send its current volumes."[116] In the Royal Society of London, "An Application being made on the part of the Electoral Society at Manheim [sic], for a mutual interchange of transactions as they are published; Resolved, that it be done accordingly."[117]

From these seemingly routine examples it is possible to infer that at all levels and whenever feasible, academies and societies established contact with one another and instituted reciprocal exchanges. Informal survey reveals, for example, that between 1750 and 1793, the Paris Academy and the Royal Society regularly received various publications from over thirty scientific societies. The Saint Petersburg Academy was regularly in receipt of publications and miscellaneous materials from over twenty other institutions; the numbers for the American Philosophical Society were in the neighborhood of a dozen, as they were for the Berlin Academy.[118] We have seen distribution lists drawn up by the Royal Society of London and the American Philosophical Society, and other institutions did likewise. These lists typically included libraries and universities as well as other academies and societies of science, so that institutional publications received an international distribution to a wider audience than merely the scientific societies, although these were clearly the mainstay institutions.[119]

Through individual exchanges, then, a whole network of inter-institutional ties was established that crisscrossed Europe and the English-speaking world and that tended to join scientific societies into a more cohesive whole. But further, these exchanges in effect created regular channels for the distribution of science and news of science. The distribution system established by the scientific societies was the most significant means for communicating science in the eighteenth century.[120] Consider, for example, the special prize sponsored by the Paris Academy of Sciences in 1775 on government initiative and with government support, concerning new ways to obtain saltpeter.[121] One might expect that this particular contest, because of its bearing on the manufacture of gunpowder, would have been limited in one way or another to France or Frenchmen. Such was not the case, however, and the Paris Academy made a special effort to contact foreign academies and to broadcast the contest abroad.[122] In this it used pre-established channels of inter-institutional communication. The records of the Royal Society of London reveal something of how this worked.

The Royal Academy of Sciences, having conveyed to this Society in a letter from their Secretary Mr. de Fouchy to Dr. Maty, S.R.S., their last Programma, offering a premium for a new & better method of making Salt Petre, to be determined at their meeting, & having desired that this notice might be made publick in the Country, that Gentlemen might have time to make experiments & prepare their observations on so universally interesting a subject; the Secretary had orders to return the Thanks of the Society to the Academy for this communication, & to express their wish to concur in their beneficent views.[123]

Another noteworthy example of the recognition of the distribution network of the scientific societies and its use as a means of disseminating science was the project undertaken in 1784 by the English Board of Longtitudes. The report appearing in the registers of the Saint Petersburg Academy gives a good account of this project.

The Secretary communicated a letter dated June 24th from the Admiralty in London and addressed to the President of the Imperial Academy of Sciences by Mr. H. Parker, Secretary of the Commissioners of Longitude, which states that the said commission has destined for the Imperial Academy a copy of all the works which it publishes and it asks [the Academy] to have them collected from the bookseller Elmsley where they will be regularly deposited. The Secretary reported having already . . . written to Mr. Magellan to have him perform this service.[124]

In letters of the same day, Parker and the board also approached the academies in Paris, Berlin, and Stockholm with the identical proposal.[125] In his letter to the Swedish Academy, Parker stated that the board was doing this "as a mark of the Respect they bear to the Learned Societies and particular Individuals in Foreign Nations."[126] The Board of Longitude apparently recognized that by late in the eighteenth century the best way to have its productions distributed throughout the learned world was precisely to offer them to the major scientific societies abroad. The episode is of small importance in and of itself, but it does indicate the extent to which the communications network of the learned societies was the foremost system for distributing the productions of science, and, by inference, how much institutional exchanges were vital to the conduct of science in the period.

The Berlin Academy was the outstanding exception to the apparently universal pattern of inter-institutional exchanges emerging in the second half of the eighteenth century. The Berlin Academy did receive publications from other institutions, but fewer than were received by other major societies. The absence of the *Mémoires* of the Paris Academy, as well as other German academies is particularly striking. More importantly, the Berlin Academy itself did not as a matter of course send its own publications in return. A regular program of exchange was not instituted with the Royal Society of London until 1769, even though the Berlin Academy had received the *Philosophical Transactions* earlier.[127] The Berlin *Mémoires* were not received in Saint Petersburg until 1782, and the exchange of publications between the Prussian and French academies of science did not begin until 1786. Incredibly, exchanges with the learned institutions in Sweden did not occur until 1810.[128]

The explanation for this peculiar state of affairs rests in the fact that, after the death of Maupertuis in 1759, the Berlin Academy came under the direct personal control of Frederick II.[129] Frederick appointed himself head of the class of speculative philosophy, and assumed the exclusive right of appointment to the academy.[130] This was his first personal involvement in the institution he then consistently referred to as "my Academy." In the period until his death in 1786, Frederick also turned to the Frenchmen d'Alembert and (then) Condorcet as the "invisible" presidents of the Berlin institution. Overall, the Berlin Academy lacked almost any semblance of autonomy in the period. As an indication of how Frederick's personal rule limited the Berlin Academy in the second half of the eighteenth century, it seems clear that he had no wish for exchanges with other scientific societies. In 1777, for reasons of simple *Kulturpolitik*, he did order the sending of the *Mémoires* of the academy to several of the smaller academies of Italy.[131] But beyond that, his leadership has to be seen as primarily responsible for the isolation of the Berlin Academy from the active pattern of institutional exchanges going on around it.

This conclusion and a suppressed desire of the academy to enter such exchanges are evident in an internal memorandum, "On the Sending of the *Mémoires* to Foreign Academies," circulated just prior to Frederick's death.[132] This very interesting document with its marginal notations (by Merian) deserves to be reproduced at least in part. In considering the sending of their proceedings, the men of the Berlin Academy first reconsidered the exchange with the Royal Society of London.

[Memorandum:] A. The [Royal] Society not having responded, either White [bookseller on Fleet Street] is a rogue who has not forwarded the copy intended for the Society, or the Society does not care about our *Mémoires*.

[Note:] The Royal Society of London, especially Mr. Magellan, its Secretary or at least one of its principal members, corresponds with Mr. Bernoulli and has sent us his own works as well as those of his colleagues. [The Society] sends us its *Transactions*, thus the exchange is advantageous for us.

[Memorandum:] B. Looking at the thing in general, I would say . . . [sic] that the rush to send our products to foreign academies is a kind of avowal of inferiority which does not at all suit.

[Note:] Yes, unless this be an exchange.

[Memorandum:] C. That the little academies of Italy to which our *Mémoires* have been sent preferentially, hardly have a right to this distinction.

[Note:] I would except the three which have replied and which are more celebrated, those of Padua, Turin especially, and Siena, by virtue of the royal order of 1777.

[Memorandum:] D. That the academies of Petersburg, Stockholm, and Paris enjoy a completely different reputation.

[Note:] An exchange is taking place between [the Academy] of Petersburg and our own; that of Paris itself has just proposed one to Mr. Bernoulli and will give us the *Description des Arts et Métiers* to boot[!]

This manuscript provides a good insight into the nature and hierarchy of exchanges and the peculiar circumstances in Berlin in the later eighteenth century. Implicit in it, too, is a sense that the Berlin Academy, above and beyond the restrictions placed on it by Frederick, was extraordinarily concerned with its position in the confraternity of scientific societies and was loath to part with its productions without getting something substantial in return. In this regard, we might recall that the academy was in the business of producing almanacs, maps, and

Mémoires for money.[133] A mercantile spirit was characteristic of the Berlin Academy at this time, and it only added to the isolation of the institution. After the death of Frederick II and the reappraisals evidenced in the memorandum above, the academy did begin a full program of exchanges not only with the major academies and societies, but also with lesser institutions such as those in Brussels, Boston, and Padua.[134] At last, the Berlin Academy, too, joined the ranks of those institutions that exchanged their works. Until then, the Berlin Academy was an important exception that, by omission, added a characteristic element to the structure of international science and the system of scientific societies in the eighteenth century.

Common Members

In addition to direct contact and exchange, the scientific societies were united through individuals who belonged to two or more societies simultaneously: their common members. We have already seen how common members as individuals often performed important services for societies and assisted in establishing and coordinating relations between them. Given his ties to the academies in Paris and Saint Petersburg, Joseph Delisle stands out as one noteworthy example for the first half of the eighteenth century. Intermediary roles continued to be exercised by a host of men in the second half of the century, and they constitute an important element underlying inter-institutional relations during the period. Charles Baër, for another example, was a Lutheran pastor attached to the Swedish embassy in Paris; he was a member of the Swedish Academy of Sciences and a correspondent of the Paris Academy of Sciences. During his thirty-year stay in Paris, Baër was a key go-between for the two institutions. He regularly attended meetings of the Paris Academy and was granted the special honor of sitting with its associate members. There, he presented the productions of the Swedish Academy and translated extracts of the *Handlingar* into French. He also assisted exchanges between the two societies and was a constant source of news.[135] His importance in making and maintaining contact between the two scientific academies in Paris and Stockholm was very great.[136]

[178]

Honorary foreign members of the Saint Petersburg Academy (man for man also members of other academies and societies) as a group stand out among the various categories and instances of membership. They were of special import for the international activities of the academy in Russia and its relations with other scientific societies. The thing that distinguished these foreign honorary members of the Saint Petersburg Academy is that they were *paid* an annual pension of two hundred rubles, and hence they were actually agents of the Saint Petersburg Academy. During the middle part of the eighteenth century, the honorary foreign members of the Russian Academy residing in Berlin—Euler and Maupertuis again—were especially important in bringing the two respective academies closer together. Euler, for example, promoted an exchange of sorts between the two institutions by privately sending the Berlin *Mémoires* to persons in Saint Petersburg for the Russian *Commentarii* destined for individuals in Berlin.[137] In part the seeming lack of direct contact between the academies in Berlin and Saint Petersburg during this and later decades was made up by these quasi-formal, indirect ties. In 1766 Catherine II (the Great) assumed the throne in Russia, and the Russian Academy received a significant boost in the level of support and financing it received from the empire. In particular, it was able to expand its class of honorary foreign members, who in the same manner as previously but more intensely, forged contacts between the Russian Academy and other European scientific societies. Inna Lubimenko describes the change.

In the early existence of the Academy, invitations to foreign savants as well as the ordering of books and instruments was accomplished through the mediation of a few of its members who were sent abroad. As some among these profited by not coming back, this occasioned a certain reluctance and cautioned prudence. The enlarged budget of the Academy permitted it to augment the number of honorary foreign members receiving a pension of 200 rubles a year, thanks to which the Academy ended up by possessing in practically every important center a paid member who concerned himself with its orders. Such was the role of Lalande in Paris, Wargentin in Stockholm, and Formey in Berlin.[138]

One need hardly point out that the men mentioned were not simply the agents of the Russian Academy, but were also important members of the scientific academies in those cities; we know that Wargentin and Formey were indeed permanent secretaries of the academies in Stockholm and Berlin; regarding Lalande's role, Pavlova, who has studied his correspondence with the Saint Petersburg Academy, likewise sees his activities as establishing closer ties between that institution and its sister academy in Paris.[139] The indirect and quasi-direct role played by the honorary foreign members of the Russian Academy clearly led to "the strengthening of its relations with foreign academies and societies."[140] This special example indicates just how important common members could be in furthering inter-institutional relations during the eighteenth century.

The significance of common members as contributors to inter-institutional relations varied tremendously, of course. On the one hand, such men as Baër, Lalande, Formey, Wargentin, and others were instrumental in establishing and maintaining ties between and among institutions. At the other extreme, there is the extraordinary case late in the century where the entire membership of the American Philosophical Society was elected en masse to the Royal Society of Valencia (presumably the *Sociedad Económica*).[141] Despite such variability, the thousands of instances of common membership in the learned scientific societies of the eighteenth century, taken together, like exchanges, created a pattern of affiliations that can reveal in a more precise way the larger structure of ties established among these institutions.

Consider, as a single example, the common members of the Paris Academy and the Royal Society of London. Figure 6 displays this portion of their respective memberships.[142] A total of 174 men belonged to both institutions at some point between 1699 and 1793. Twenty-nine men who were based around the London society became members of the Paris institution; conversely, sixty-three Paris academicians were elected F.R.S. In this case, not at all surprisingly, the Royal Society shows itself the more open institution. It was easier for the men of Paris to get elected to the Royal Society than it was for members of the Royal Society to get elected to the Paris Academy. More significantly, the chart displays the geographical distribution of this instance of common membership.[143] The nonresident membership of any institution was not limited to a single geographical locale, of course, but was spread out. These third-party members of any pairs of societies themselves were not unaffiliated, but belonged to and many times led other scientific societies. For example, the common member of the

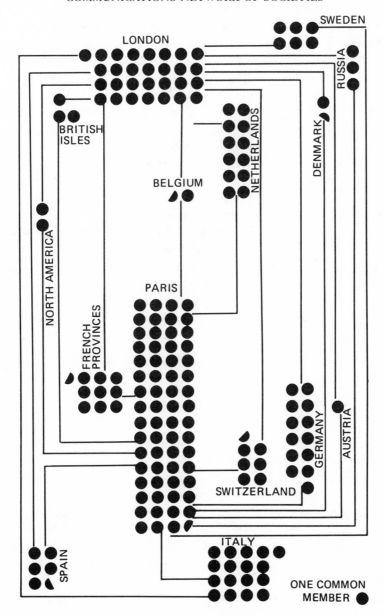

Figure 6: Common Members of the Paris Academy and the
Royal Society.

Paris Academy and the Royal Society from the United States was Franklin, naturally, and his common membership in both institutions reinforced and strengthened the ties already and otherwise established between the London and Paris societies and the American Philosophical Society. Similarly, Euler sounded a reinforcing note for the Saint Petersburg and Berlin academies, Haller for the Göttingen society, Wargentin for the Swedish academy, Jablonowski for the Leipzig group; Zanotti, Bologna; Lorgna, Verona, and so on. One sees in figure 6 only one slice through the international scientific community of the eighteenth century, and considering that such patterns of direct and indirect common membership arose between each pair of scientific societies—or, put another way, that each society took a stance vis-à-vis the others in a unique way—one is led to the realization that the scientific societies (and other notable institutions of science) were tied inextricably together through complex, interwoven sets of common members.[144] In this way the international scientific community of the eighteenth century was largely institutionalized in the scientific societies; its top members in top institutions. And, in this way also, the system of scientific societies achieved its reality through a human dimension.

Movements Toward Union

We have been making the case that eighteenth-century scientific societies collectively constituted a linked system of institutions. If such a system existed, and given what has already been encountered concerning the common historical development of scientific societies, contacts and exchanges among institutions, and their common members, one would expect to find evidence of efforts to unite formally individual societies at higher levels of organization. In this, we are not disappointed, and, while of limited success, plans to organize societies into various national and international groupings again testify to the reality of the network and system of the scientific societies in the later eighteenth century.

France was a natural place for such efforts. There were more learned societies scientific and otherwise there than anywhere else, and France fell under the umbrella of one national government and culture. One sees that several efforts, in fact, were made to unite the activities and administration of the French academies.

To begin with, as the number of French provincial academies continued to grow through the eighteenth century and as their places in local contexts became more secure, provincial academies themselves began to establish more formal and systematic contacts. Roger Tisserand mentions special institutional ties established in the period 1765–1770 between the Dijon Academy and those in Besançon, Metz, Lyon, and Montpellier. Pierre Barrière speaks of the same for the academies in Nancy and Lyon and those in Bordeaux and Marseille, likewise in the 1760s; more instances could probably be added to the list.[145] As a result of such contacts, inter-institutional correspondence and exchanges (more often of prize circulars and notices of public meetings than of published *Mémoires*) became commonplace among the academies of provincial France as they did elsewhere in Europe. What distinguishes French provincial academies and their interconnections from other groups of scientific societies in the same period is that further efforts were made to join them into a more cohesive whole. Already in the late 1750s, the Paris-based periodicals *Le Mercure* and *La France Littéraire* began to coordinate the flow of information among provincial academies.[146] In the 1760s several other like proposals were put forward by the academies themselves, with the result that smaller groups of institutions (such as the academies in Lyon, Toulouse, and Bordeaux) became closely associated with one another.[147] The most outstanding plan of this sort was that put forward by the Abbé Yart of the Rouen Academy in 1774 for a central office for provincial academies in Paris to be staffed by a paid journalist and to publish both a monthly bulletin and an annual collection of memoirs. Yart defined the problem as follows:

The utility of academies is so largely recognized that we would do well to take every means of extending and augmenting it. No way is surer than that of a general communication among their scientific or literary members. They have a pressing need of each other to enlighten and perfect themselves by mutual aid. Yet, our academies hardly know one another. The success of some are lost for others, and while they should form but a single body of light, hardly a few rays escape from them to reflect on lands near or far. The capital takes little benefit from the knowledge in its provinces as do these from their reciprocal works.[148]

Nothing came of Yart's proposal, but, as Barrière puts it, the idea of creating a federation of learned societies was common currency in the French provinces in the 1760s and 1770s.[149]

The next project of this nature likewise dates from 1774, but was initiated from Paris and not from the provincial level. The relations between the provincial academies and the Paris Academy of Sciences in particular were not strong in the period. There were, of course, the special ties between the science academies in Paris and Montpellier, and several permanent secretaries of the provinces were in contact with the Paris Academy over details of academic organization and administration.[150] In addition, efforts were made to formalize a union between the Paris Academy and the Marine Academy at Brest, unsuccessfully in 1753 and then successfully in 1771.[151] It was with a view toward strengthening the position of the Paris Academy and alleviating the problems of communication between the capital and the provinces (to which Yart refers) that Condorcet offered his plan to link all the scientific academies into a single administration.[152]

Condorcet framed his proposal in various forms. It was different for presentation to the royal ministers, Maurepas and Malesherbes, and so also when it was forwarded to the various provincial academies.[153] Essentially what Condorcet had in mind, however, was the fusion of all French academies into a single national organization, the elimination of local customs and procedures, the regularization of membership and organizational forms, the standardization of equipment and observations, and the creation of a central administration in Paris, presumably headed by the Academy of Sciences, to oversee the whole thing. This was an ambitious proposal, the possible realization of which bespoke a new level of organizational sophistication among French academies.

Between 1774 and 1776 Condorcet forwarded versions of his proposal to the academies in Dijon, Lyon, Rouen, Bordeaux, and possibly elsewhere.[154] The response was mixed, but generally negative. The Dijon Academy, seemingly always on Paris' side, was impressed by the emphasis Condorcet placed on the availability of a new publication forum for provincial academies, and, despite a letter from Buffon who suggested that "the moment has not yet arrived when this project conceived by M. Condorcet can be executed," the plan received a favorable report by a special committee.[155] Guyton de Morveau wrote the acceptance of the Dijon Academy: "Your project can only be very advantageous and very honorable for the provincial academies. . . . You can therefore be assured, Sir, that once this association is established, our academy will willingly engage itself in the fulfillment of its conditions."[156]

The response of Séguier, permanent secretary of the academy at Nîmes, revealed a mass of contradictions and problems with Condorcet's plan, however.[157] Séguier suggested that the level of intellectual and scientific activity in the provinces was not that great, that papers for the proposed publication would be hard to come by, that correspondents of the Paris Academy already enjoyed all the perquisites suggested in the plan, and that Condorcet was asking a lot of permanent secretaries. Séguier emphasized that provincial academies also concerned themselves with literature, eloquence, poetry, and historical studies, and hence that coordination with the Paris Academy of Sciences would be difficult. About what was diplomatically phrased as the advantage stemming from an increased protection from the government, Séguier concluded, "I don't believe that this last reflection can be read in full assembly. It must be modified . . . you must be more prudent."[158] The academies in Lyon and Bordeaux rejected Condorcet's proposal on much the same grounds.[159]

Indeed, Condorcet's plan for a general administration of provincial academies was heavily nationalistic, bureaucratic, and "centralist" in its overtones and clearly something associated with the new regime of Louis XVI and Turgot. In this, Condorcet overlooked the strong grain of particularism and egalitarianism affecting the academies of the provinces, and these feelings were primarily responsible for the failure of his plan. But with the fall of Turgot's ministry in 1776, the plan was a dead letter anyway.[160]

That did not mean, however, that the idea of a union among provincial academies did not continue to receive active support. In 1777 the academies in Lyon and Bordeaux were again in contact over the question, and in the early 1780s *La France Littéraire* revived its correspondence project.[161] Until 1786, these various efforts to unite provincial academies in one form or another had made little headway. At that time Dubois de Fosseux, secretary of the academy at Arras, established his successful and impressive bureau of correspondence.

Dubois de Fosseux was elected secretary of the Arras Academy in 1785, and from the beginning of his tenure, it was clear that he was interested in expanding the dimensions of his office and his academy, as seen in his successful efforts to elect the first nonresident correspondents to the academy. His vision was larger, however, and intimately connected with the (by then traditional) problem of securing coordination among the provincial academies of France. Shortly after his election, de Fosseux wrote to the secretaries of several dozen other academies to propose active relations with the academy of Arras. Between 1785 and 1792 he made many efforts to establish and enlarge a list of individual correspondents, sending out nearly 14,000 letters during the period.[162] In the end Fosseux was successful in creating an immense network of correspondence that encompassed some fifty-four learned societies and hundreds of individual correspondents.[163] There were two keys to Fosseux's success. On the one hand, he was able to obtain free postal service from the government. After several preliminary steps and a petition to Paris, in 1787 regional intendants became empowered to provide a postal cover for inter-academic correspondence, and thus a single correspondence network covering the whole of France and linking all the academies came into being.[164] The second element involved in Fosseux's bureau and its success was his development of the collective letter. Rather than writing separate letters to each of his correspondents, Fosseux coded paragraphs according to subject and assembled the appropriate paragraphs so that each correspondent received the individualized communication that best matched his own interests, and Fosseux was spared considerable time and effort. Three hundred collective letters were expedited in 1787; over twenty-five hundred were sent in 1788.[165] From these figures, the scope and magnitude of the Fosseux enterprise are evident. Here, at last, was the emergence of a national organization for science in France and what has been termed an "academy the size of the kingdom."[166]

Because French academies outnumbered the learned societies of other countries and because they shared a single national identity and culture, the development of an informal national organization of academies in France in the second half of the eighteenth century was clearly of great moment for the whole system of scientific societies and the overall organization of science on the international level. Yet efforts to establish larger, national organizations were not limited to France. In the United States, for example, there were two proposals in the American Philosophical Society—one in 1780, the other in 1790—to create committees of correspondence covering the American continent to communicate with the society in Philadelphia.[167] The second of these was written into the bylaws of the society, which, admittedly, on a much lesser scale than in France, tended to enforce a modicum of national organization for science in the United States. Similarly, from Italy, Antonio Lorgna of the *Società Italiana* wrote to Sir Joseph Banks in 1783 of his idea for a general learned society to unite that country.[168] Then, against the backdrop of many plans and trials to combine German language academies (the *Deutsches Gesellschaften* movement) and to form an overarching "Deutsche Gelehrtenrepublik" for the German-speaking world, in 1793 one R. Z. Becker made an important proposal for the union of all German scientific, economic, and industrial societies.[169] This plan was successful, and in 1794 some fourteen different German societies, including all the German scientific societies mentioned earlier, became united in a loose confederation.[170] The foundation of this confederation occurs somewhat late in the chronology of this tale, but it indicates that in Germany, too, the same tendency toward unions of institutions was operating.[171] Later in 1793 Becker made another proposal for uniting all European learned societies.[172] Nothing came of this plan, but it evidences the beginning of a movement to formally organize science and scientific societies on the international level.[173]

The Problem of Communications

The long trail of evidence and history encountered thus far strongly suggests that old style scientific societies of the period 1750 to 1790 collectively constituted the most successful and imposing form for organized science of that day or any preceding day. The course of the scientific society movement from its origins in the seventeenth century, the establishment of academies and societies of science as the dominant scientific institutions of the period, and the rise of a real and working network and system of scientific societies linked internationally through patterns of exchange, common members, and an impressive series of common and cooperative endeavors indicate that eighteenth-century scientific societies lent a structure and coherence to the organization of science that was unmatched by any other potential mechanism, institutional or otherwise, for uniting and coordinating the goals and processes of science on local, national, and international levels.

As successful as they were, academies and societies of science of the eighteenth century and the unified institutional network they created inevitably turned out to be limited and imperfect. Many forces, each deserving particular attention, operated independently against further international cooperation in eighteenth-century science and against strengthening the system of scientific societies: the weather and other material factors, personal rivalries, war, nationalism, and so on. Each of these limits undercut the strength and vitality of the system of eighteenth-century scientific societies and in the end reduced the whole structure of internationally organized science as it was fashioned by these institutions to something less than it might have been. While real and not a figment of our historical imaginations, the international system of scientific societies so far documented was in many ways weak and tenuously based.

By far, most of the problems affecting the societies were not directly their fault. But there was one major and particular problem facing organized science in the last third of the eighteenth century for which the scientific societies were largely responsible: the lack of swift and efficient communications in science. The publications of the scientific societies were the major forums for the presentation of new research in science in the eighteenth century, and through contact and exchange, the scientific societies maintained real and important channels for communicating and disseminating science. But these publications and channels served only the scientific societies and their closest members. Apart from the institutional mechanism offered by the learned societies, quick and efficient communications in science were available only to persons able to invest time in a large-scale personal correspondence or those willing to follow developments in science at several removes in the popular press. The main problem in scientific communications in the eighteenth century involved lengthy delays in the publication of scientific research in the proceedings of the scientific societies, delays raging up to a decade or more. In this regard, one might even question how much scientific societies, in publishing their proceedings, were actively seeking to promote scientific communications, as opposed to establishing archival or merely symbolic records of their accomplishments.[174]

Improved dissemination of the work of the scientific societies to and for a larger community was already an issue in the early 1750s, when the *Collection Académique* first appeared.[175] That there were problems in quickly disseminating the products of the scientific societies or that the *Collection Académique* filled such a gap is evident in the following communication of 1774 sent by Lalande in Paris to Wargentin in Stockholm.

> The eighth volume from Bologna, which is the only one that you lack, is as rare here as it is in your neighborhood, and I don't know if I will find it. But if you want the French volume of the *Collection Académique* containing extracts from Bologna, or even the volume of the Academy of Sweden [in the *Collection Académique*], I can send them to you. I don't know either if the fourth volume from Turin can be found in Paris.[176]

The *Collection Académique* is an interesting development, since it suggests that not only were academies and societies reaching a new level of maturity and visibility at the mid-century point, but also that there were scientific shortcomings to the ways in which scientific societies by themselves institutionalized and communicated science.

Another indication that scientific societies were not satisfactorily handling the task of communications in eighteenth-century science can be seen in calls made for better communication of the results of prize contests. Comprehensive lists of contests and winners were compiled by de Ruffey of the Dijon Academy in 1769 and by Delandine in 1787. These lists are potent manifestations of a felt need for improved communications regarding prizes.[177] Similarly, the projects of Condorcet, Dubois de Fosseux, and others for greater contact among French provincial academies likewise incorporated plans for improved scientific communications, and they indicate that, to an important extent, the scientific societies themselves strove to shore up this fundamental shortcoming in the way they organized and communicated science.

One feature of the problem was that in academies generally only resident academicians were allowed to publish in an institution's

Mémoires. The Paris Academy made a significant effort to rectify this situation. In 1747 papers of foreign men of science sent to the academy were assembled, and in 1751 appeared the first volume of the *Mémoires de mathématiques et de physique présentés à l'Académie Royale des Sciences par divers sçavans & lûs dans ses Assemblées,* or *Savants Étrangers* as it was known.[178] Eleven volumes of this series were published until 1786 and, therewith, the learned world outside of Paris secured not only an additional forum for the publication and dissemination of its scientific productions but also the recognition that its works were an integral part of the activities and purposes of the Paris institution.[179]

In the final analysis, however, as a mature form for the organization of science, academies and societies themselves were unable to completely effectively or satisfactorily handle communications in eighteenth-century science. The best evidence for this judgment (and, indeed, that there was such a problem) is provided by the several projects of the 1770s and 1780s to improve communications in science which arose outside established institutional and learned society channels and which sought to build upon the flawed organization of science inherited from the scientific societies. The three projects were Rozier's journal, the correspondence bureau of the Patriotic Society of Hesse-Hamburg, and the Paris assemblies of Pahin de la Blancherie.

The Abbé François Rozier (1734–93), the prime mover behind the first and most successful of these projects, in 1773 gave the most forceful expression of this problem of communications and its roots in the failure of the learned societies.

The learned societies succcessively established by their sovereigns made known great numbers of discoveries by the publication of the collections of their memoirs. However, most academical collections are written in national languages and printed several years after the memoirs have been read. During this time we remain ignorant of facts that could be of great utility for the sciences. And then, these collections have become very numerous and thus very expensive and [they] are often beyond the means of those who could best profit from them.

It seems that as the number of savants has grown, correspondence between nations has slowed down. Everyone doubtless believed that the national academies would be sufficient and that all the necessary support would be gained from them. The constitutions of these companies, formed by sovereigns, having admitted foreign correspondents, seemed to remedy the problem. But this precaution, so wisely undertaken, has not been justified by success.

It results from this poverty of communication that the progress of the sciences is very slow, that the learned of different nations work for a long time on the same subjects, and that they lose precious time in acquiring a glory that in the end is problematical. This drawback is even less than that of work on a subject already enlightened by published works of which one is ignorant. An author wastes time he could better employ for the good and glory of his country, if, in entering upon the career he choses to follow, he had beneath his eyes the [full] picture of the current state of physical knowledge and the term at which it had arrived.[180]

For science nothing could be more reasonable. The means by which Rozier sought to facilitate communications and to secure a better flow of information for science at large was through the publication of his own journal, the *Observations sur la physique, sur l'histoire naturelle et sur les arts* (Paris, 1771/1773–1826).[181] The *Observations* (or Rozier's Journal, as it is commonly known) was a novel and important periodical for a number of reasons. On its own, Rozier's Journal was exclusively a scientific journal and was limited to publishing original material in science. (These characteristics distinguish the *Observations* from the rest of the loosely scientific press of the period and place it alongside the proceedings of the scientific societies as a major forum for publication in science in the late eighteenth century.) But speed of publication was the key characteristic of the *Observations*. It was published monthly and got material into print with a promptness "that would be considered exemplary even today."[182] All things considered, it proved that day's best means of spreading news of science. Speed of publication also separated the *Observations* from proceedings of the scientific societies. Because publication of news and memoirs in it was so quick and efficient and because delays in the publication of society proceedings were so widespread, Rozier's Journal "filled a lamentable gap in the scientific periodical literature" and provided a stong alternative medium for publication in science in the 1770s and 1780s.[183] Yet another important feature of the *Observations* was its self-proclaimed audience of the serious, professional man of science. As Rozier put it in 1773:

We believe ourselves obliged to circumscribe our limits in order to make this work even more worthy of men of science [*savants*]. . . . We will not offer to idle amateurs purely agreeable works or the sweet illusion of believing themselves to be initiated into sciences of which they know nothing. . . . We offer this collection to true scientists [*les vrais savants.*][184]

In addition, while Rozier's Journal was clearly a new and important means of hastening communications in science, not the least of the journal's significance derives from its connections to the scientific societies. Even though, in a sense, Rozier's Journal competed with the productions of the societies, academies and societies of science loomed large in Rozier's program, and they were a firm resource behind his success. Rozier maintained regular and fairly close contact with a number of scientific societies, he actively and systematically sought their cooperation and participation, and he cast the *Observations* in the mainstream of their work and fully the successor to the *Collection Académique*. As Rozier himself put it in the *Observations* in 1779: "This journal bears no relation to periodical works that are distributed in France or in foreign countries. Its goal is to make known all the discoveries that are made daily in the sciences and to serve as the sequel or supplement to the volumes of the academies."[185] About his activities vis-à-vis the societies he wrote more directly:

My goal in sending the *Journal de physique* to all the learned companies of Europe was to engage them to cooperate with me in the progress of the sciences. The slowness with which they are propagated, the little communication that there is between one kingdom and another, in a word, everything conspires to leave us ignorant often for several years of the most useful discoveries. I therefore sought to bring the sciences together in a common center, a general depository, where everyone should have the opportunity to take credit for his work and make himself known.[186]

Summarizing what has been said elsewhere, Rozier *systematically* contacted the scientific societies with circulars, and he sent them his journal gratis.[187] Rozier came to be well connected to the Paris Academy in particular. His journal received a favorable committee report, he became a corresponding member of the academy, and he prepared special tables of their *Mémoires*. There was even a small move to have his journal taken on as an official publication of the academy. Rozier was also well received throughout the French provinces: his "home" academy in Lyon and those in Montpellier, Dijon, Bordeaux, Marseille, and elsewhere welcomed him. Rozier was particularly interested in contacting foreign academies and societies of science, and ultimately those in London, Berlin, Saint Petersburg, Stockholm, Philadelphia, and other locales were receiving his journal regularly and were in contact with him.

The story and success of Rozier's program and his journal in the period of the 1770s testify clearly enough to the reality of the problem of communications and especially international communications in late eighteenth-century science and, by comparison, to the poor job the scientific societies were doing in coping with the problem. The episode also reveals that, in launching his drive to upgrade and improve communications in the contemporary world of science, Rozier did not seek to challenge or replace the dominant scientific societies or the characteristic pattern for the international organization of scientific communications they had created. Rather, and not without reason, Rozier based his project directly on the existing system of the scientific societies, and, all things considered, he was positively received and generally accepted by them.[188] His enterprise was thus closely and concretely intertwined with the scientific societies, and in the end it formed a stopgap addition to the system of such societies and their established practices. But Rozier was not alone in this.

The Patriotic Society of Hesse-Hamburg (1775–1781) was responsible for another ambitious plan for improving communications among scientists (and others) which involved the scientific societies.[189] What distinguished this second major assault on the problem of communications among the learned in the late eighteenth century was the idea of establishing a wholly new network of corresponding bureaus, linked by a central committee, to handle the mechanics of information flow for other institutions and men. Somewhat surprisingly, the Hesse-Hamburg Society, backed by the Landgrave of Hesse, achieved a remarkable degree of success. Very soon after its foundation in 1775 it succeeded in forming over 100 corresponding committees throughout Europe and building a membership of at least 150 individuals (though the number 600 is spoken of).[190] In its first stage through 1779 the society (in conjunction with its bureaus in Paris and Zweibrücken) also published several volumes of the *Bibliothèque du Nord* as part of its overall effort.[191] The society was dedicated to a patriotic utility, and it was in close contact and formed bureaus with many economic and patriotic societies of north and central Europe.[192] While decidedly more a patriotic society, the Hesse-Hamburg organization was also in contact with the science academies in Dijon, Nancy, Besançon, Grenoble, Marseille, Montpellier, and possibly Toulouse in this early phase.[193]

In 1779, after initial organizational problems, the society reformed its activities, and in particular (like Rozier before) systematically sought to enlist scientific societies in its project.[194] One sense of the society's new focus can be seen in the letters it addressed to Samuel Formey and the Berlin Academy.

Sir: I have the honor to send to you on the part of the directorate of the Patriotic Society of Hesse Hamburg and His Highness the Landgrave of Hesse, Head and Protector of the said Society, a correspondence project he wishes to establish with the Royal Academy of Sciences and Belles Lettres of Berlin. . . . Its goal is to devote itself to the service of all the other societies, to link them by an easy and rapid correspondence, to gather all knowledge in a common depot in order to then more quickly and fairly disperse it back to the core of all nations. . . . It is certain that distance and the lack of communication among savants and the associations they have formed in the different states of Europe hurt the progress of the sciences and arts which they cultivate. Uniting them through a regular correspondence and circulating their discoveries would be to put their works to, and purify them in the crucible of discussion, experience, and tastes, and it would allow all nations to participate equally in the advantages which should result. Thus knowledge and happiness should be increased. Such has been the basis of our institute and the motive which engages us to extend its branches into all parts of Europe. Now that this extension is taking place, we desire that the academies and other societies would wish to aid us in perfecting an enterprise undertaken for their own benefit.[195]

The emphasis here on the problem of communication and the desire of the Society to enlist "academies and other societies" is noteworthy in the present context. In addition to the Berlin Academy, the Hesse-Hamburg Society made contact with the scientific societies in Stockholm, Saint Petersburg, Mannheim, Leipzig, Göttingen and possibly others.[196] The response of the societies is hard to judge. Despite the fact that the oversight committee in Paris was of overwhelming importance in the first phases of the Hesse-Hamburg Society's existence, there does not seem to have been any contact with the Paris Academy.[197] In Saint Petersburg, contrary to its established practices with regard to economic societies and questions of interest to them, the Saint Petersburg Academy accepted the proposal of the Hesse-Hamburg Society and charged the younger Euler with maintaining the correspondence.[198] The Mannheim Academy said it would join in when everybody else did.[199] The response of Besançon was lukewarm.[200] The rest remains a mystery. Nevertheless, in setting out to create and lead an auxilliary system of communications to augment existing networks of inter-institutional and personal contacts (including those of the scientific societies), the Hesse-Hamburg Society cut out a large and idiosyncratic role for itself, one that testifies further to problems in the existing systems of communications in the eighteenth century.

The third attempt to improve communications in the late eighteenth century was seated in Paris. As he outlined it in his pamphlet, *Correspondence générale sur les sciences et les arts* (1779), Pahin de la Blancherie's program was obviously related to the efforts of Rozier and the Hesse-Hamburg Society. But his appreciation of the problem was hardly as explicit or self-conscious as was theirs, and the solutions he proposed were hardly as innovative or avant-garde.

The first of the three main components in de la Blancherie's peculiar effort was the (reputedly) vast network of personal correspondence that he had built up over the years.[201] In attempting to fashion a personal and private network of systematic correspondence, de la Blancherie was actually regressing to seventeenth-century practices, before the scientific societies had even appeared on the scene. The second pillar upon which he rested his project was the establishment of a weekly reunion or conference in Paris, called the *Assemblée ordinaire des savants et des artistes*. This assembly was a strange affair, being organized and conducted outside normal institutional channels and involving a hodge-podge of savants, artists, amateurs, and women (who could attend the afternoon exhibition associated with the assembly).[202] The creation of such an extraordinary and semi-public reunion was again more like earlier, seventeenth-century practices and the conferences of Théophraste Renaudot's *Bureau d'adresse* than it was a significant innovation in the ways science was organized in the eighteenth century. Finally, to subsidize and extend his correspondence and his conferences, de la Blancherie proposed the publication of a new journal to be issued twice monthly and entitled, after Bayle, *Nouvelles de la République des Lettres et des Arts*. This part of his program (and especially the speed of the publication of the proposed *Nouvelles*) seems more serious and productive than the others. But the *Nouvelles* did not share the same commitment to the quick dissemination of original science, but, rather, was envisioned by de la Blancherie more as a means to stimulate further correspondence.[203]

It is evident from his tract that de la Blancherie was essentially setting himself up as an independent agent to promote communications for a broad range of subjects, and he seems more concerned about his personal role and the general standing of his antiquated program than he does about its possible effectiveness or success. It is also evident that the sciences and better scientific communication played only a small part in his program. Even more than was the case with the Hesse-Hamburg Society and indicative of the changing cultural climate in pre-Revolutionary France, de la Blancherie addressed himself to "the citizens of this capital" and "the artisans of all classes," not to speak of writers, painters, sculptors, and the like.[204] The bulk of his enterprise, in other words, seems directed to those who were disenfranchised by, rather than representative of the official and institutionalized science in the period.

By the same token, science and the international network of scientific societies did attract a portion of his attention. In his pamphlet he says that in the course of establishing his correspondence, "I have found myself postman for the academies. . . . I cannot terminate . . . without again offering the resource of my correspondence to all the companies and all the individuals who cultivate the science and the arts."[205] In addition, while it would seem that de la Blancherie ran his correspondence bureau largely outside established institutional channels, he did make a real effort to contact and enlist at least a few academies of science. Surprisingly, the Paris Academy of Sciences gave him a favorable report:

We have attended his weekly assemblies, called rendezvous of the republic of letters. There we have seen savants, artists, and amateurs from almost all parts of Europe. We have seen in his registers proofs of a correspondence he could only have formed with a lot of time and trouble, and we have been witness to a very rare activity and zeal that can only be very useful to the progress of the sciences and the arts. . . . The more he is encouraged, the more he will become useful both to Frenchmen and foreigners, whom he wishes to spare the trouble of a correspondence for which many are very ill suited, which fatigues many others, and which makes them lose a lot of time for lack of having at their disposal the means, contacts, and sources that M. de la Blancherie has been able to procure. One cannot overly favor correspondences, which are one of the major means of accelerating the progress of human knowledge. Consequently, we believe that the project of M. de la Blancherie merits encouragement and that the Academy can only see with pleasure the success of this establishment.[206]

De la Blancherie made a great deal of this bit of encouragement and sent his *Nouvelles* to the Paris Academy during the period 1779–1781 while his assembly was in operation.[207] Similarly, he was in contact with the Saint Petersburg Academy on several occasions, pressing them for news of that establishment and their support for his enterprise.[208] Initially his project was received cooly ("the Conference seemed little disposed to satisfy all these demands and questions"), but later copies of the *Acta* of the Saint Petersburg Academy were officially sent to de la Blancherie in Paris.[209] The Dijon Academy, too, entered into official contact with de la Blancherie, and it is probable other institutions did too.[210]

The initiatives undertaken by de la Blancherie indicate that, while attempting a personal and more global assault on the problem of communications in the late eighteenth century, his program still made room for its more purely scientific component and that, to that extent, the scientific societies had to be taken into account. Conversely, the positive reception by the three scientific academies in Paris, Saint Petersburg, and Dijon indicate that, while de la Blancherie's orientation was slightly different from theirs, it was sufficiently in line with their purposes to be viewed as a constructive means for augmenting and improving their way of handling scientific communications.

Taken together, the three programs of Rozier, Hesse-Hamburg, and de la Blancherie demonstrate that, having reached its limits in the 1770s, the international system of scientific societies was proving inadequate to meet further demands for better communications that the needs of science and the times placed upon it. The scientific societies earlier were progressive forces improving communications and the dissemination of science through their journals and their mutual contact and exchange. Their very success caused the problems of the 1770s and 1780s. A key point about these alternative projects is that they did not reject the scientific societies as the major, established means of organizing science and scientific communications. Rather, to varying degrees, they sought to alleviate the problem in conjunction with these other institutions. These alternative programs did not especially presage a new means or form for the organization or communication of science so much as they constituted ad hoc additions to the existing network of the scientific societies. Although disciplinary specialization in science was making itself felt for the first time in the last decades of the eighteenth century, it would be a while before the specialized institution and the specialized journal emerged to effectively challenge and replace the journals of the scientific societies or their channels as the primary means for men of science to communicate among themselves and exchange information.

CHAPTER VI

A Record of
Common Endeavors

Having explored the indirect ties that bound the scientific societies into a larger structure of international scientific organization in the eighteenth century, we can turn at last to the record of their common and cooperative endeavors. We observe and this chapter seeks to detail that after 1750 the leading scientific societies engaged in several cooperative projects, some of truly spectacular proportion. All of these were projects that no one institution or group of individuals could possibly have undertaken. They involved advanced planning, a division of labor between and among the scientific societies (as well as other participants), actual work, and a sharing of results (using the communications network). These projects represent the highest form of inter-institutional contact and action, and they provide the best evidence for considering the scientific societies collectively as a systematically functioning network of institutions. That the scientific societies should have begun to cooperate formally at the mid-eighteenth-century mark is also important further evidence of their transition at that date.

From the beginning in the 1660s one thing scientific societies offered as separate, individual institutions was the possibility of harnessing resources beyond the capabilities of individual researchers. That is to say, academies and societies could and did undertake large projects separately on a single institution basis. We have seen quite a great deal of this already. Recall the expedition to Lapland sent out by the *Collegium Curiosorum* from Uppsala; the Lewis-and-Clark kind of explorations and the cartological projects of the Saint Petersburg Academy; the Cassini map of France produced under the auspices of the Paris

Academy; the natural history surveys of the Bordeaux, Montpellier, and other academies. The Royal Society of London, too, sponsored institutional projects no individual could have done. The meteorological project of the Royal Society of the 1720s is one example; Halley's earlier voyage to Cayenne to test chronometers is another.[1]

The power of institutions to do science in these extended, institutional capacities is certainly something, and science and organized science were thus considerably enriched when the scientific societies came on the scene. But projects undertaken and pursued by individual institutions, as interesting and important as they were and are, are not really the concern here. Rather, it is projects undertaken by groups of institutions that demand our attention.

Regarding the Royal Society's meteorological project of the 1720s, while the invitation for observations was directed in the first instance toward individuals and the network of Fellows of the Royal Society, other institutions (particularly the Berlin Academy) became involved. The case marks the first tentative effort to enlist the collective powers of the scientific societies. To understand and appreciate the collective efforts of eighteenth-century scientific societies, however, a last single-institution project—the Paris Academy's expeditions to Lapland and Peru—needs to be considered. It was this project that set the stage for the series of great common enterprises taken up by the scientific societies in the second half of the eighteenth century.

Lapland and Peru

For a number of reasons suggested earlier (its quality staff, its *Mémoires*, its prize competitions), the reputation of the Paris Academy of Sciences grew considerably after 1699. Nothing quite so solidified the position and the visibility of the Paris Academy as the world's leading scientific institution in the first half of the eighteenth century than the expeditions it sent out in 1735 to measure the length of a celestial arc in Lapland and Peru.[2] The *reason* for sending these expeditions had to do with the biggest theoretical question in science of the day (outside of England, at least), and that was choosing between Newton and Descartes and their differing world views and scientific approaches. In particular, according to Newton and his champion at the academy, Maupertuis, the rotation of the mass of the earth should cause it to flatten at the poles. According to Descartes and his academic supporters, notably Cassini, the spin of the ethereal vortex should cause the earth to flatten along the equator. By sighting on a fixed star at one point, moving on the surface of the earth to another point where the apparent position of the star has changed one degree, and then measuring the distance between the two points, one has measured one degree of arc. What the Paris Academy did was to set out to compare two of these measurements, one at the pole, the other at the equator. If they were the same, the earth is a sphere. If the one at the pole is larger, then the earth is flattened there, and Newton is right. If the one at the equator is larger, then it is flattened there, and Descartes is right. To settle the question in general and in particular, the Paris Academy sent Maupertuis, Clairaut, Le Monnier, and Celsius to Lapland, and La Condamine, Godin, and Bouguer to Peru.[3]

The question and the event interested and affected the entire international community, of course. Indeed, one of Mortimer's communications from the Royal Society to the Paris Academy at just this time was a paper concerning an experiment conducted in London and Jamaica by Graham and a certain Colin Campbell regarding differing beats of pendulums in those places.[4] This seems to have been a move on the part of the English to assert not only their Newtonianism but also the priority of the Royal Society over the Paris Academy in dealing with the question. The exchange of weights and measures between the Royal Society and the Paris Academy (begun in 1735) was clearly an outgrowth of the controversy over the shape of the earth. The strong impact of the French expedition on the Swedish scene has been mentioned, and the participation of Celsius lent that expedition something of a more international air. In Russia in 1737, Delisle proposed and submitted to the Academic Chancellory of the Saint Petersburg Academy a "memoir on things necessary for undertaking the measurement of the earth in Russia," and later in the same year he was already detailing an "abridged report on the first operations made to measure the earth in Russia."[5] And, while the expeditions were in the field and after they returned, reports were communicated from Paris to the Royal

Society, the Saint Petersburg Academy, Swedish scientists, and almost necessarily elsewhere.[6]

The expeditions to Lapland and Peru were the largest scientific projects in the first half of the eighteenth century. Formally, they were organized and carried out by a single institution, but because of the scope and significance of the effort they attracted and affected larger communities of men and institutions. The endeavor must also have made clear that even larger projects would benefit from the mutual cooperation of several institutions.

LaCaille and the Transit of Mercury

In 1750 the Abbé N. L. de LaCaille was sent to the Cape of Good Hope under the auspices of the Paris Academy to catalogue stars in the southern hemisphere. Coincidentally, the planet Mercury was to transit the face of the sun in 1753, and LaCaille's voyage offered the further opportunity to conduct solar parallax experiments. This was the occasion for the first truly collective endeavor undertaken by the scientific societies.

In this connection a word needs be said about solar parallax, for it, too, was a pressing question in science. With the accurate determination of solar parallax eighteenth-century astronomers could arrive at a better notion of the astronomical unit (the distance between the earth and the sun) and hence the real dimensions of the solar system. The subject of solar parallax already had a history in contemporary astronomy stretching back to the seventeenth century, and it had captured the attention of such men as Gassendi, Halley, and Delisle. But the key value they were all seeking was still not known precisely by the middle of the eighteenth century.[7] Not only was the question of universal interest to astronomers, but the techniques then available to resolve it *depended* on simultaneous observations (of a planet's passage over the sun, of the moon, of the satellites of Jupiter) taken at more than one station. The astronomical unit was not to be determined without some international effort.

Harry Woolf has treated the LaCaille expedition and the similar and subsequent projects centering on the transits of Venus in 1761 and 1769 in his book *The Transits of Venus, A Study of Eighteenth-Century Science*.[8]

Regarding the effort of the early 1750s, Woolf describes the promotional activities of LaCaille and Delisle, the various *Avis aux astronomes* and *mappemondes* produced and distributed by these men for the occasions, and the important role played by private correspondence in marshalling the concerted effort needed to exploit fully the opportunities afforded by LaCaille's trip and the Mercury transit.[9] Prominent in Woolf's tale, too, is the role played by the scientific societies. He mentions the sponsorship of the Paris Academy of Sciences for LaCaille and points to the involvements of the Berlin Academy, the Bologna Academy, and the Royal Society of London.[10] Still, the full role of the scientific societies in the LaCaille affair has yet to be detailed.

At the heart of the plan to send LaCaille to the Cape of Good Hope stood the notion that corresponding observations should be made simultaneously in the northern hemisphere in order to provide a basis of comparison for his parallax experiments and to harmonize LaCaille's findings in the south with known stellar positions in the north.[11] In an important innovation LaCaille, Delisle, and the Paris Academy also worked directly with other scientific societies to further the project.

The most outstanding case in point was the new relation contracted between the academies in Paris and Stockholm for the occasion.[12] We have mentioned how Wargentin came to be associated with the Paris Academy in 1748 and how in 1750 the two institutions began their exchange of publications. But the status of their ties did not long remain at that level. Delisle, back in Paris and apparently wanting to participate more fully in the LaCaille expedition, applied to the French minister d'Argenson for permission to undertake his own corresponding expedition to the north.[13] Whereas twenty years earlier (repeating the Peru and Lapland expeditions) something like this would have been the order of the day, circumstances had changed, and now it was easier and cheaper to get Swedes to do it. Delisle was not to be the second Maupertuis. As reported by Grandjean de Fouchy in the registers of the Paris Academy: "The proposed voyage of M. de l'Isle was deliberated upon. The academy said that it believed it sufficient to ask the academies of Stockholm and Uppsala to undertake the observations corresponding to those of M. the Abbé de la Caille, and it was resolved that I would write to them as a consequence."[14]

De Fouchy wrote to Wargentin the following month: "The [Paris]

Academy flatters itself that the astronomers of Stockholm and Uppsala would wish to take part in so useful a work in making their own corresponding observations to those of the Abbé de la Caille."[15] There was, apparently, some uncertainty on the part of the academicians of Paris as to whether the Swedish academies could successfully perform their part in the enterprise, for yet a month later the Paris Academy decided to write another letter to Sweden, "to exhort them to apply themselves with all possible diligence to the observations corresponding to those of the Abbé de la Caille."[16] Wargentin wrote a forceful letter back promising in the name of the Swedish Academy to have the corresponding observations made.[17] Delisle himself wrote Wargentin that the Paris Academy was "delighted with the arrangements you have made to procure for us observations corresponding to those of Mr. de la Caille from the greatest possible number of locations in Sweden."[18] In the end, Wargentin got special Swedish government subsidies for his academy and produced seven sets of corresponding observations for the southern stars project and four sets for the transits of Mercury work. Sten Lindroth is clearly right when he says that this effort on the part of the Swedish Academy and Swedish astronomy was conducted "on assignment from the French Academy of Science."[19]

Inter-institutional contact and cooperation over the LaCaille observations and the transit of Mercury were not limited to the Paris-Stockholm axis, however. In a similar, but less directly formal way, the Paris Academy and the Saint Petersburg Academy also cooperated in the project. The academician A. N. Grischov accepted the task of undertaking corresponding observations in Russia, and between 1751 and 1756, he made three extended trips to the northern island of Oesel in Estonia to perform observations.[20] Something of the ties between the Paris and Saint Petersburg academies over this work can be seen in a letter of 1752 from Grischov at Oesel to LaCondamine in Paris.

My duty toward your illustrious Academy and toward you, Sir, is presently to inform you that I have continued up until now with great care and exactitude the corresponding observations of Saint Petersburg. . . . The Imperial Academy of Saint Petersburg [is] infinitely indebted to your illustrious Academy of Sciences for deciding to prolong the stay of the Abbé de la Caille at the Cape. I must communicate to you the observations that I am going to make here and ask you to present them to your illustrious Company. Being presently in a position to continue the corresponding observations here with the greatest exactitude as long as it pleases your illustrious Academy to leave the Abbé de la Caille at the Cape of Good Hope, I beg you, Sir, to show me the kindness of informing me of the approximate date up until which you believe the Abbé de la Caille will continue his observations.[21]

In all, the Russian Academy of Sciences mounted four stations: Oesel, Saint Petersburg, Moscow, and Archangel.[22]

The Berlin Academy was involved in the observations and dealt with the Paris Academy, too. Berlin falls on about the same longitude as the Cape of Good Hope, a situation which made it the ideal spot from which to make observations in conjunction with LaCaille. Maupertuis in Berlin extended an invitation to the young Lalande to come to Prussia from the Paris Academy to perform this task.[23] The report appearing in the registers of the Paris Academy concerning the Lalande trip to Berlin is revealing of the situation. "M. le Comte de Maillebois, President [of the academy], said that the King had accorded 2,500 livres to M. de la Lande, student of Mr. de l'Isle, to go to Berlin to make observations corresponding to those of the Abbe de la Caille. He also invited the Academy to take steps to have them made at Malta and Tripoli"[!][24]

The Paris Academy was clearly being given abundant support for this project. In addition to rallying scientific societies in Sweden, Russia, and Prussia, the Paris Academy also contacted academies and societies in Bologna, Montpellier, and London, if not still others, to secure an even greater number of observations.[25] With regard to the Royal Society of London, the Astronomer Royal and an important member of the society, James Bradley, fully participated in the project and sent his own corresponding observations to the Paris Academy in 1752 and again in 1754.[26] In addition, for the Mercury transit, Delisle wrote to Thomas Birch, S.R.S., asking him "to propose to the Royal Society and to employ all your credit to procure for us from your English colonies the most exact observations possible."[27] Yet the Paris Academy was not the sole motive force behind the operation. Wargentin and the Swedish Academy were especially active, being in touch with Grischov in Russia, Zanotti in Bologna, and Mortimer in England.[28] Euler also informs us in a letter to Saint Petersburg from Berlin that the Royal Society of London came through in a traditional capacity by supplying astronomical instruments for Grischov in Oesel.[29]

The LaCaille and Mercury transit observations of the early 1750s clearly represent something new for the learned societies and for organized science. The occasion certainly evidences the changed circumstances of the scientific societies at mid-century, and demonstrates how the societies came to function collectively as a system of institutions. Again, the challenge of the LaCaille and Mercury observations was larger than any one institution could support. With the Paris Academy at the helm, the effort implied and embodied a real division and subordination of labor among the component parts of the emergent system of scientific societies. The contrast between the level and quality of organized science, institutionalized science, or international science in the 1660s or even the 1730s and the achievement of the scientific societies in this effort of the 1750s is dramatic and remarkable.

The Lauragais Prize

Not all of the common endeavors of the scientific societies in the second half of the eighteenth century involved large-scale, "Big Science" projects. After the LaCaille and Mercury transit observations came a proposal, unfortunately aborted, for more modest cooperation among the academies and societies. In 1758 the French Count de Lauragais proposed an exceptional prize contest to be judged in common by the Paris Academy, the Berlin Academy, and the Royal Society of London.[30] As his proposal was of special interest to the resident mathematicians of the Paris institution, this group stayed behind after the regular meeting to discuss the matter. The minutes of this meeting inform us of the details of the prize.

The question at hand concerned the proposition made by M. le Comte de Lauragais for a prize of 2,400 livres to be awarded to the person who would give the most elegant solution (without any approximation) of the three-body problem. As it is the intention of M. de Lauragais not to exclude any academician from participating, this prize cannot be adjudged in the ordinary fashion. The manner in which entries should be judged to avoid any suspicion of partiality was deliberated upon, and it was proposed to engage the Royal Society of London and the Academy of Berlin to each name three judges who, with an equal number from the [Paris] Academy, would form a committee of nine persons. If any of these judges wishes to work for the prize, he will be required to so declare in order that another can be named to his place. The project was approved, but it was decided that before this deliberation was recorded . . . [one would wait] the acceptance of the two foreign companies.[31]

The exciting points about the Lauragais proposal were its subject and the form of the competition. With regard to the subject of the prize, Lauragais's initiative followed on the heels of the Saint Petersburg astronomy contest of 1751 (about whether the observed irregularities in the motion of the moon fit Newton's theory) and Clairaut's famous contradictory statements in public meetings of the Paris Academy in the late 1740s and early 1750s that Newtonian theory first did not and then did account for the observed phenomena.[32] As well as being a strenuous exercise in mathematics, the Lauragais prize was thus aimed at the recognized heart of contemporary celestial mechanics, and had the competition come off as proposed, its results would have marked an important milestone in the history of that science.

With regard to the form of the competition, it is evident that the Lauragais proposal sought to alleviate two problems in the way academic prize contests were normally administered. First, because only a small group of persons were qualified to compete in this contest, it was important for its success that all those who were qualified and willing to enter should be able to do so. Because Euler, for example, although in Berlin, was one of the judges of the Petersburg contest, he was not eligible to enter, and that prize contest was denied one of its most able potential candidates.[33] That the three-body contest was to be sponsored by more than one institution and subject to an international jury, it followed that no one would be excluded on account of his membership in a sponsoring institution and that at the same time competent judges could be secured. Second, it is evident in the above report and other documents relating to this proposal that academic prize contests were not always felt to be impartial, even though all entries were in theory anonymous. Here again, the possibility of a multi-society prize contest was seen as a mechanism to avoid charges of partiality in determining winners. It need hardly be pointed out that these innovations in the administration of this prize proposal would have been inconceivable without the presence and the perceived abilities of the Paris Academy, the Berlin Academy, and the Royal Society of London to decide the question collectively.

[207]

Unfortunately, as suggested, nothing resulted from the Lauragais proposal. In July of 1759, Grandjean de Fouchy did prepare letters to the Royal Society and the Berlin Academy, but because of the Seven Years War (then in progress), he sought official permission before actually sending them.[34] The response of the minister Florentin put an end to the project:

I have, Sir, informed the King of the proposed letter to be written to the secretaries of the academies in London and Berlin. The King has strongly approved of the zeal of the Count of Lauragais to found a prize that can only contribute to emulation and the progress of the sciences. But His Majesty thinks that it would be better to put off the writing of the proposed letters to the secretaries of these two foreign academies until peace comes.[35]

The 1761 Transit of Venus

The next project to engage the collective attentions and energies of eighteenth-century scientific societies was eminently more successful, large-scale, and indicative of a coordinated functioning of a system of institutions. That project was the observation of the passage of Venus across the face of the sun in 1761, the goal of which, like the 1753 Mercury transit, was to determine the real distance between the earth and the sun by parallax. As this story, like the earlier LaCaille and Mercury transit observations, has received solid treatment by Harry Woolf, the reader is referred to his more general account.[36] Here the essential activities and interactions of the scientific societies will be highlighted.

There is no doubt that the observation of the 1761 Venus transit was a truly international event or that the scientific societies of the day provided an institutional sine qua non for its success. Of a striking total of some one hundred and twenty observations taken at sixty different places on the globe, most were made either directly or indirectly under the auspices of the scientific societies, and nearly all were reported to and disseminated through these institutions.[37] The French and the Paris Academy of Sciences constitute the prime movers behind the plans for observing the transit in 1761. As he had done earlier, Delisle prepared a *mappemonde* and extensive instructions for the proposed observation. Two hundred copies of this advertisement and program for the event were distributed "virtually as wide as the civilized world."[38] With such a document, it was easy to consider the choice of proposed stations for the observation, and, although the subject had been raised earlier, it was after Delisle's *mappemonde* and memoir had been submitted to the Paris Academy and approved by committee, that discussion of the 1761 transit was taken up seriously by that institution.[39] In addition to numerous local Paris observations and those of Cassini de Thury made at the Imperial Observatory in Vienna, the Paris Academy and the French government sponsored three major expeditions for their part in the overall undertaking in 1761: Chappe to Tobolsk in Siberia, Le Gentil to Pondicherry in India, and Pingré to Isle Rodrique off Madagascar.[40]

The expedition of Chappe to Siberia was particularly noteworthy because it involved active cooperation between the Paris Academy and the Saint Petersburg Academy. In April of 1760, LaCaille informed the Paris Academy of a letter "from the Secretary of the Academy of Petersburg in which he asks on the part of that Company if someone of the astronomers of the [Paris] Academy would wish to go to Siberia to observe the passage of Venus over the sun which should arrive in 1761."[41] Charles Baër, writing to Wargentin in Stockholm, saw in this request an indication of lack of confidence on the part of the Saint Petersburg Academy in the capacity of its members, but Grandjean de Fouchy was quick to reply to Saint Petersburg that the Paris Academy "would believe to fall short of its purposes if it hesitated to concur in the zeal you evidence for the advancement of the sciences."[42] Chappe was duly appointed and enjoyed the seemingly unrecognized advantage of being paid by both the French and Russian crowns for his work.[43] Chappe was well received at the academy in Saint Petersburg, and special arrangements were made, once he was on his way to Tobolsk, to speed his results the eight hundred leagues back to Saint Petersburg and thence to Paris.[44] On its own, the Russian Academy sponsored other expeditions to Irkutsk and Selenginsk, as well as several observations in the vicinity of Saint Petersburg.[45]

The Swedish Academy, informed of events through Baër and again pressed by Delisle as it had been earlier for the LaCaille observations, ended up with government support sponsoring eight observations in various places in Scandinavia.[46] With a total of twenty-one sets of observations, the Swedish were the second largest national group, after the French, to participate in the observations of 1761. Woolf finds this "surprising," but from what has been seen of the vitality of the Swedish Academy and Swedish science in the international scientific community of the eighteenth-century the surprise is moderated.[47]

British preparations and those of the Royal Society of London were, comparatively, late and not as thorough as they might have been. There was no discussion of the 1761 transit within the Royal Society until June of 1760, when it received a copy of Delisle's *mappemonde* and a letter from Roger Boscovitch who had been in Paris.[48] The Boscovitch letter is interesting because it reveals something of the spirit that motivated the Royal Society in its participation. Boscovitch discussed the Venus transit and the necessity of proper instrumentation, and then he added:

May I presume in the third place just to mention the emulation of a neighbouring nation? The Royal Academy of Sciences at Paris, before I left the place, had sent one of their Members as far as the East Indies; another was appointed to Siberia; and the day of my departure from thence, they were debating about fitting out a third for the Cape of Good Hope: Besides these a fourth was solliciting orders to the Island of Cyprus. Surely it can never enter into my mind that so flourishing a Nation [as England], whose ships ride the ocean without controul, will endure to be outrival'd in literary Expeditions![49]

This exhortation must have had its effect, because at just this time, in an extraordinary resolution passed by the regular assembly of the Royal Society, the council of the society was directed "to think of, treat with, and at their Discretion to conclude agreements with proper persons to go to the proper places, in order to observe the expected transit of Venus over the sun."[50] After applying to the English government for eight hundred pounds, the Royal Society sponsored two expeditions, one led by Mason and Dixon to the Cape of Good Hope, and another by Maskelyne to the island of Saint Helena.[51] The trip to Saint Helena is interesting because it occasioned a special instance of cooperation and exchange within the overall program, with Maskelyne of the Royal Society and LaCaille of the Paris Academy agreeing to carry out further coordinated observations of the moon to be taken at Saint Helena, Paris, and Greenwich.[52] There were also British observations in London, Greenwich, and other places in England.[53] In America, although there was no learned society there in 1761, John Winthrop of Harvard, guided by Delisle's *mappemonde*, made an important observation at Saint Johns, Newfoundland.[54]

In addition to these major expeditions and observations, other academies and societies contributed to the effort. The Bologna Academy made six observations; the Montpellier Society made three; the Danish Academy made two, including the expedition it partially sponsored to the northern location of Trondheim.[55] The Berlin Academy, because of its insulated nature and the Seven Years War, did not take part in the 1761 transit observations.[56]

The extent of institutional backing and cooperation in observing the 1761 Venus transit hardly needs emphasis. On the one hand, as indicated by the distribution of Delisle's *mappemonde* or the above letters of Baër and Boscovitch, institutions were kept fully informed of one another's plans.[57] In addition, it is evident that more direct institutional cooperation and outright division of labor among institutions were facets of plans and preparations for the 1761 transit. The cooperation between the academies in Paris and Saint Petersburg over sending Chappe to Siberia is one important instance; the agreement between Maskelyne and LaCaille is another. Baër in his letter to Wargentin about this latter matter, made clear that observational responsibilities were being shared between institutions (the Royal Society and the Paris Academy, at least) and that duplication of resources was to be avoided: "Father Pingré is also leaving at the first opportunity for the western shores of Africa for the same purpose. He was going to go to Saint Helena but it was learned that the Society in London is sending someone there, and that's sufficient."[58] Similarly, in discussing a proposed French station at Hudson Bay, Le Gentil remarked that such would be fine, "unless the English take that charge upon themselves."[59] He continued, "One must rest content with the observations we are going to make in the Indies and those that can be made in Europe, be that at Stockholm, Petersburg, London, Paris, or finally Bologna, in Italy, places where competent observers are not lacking." Not coincidentally, these were also places where scientific societies were most firmly established. From these remarks and other examples that could be cited, it is evident that no one institution took on a sole or separate responsibility for observing the 1761 transit, but that each participated according to its own lights and as part of a loosely coordinated whole.

This was true even for the Royal Society of London, where feelings of national pride and rivalry strongly motivated its involvement. A letter written by Lord Macclesfield, P.R.S., to the Duke of Newcastle, to accompany the Royal Society's solicitation of government funds, is revealing of the situation. Maccelesfield begins by emphasizing nationalism and competition.

The [accompanying] Memorial itself plainly Shews, that the Motives on which it is founded, are the Improvement of Astronomy, and the honour of this Nation; which Seems to be more particularly concerned in the exact observation of this rare Phaenomenon. . . . And, it might afford too just ground to Foreigners for reproaching this Nation in general (not inferier to any other in every branch of Learning and more especially in Astronomy); if, while the French King is Sending observers for this purpose, not only to Pondicherie and the Cape of Good Hope, but also to the Northern parts of Siberia, and the Court of Russia are doing the Same to the most Eastern Confines of the Greater Tartary; not to mention the Several observers who are going to various places, on the Same Errand from different parts of Europe; England should neglect to Send observers to Such places as are most proper for that purpose and Subject to the Crown of Great Britain.[60]

That said and felt, however, Macclesfield goes on to point out the universal and international needs, aims, and expectations for the transit observation and the desire of the Royal Society to participate along with other scientific societies in the larger, common project:

This is by Foreign Countries in general expected from us; Because the use that may be derived from this Phaenomenon, will be proportionate to the Number of distant places where proper observations can and Shall be made of it; And the Royal Society, being extremely desirous of Satisfying the Universal expectations of the World in this respect, have thought it incumbent upon them to lay this matter before your Grace, who is so great a Patron of Learning, and to request your effectual intercession with his Majesty, that he would be graciously placed to *enable them*, in Such manner as his Majesty Shall think fit, to *Accomplish this their desire and to answer the expectation of the world.*[61]

Motivated by national pride, yes, the Royal Society participated as only one part of a larger whole.

A related aspect of the collective role of academies and societies in the 1761 transit observations was the impressive exchange of results that took place through their channels afterward. The Paris Academy, for example, received forty-five sets of observations, some communicated by individuals, most by institutions and individuals closely associated with institutions.[62] The Royal Society of London received results from France, Sweden, Bologna, Mannheim, Naples, and Cambridge, Massachusetts, and it published twenty-six different observations on its own.[63] In Montpellier, de Ratte wrote an analysis of the results based on his own observations plus a dozen others from the Paris Academy, the Royal Society of London, and the Swedish Academy of Sciences.[64] Finally in this connection, in sending his observations in a letter to Wargentin, Delisle chided the Swedish secretary for not yet having forwarded the northern observations to him; fortunately, Delisle continued, he was able to obtain them through England and the Royal Society.[65] Such was the extent of observational exchanges post 1761.

It might also be pointed out that the 1761 observation took place during the Seven Years War, a state of affairs that caused the failure of the Lauragais prize proposal and that made the Venus enterprise that much more remarkable. Mason and Dixon were disturbed by the French while in the Channel, and Pingré was later harassed by the English in the Indian Ocean despite his British passport.[66] These minor setbacks aside, the observation proceeded despite hostile relations between nations. Dortous de Mairan of the Paris Academy wondered at this:

it is satisfying for those who are interested in the progress of the sciences to learn that in a time of war, when the enormous expenses involved would seem to overshadow those needed by knowledge that might seem only a pure curiosity, all the astronomers of Europe are active in getting themselves each to the most advantageous stations to undertake a most courageous, most useful, and most subtle enterprise; . . . What a glory for astronomers to undertake voyages for the discovery of a truth that others undertake solely to amass treasures.[67]

These remarks and the event itself are impressive testimony to how much the Republic of the Sciences and the international system of scientific societies indeed were operating with a measure of independence outside the restrictions of nationalism.

The results of the 1761 transit observation were inconclusive. As Lalande put it after his visit to London and the Royal Society in 1763: "Everyone in England is persuaded that the parallax of the sun is 8″ ½, but M. Pingré still claims 10″ ½ here [Paris]. I think the question will not be decided until 1769."[68] That would be the second transit of

Venus, scheduled to appear twice every 110 years or so. The world waited, and academies and societies prepared to gear up again.

Interlude: Harrison and Chabert

Academies, societies, and internationally organized science were not quiescent between the first transit of Venus in 1761 and the second in 1769. On the one hand, as Woolf indicates, the two events merged into one long train where discussion of results of the first observations occasioned discussion of plans for the second.[69] On the other hand, there were two instances in the intervening eight years that, on a lesser scale than the 1761 transit observation, continued the tradition of academies and societies of science working together on common projects. The first of these was the judgment of Harrison's chronometer.

The development of accurate time-keeping devices had long been a desideratum of science, technology, and government.[70] With a clock of sufficient accuracy and durability, the problem of finding one's longitude at sea or on land could be solved, and navigation and trade made that much more precise and reliable. Both the French and British crowns had offered substantial premiums from early in the eighteenth-century for the development of such clocks, and precision instrument makers had worked on the problem for years.[71] In 1761 the Englishman John Harrison thought he had perfected his chronometer and demanded that it be judged by the Board of Longitude.[72] The board did not proceed to do this on its own account, but wrote to the Royal Society of London, as "the most Competent Judges" in this matter, and asked that that body name persons to test Harrison's clock and specify the tests to be employed.[73] The Royal Society went further and approached the Paris Academy about sending a delegation to the trials.[74] The Paris Academy responded positively to the offer to participate, and the mathematician Camus and the clockmaker Berthoud duly went to England in April of 1763 and reported back to the academy in June of the same year.[75] This was followed by an official report to Paris from the Royal Society approving Harrison's invention; the Saint Petersburg Academy was also kept informed by the Royal Society.[76] After a sea voyage of 147 days, Harrison's clock was off by less than two minutes, and he was awarded a total of twenty thousand pounds by the Board of Longitude.[77]

Firm details of the second instance of inter-society cooperation occurring in the interim between the first and second transits of Venus are not to be had. The story concerns another set of coordinated astronomical and geographical observations, reminiscent of those associated with the LaCaille observations at the Cape of Good Hope a decade and more previously. This time the Paris Academy, again, initiated a project to have other institutions and individuals join it in preparing corresponding observations to those of one of its members, J. B. Chabert.[78] The Paris Academy and Grandjean de Fouchy launched the project in 1765, seeking the cooperation of Royal Society of London, the Swedish Academy, and Saint Petersburg Academy, the Bolognese Academy, and possibly other learned societies.[79] As best can be made out, the goal of this project was to compile comparative data on longitudes and hydrographical conditions at various coastal points in the Mediterranean with an eye toward improving maps.[80] The response to this initiative was slight, however. Lord Morton, president of the Royal Society of London, wrote back saying that the Paris Academy ought to contact Maty, Bird, or Charles Morton of the society; Zanotti replied from Bologna that he had no observations; there is no evidence of Swedish participation.[81] The Saint Petersburg Academy did debate the matter and decided a positive response should be made to Fouchy.[82] What this entailed is likewise unclear. In only one instance, a reply from a certain M. D'Arquier of the Toulouse Academy, did the Paris Academy receive hard data back from its requests.[83]

This project was of larger scale than the episode concerning Harrison's chronometer, but it was also less successful. Judging from the evidence available, one might infer that the Paris Academy, riding the crest of events between 1761 and 1769, was seeking to maintain its vanguard position and the momentum of institutional cooperation until that more attractive date arrived.

The 1769 Transit of Venus

With the experience of 1761 behind them, men and institutions of science were in a better position to plan for, coordinate, and observe the second transit of Venus in 1769.[84] This they did on an even larger scale than the first time. Compared with the one hundred and twenty observations from sixty different stations that constituted the 1761 transit observation, the 1769 transit was observed by at least one hundred and fifty-one observers at seventy-seven different stations.[85] To give another indication of this difference in scale, whereas the Royal Society solicited eight hundred pounds from the English government in 1761, it requested and received four thousand pounds for 1769.[86]

Once again academies and societies of science provided the primary basis and the main initiative for the transit observation, but this time the Royal Society of London and the English led the way. Knowing the importance of the event and not needing any stimulus from the regular body of the society, the council of the Royal Society began plans for the 1769 transit as early as 1765.[87] In 1767 a committee was formed to formulate specific proposals, and out of this the Royal Society ended up sponsoring four expeditions to observe the transit.[88] Two of them went north, one to the North Cape, and the other (led by Dixon) to Hammerfest in Norway. The third was outfitted with a special portable observatory and sent to Hudson Bay, and the fourth and most famous of all the Venus transit expeditions was sent to the South Seas. This expedition was led by Captain Cook (appointed an official observer by the Royal Society) and included Banks and the Swede Solander among its personnel. It was Cook's first voyage to the Pacific. This historic voyage, as well as the other expeditions sent out by the Royal Society, required close coordination among the Royal Society, the Admiralty, and Hudson Bay and East India companies. As well as its scientific interest, the trip to Tahiti, where the Venus transit was observed, promised the possibility of contact and trade with an unknown but presumably populous southern continent.[89] In all, the Cook expedition of the Royal Society was possibly the most impressive and important of all eighteenth-century scientific expeditions; it was a major event in the history of Pacific exploration, and the observatory erected at Fort Tahiti was certainly the most far-flung outpost of organized science of the age.[90] Nevertheless, this spectacular expedition to the South Seas was only part of the Royal Society's participation and that of the English in general. Including the expeditions named, the British mounted a total of sixty-nine different observations, which, as Woolf informs us, made them the leading nation to participate in the event.[91]

[216]

Yet the French and the Paris Academy of Sciences in no way diminished their efforts from what they had done as leaders in 1761. In addition to the again numerous observations in and around Paris, the Paris Academy sponsored three voyages. Le Gentil was still in the East Indies, having failed in his attempt to observe the 1761 transit; although, as bad luck would have it, he would fail again; he was directed to observe either at Manila or Pondicherry. Chappe, returned from Siberia, was sent to southern California with the aid and blessing of the Spanish court. The third academy-sponsored expedition went to Cap-Français at Santo Domingo where Pingré observed the transit. This voyage was actually sent out to test French chronometers at sea, the occasion being used additionally for the purposes of the transit.[92]

Stations in the far north and the southern Pacific were of particular importance to the success of the 1769 transit observation. Located in the north, the Swedish Academy of Sciences was aware of its duties toward the learned world and sponsored a total of ten different observations, including two expeditions sent to extreme northern latitudes, again with Swedish government support.[93] At least fifteen different Swedes observed and recorded the 1769 transit, maintaining the example set in 1761 of strong participation by that nation.[94]

The Russian Academy of Sciences also began its preparations early.[95] Between 1764 and 1769 there was a steady stream of discussions and plans, largely involving contact and coordination with foreign academies and societies. In all, the Russian Academy sent out eight expeditions, including one to Kamchatka. A total of thirteen Russian observers witnessed and recorded the 1769 transit.[96]

The abortive attempt of the Berlin Academy to join its sister institutions in the observation is interesting. After Euler's return to Saint Petersburg in 1766, Lagrange came from Turin to replace him. Lagrange, a man with strong internationalist sympathies, was unwilling to let the occasion pass unheralded in Berlin. In 1767 he addressed the Berlin Academy, saying, "Sirs, you realize that the principal academies of Europe are making great preparations for the observation of the next transit of Venus. It is appropriate, it seems to me, that our Academy take some part in it."[97] He went on to propose an expedition to the north, and, incidentally, that the academy employ an additional astronomer to ease the burden on the regular astronomer who was limited to taking observations for the academy's almanacs. Lagrange's proposition was taken up by a committee of the academy, papers were read on the subject, and in general it looked as if something might come of it.[98] Frederick was not to be moved, however. The last entry in the records of the Berlin Academy relating to this matter reads: "The secretary reported that His Majesty has not responded to the Academy's letter concerning the voyage to observe the transit of Venus."[99]

If the project to observe the 1769 Venus transit made no headway in Berlin, just the opposite was the case on the American continent. Organizing itself in 1768 and 1769, the American Philosophical Society made the observation of the 1769 transit one of its first orders of business.[100] It, too, proposed an expedition to Hudson Bay, but in the end, the American society could officially sponsor only three observations, two in Philadelphia and one in Delaware.[101] It received a total of four hundred pounds from the Pennsylvania Assembly for the purchase of instruments and the conversion of Assembly buildings in Philadelphia into a serviceable observatory.[102] There were a total of twenty-two American observations of the 1769 transit, the bulk of which were coordinated through the American Philosophical Society and its special transit committee.[103]

Participation of other scientific societies in the 1769 transit seems somewhat less than it was in 1761. The Danish Academy sponsored observations at Vardo, including those of the Viennese astronomer Father Hell.[104] Italian participation—Bologna excepted—was almost completely lacking; the Montpellier Society did not observe on account of a dispute with a local religious order over the height of its observing tower.[105] The tables presented by Woolf summarizing the 1769 observations indicate that there were stations in Toulouse, Rouen, Bordeaux, and Brest, and these may well have been associated with academies in those cities.[106]

As was the case in the 1761 transit observation, institutional cooperation and a division of labor among the participating academies and societies was a key element of the 1769 operation. Again, each institution knew what the others were planning. For example, Razumovskii, director of the Saint Petersburg Academy, wrote to the Royal Society in London informing them of Russian preparations, mentioning at the same time details of Swedish plans conveyed to him by Wargentin.[107] More clearly than in 1761, tasks involved in the observations were divided among institutions. The Royal Society, more or less taking charge, was instrumental in this. In the report made by Maskelyne and the transit committee of the Royal Society, great attention was paid to the most desirable locations for observations and the preparation of other nations. In particular, regarding observations in Russia, the committee concluded that, given the plans under way there, "It will be needless to send any English Astronomers into those parts." Acting on the committee report, the council resolved in addition, "That the Astronomer royal, be desired to write to the Swedish Astronomers, to know what places they intend to observe the ensuing Transit of Venus, and to acquaint them with the kind and size of Instruments, which are to be used by the English Astronomers."[108]

On behalf of the Royal Society, its president, Lord Morton, wrote to the Saint Petersburg Academy informing them of English plans and expressing "the wish and the hope" that the Saint Petersburg Academy would undertake observations in Kamchatka and other places in the Russian empire where the transit would be most visible. Morton also offered English instruments to the Russians, and there were numerous communications back and forth over this question.[109] The Russian Academy itself again requested an astronomer from the Paris Academy to aid them in their observations, but this time nothing came of it.[110] Still, J. A. Euler and Lalande of those academies were in contact over coordinated plans for the observations, and the Russian Academy was able to obtain the services of Lexell from the Swedish Academy, Christian Mayer from the Mannheim Academy, and the Swiss, Mallet and Pictet, to augment its own corps of observers.[111] (It seems that Mayer went from Mannheim to Saint Petersburg only because the Royal Society in London ignored the request of the Palatine Elector for his astronomer to join the English.[112] The Royal Society and the Paris Academy likewise kept in touch over their plans, and the 1769 transit observation was the occasion for a special exchange between Maskelyne and Lalande.[113] The American Philosophical Society was in contact with Maskelyne and the British regarding its observations and was able to secure instruments from England.[114] No doubt a more thorough search of the records of these and other societies would reveal further instances of cooperation and coordination among them.

Finally, after June 1769, the month of the transit, another massive exchange of results took place. Again institutions exchanged their own results and served as clearing houses for those sent in by individuals. To single out the Saint Petersburg Academy as an example, one discovers it received results from the Royal Society of London, the Paris Academy, the Berlin Academy (local observations), the Danish Academy, the Göttingen Society, the Munich Academy, the Mannheim Academy, the Bolognese Institute, and the American Philosophical Society, as well as from numerous individuals.[115] In fact, so thorough was the flow of information to Saint Petersburg that the Russian Academy received duplicate sets of the American observations, one communicated through Mayer in Mannheim, the other through Morton and the Royal Society.[116]

The formulated results of the 1769 transit observation still varied significantly, but they were closer to the modern value of solar parallax than those obtained after 1761.[117] Somehow, the fact that astronomers were unable to agree on a final figure pales before other aspects of this great enterprise.[118] Woolf points to new knowledge in botany and natural history that grew out of the expeditions and to increased contact between science and governments as two other benefits forthcoming from the project. More than that, one can only be struck by the sheer size and purposefulness of so large a scientific undertaking. The 1769 transit of Venus observation is formidable, indeed overwhelming testimony to the progress of science, its acculturation over the preceding century, and the absolutely central position of academies and societies for organized science at the time. Concludes Woolf,

The transit story increased the prestige of scientific societies everywhere and lent additional weight to their future claims for a larger share of national wealth and attention . . . It brought to a common focus men of almost every national background with an abiding concern for the advancement of knowledge. In so doing, it helped to shape the growing international community of science and to demonstrate with striking clarity what cooperation and good will might achieve in the peaceful pursuit of truth.[119]

Meteorology and Mannheim

Systematic attempts on the part of scientific societies to tackle the problem of understanding the weather was a part of the common history of academies and societies from the 1720s, when the Royal Society of London called on its international network of F.R.S., the Berlin *Societas*, and the Saint Petersburg Academy to join it in a campaign of data gathering and analysis. As late as 1755, the call issued within the Royal Society for "such members of the Society as reside in different parts of the world to transmit their meteorological observations to the Secretary at the end of every year."[120] On the Continent, the Saint Petersburg and Berlin academies each made weather observations on a regular basis. In 1768 J. A. Euler took up an unbroken chain of Petersburg observations that stretched back to the 1740s.[121] And in 1771, an official, monthly exchange of weather data between secretaries Euler and Formey was begun.[122] In France in the 1770s, meteorologist and member of the Paris Academy of Sciences Father Louis Cotte engaged in a personal effort to collect and coordinate observations being taken by scientific societies throughout France and Europe.[123] Cotte tried contacting some of the French provincial academies, and he received weather data from Wargentin and the Swedish Academy as forwarded to Cotte through Lalande.[124] More significantly Cotte took over and coordinated the meteorological part of a separate program of meteorological and epidemiological research undertaken by the Paris Royal Society of Medicine. This program is deserving of at least brief separate mention.

The *Société royale de médecine* was chartered in 1776 as the French learned medical society.[125] It was a typical academy in its organization, and it arose in opposition to the established medical faculty in Paris. Its medical doctrine was different, in that it stressed community medicine and "environmental" causes of disease. It therefore supported a Baconian program of investigation to uncover the relations between environment and health, and here it made contact with the science of meteorology.[126] With considerable support from the French government, it systematically established a network of some 100 to 200 stations reporting weather data (temperatures, pressures, conditions); these were funneled through Cotte and duly reported in the *Mémoires* of the *Société* (9 vols., 1776–1789).[127] The *Société* made no special provision for instruments, except to recommend the Paris instrument-maker Mossy for off-price and presumably comparable instruments.[128] All things considered the *Société* did not rely on other learned societies to carry out its program, which was effectively run through individual observers and the national system of intendants. But to an extent deserving our notice, the meteorological and medical program of the *Société* did make contact with the system of scientific societies as it has been depicted here. The *Académie* at Dijon was early and officially affiliated with the program.[129] Besançon and other French provincial academies may have been involved; Bologna and a few northern Italian academies reported back; and while there seemingly was nothing official with the Paris Academy of Sciences, twelve of the members of the *Société royale de médecine* were already *pensionnaires* of the Academy of Sciences.[130] In all, the episode surrounding the *Société* is somewhat tangential to the main thrust here, but it indicates another facet of the "academic movement" in the eighteenth-century and how much meteorology was the subject of practical concern and research.[131]

By the 1780s, then, there was what amounted to a tradition of attempts to coordinate the power of the scientific societies in a directed assault on the natural forces of the air. In a review of Cotte's *Traité de Météorologie* in 1774, the Abbé Rozier summarized the state of meteorological science:

For a long time, *physiciens* have desired a body of doctrine in which all the related parts of meteorology would be included. The work of Father Cotte fulfills this goal and it is the only one of its type published. He has cleared a route which [further] observations continued with care will render more certain and more useful for agriculturalists and *physiciens*. Those that are presented today are the first foundations for an immense edifice. When will it be completed? Who knows! Far from knowing the principles of nature, we see only her effects and her vicissitudes; there must, however, be a constant order in this apparent disorder.[132]

Rozier's estimation makes clear that there was still a long way to go and that felt uncertainties about the weather made meteorology an attractive area for further scientific research. Several problems still hindered extensive international cooperation in meteorology, however: procedures were not standardized and observations generally were taken on instruments with different scales, so that results were practically impossible to compare.[133] Then, which was better, air, water, alcohol, or mercury thermometers? How to guarantee tubes with like diameters? Does one read from the top or bottom of the meniscus? And so on. Further, the needs of science in these regards could only be met by corresponding skill in artisanal manufacture, a conjuncture achieved only in the last decades of the eighteenth century.[134]

The earlier attempts at systematic collection of weather data from a large area do seem slight and pitifully inadequate compared to the magnitude of the task, but all that changed in 1780 with the creation of the Meteorological Society of Mannheim (*Societas Meteorologicae Palatinae*). The Meteorological Society was a separate adjunct of the Academy of Sciences which existed in Mannheim from 1763.[135] The Meteorological Society was created explicitly as a specialized society to do what needed to be done in meteorological research, and it adopted two novel approaches. The Meteorological Society of Mannheim directly and primarily enlisted other scientific societies, and it sent out calibrated, standard sets of instruments to insure uniform results. The Meteorological Society thus initiated and led one of the great cooperative projects of eighteenth-century science.

The full program and panoply of the Mannheim Society are laid out in its first volume of proceedings, the *Ephemerides Societatis Meteorologicae Palatinae*.[136] The formalities of its creation, its climatological, medical, and cameralist philosophy, lengthy and specific instructions to observers, and discussion of instruments and reporting channels are fully detailed. Underscored, too, in model letters, responses, and lists of institutions contacted is the extent to which the Mannheim Society was specifically directing its effort through the existing system of scientific societies.[137] A later library "Avertissement" for the volume of *Ephemerides* for 1785 likewise makes clear the primacy of institutions in the project.

The Society received particular orders to exercise care in the choice of stations and learned observers. As a result it expedited letters of cooperation in this great and vast institute to various academies, universities, and literary bodies and offered them, according to the desire of his supreme Highness, all the instruments free, with the condition, however, that all of these bodies engage themselves to carefully conserve them, to choose good observers, to continue the observations without interruption, and annually to send their work to the Academy.[138]

The program of the Meteorological Society of Mannheim thus was directed primarily at other institutions, as distinct from the earlier meteorological project of the Royal Society, which was geared more toward individuals. Large scale meteorological research was clearly beyond the capabilities of single individuals, and institutions—meaning other scientific societies, largely—could guarantee more reliable and continuous observations. Motivating institutions to join the project was facilitated by offering the instruments gratis (how could they be turned down?), and standard instruments and procedures insured standard observations. Coordination of results was insured by the conditions imposed upon the recipients. And the published *Ephemerides* were the vehicles for reporting back data and results. The success of this program required the Mannheim Society as a central administration, the munificent support of the Palatine Elector Carl-Theodore, and the industry of the Italian instrument-maker Carlo Artaria hired for this purpose.[139] Clearly, in the history of the involvement of learned societies with the study of meteorology, a new stage had been reached.

The contact between the Meteorological Society in Mannheim and the Academy of Sciences in Saint Petersburg was the first and probably the most important of all the institutional ties established in this project. The Saint Petersburg Academy got early word of the Mannheim Society and its notions through the academician Lexell, who was traveling through Mannheim on leave of absence. The plan was reported straightforwardly as follows:

The Palatine elector is having analogous instruments constructed for meteorological observations, notably barometers, thermometers, and hygrometers which His Electoral Highness proposes to have distributed everywhere in Europe in order to obtain corresponding observations made with instruments of the same construction. [The Meteorological Society] consequently wishes to learn if the Imperial Academy of Sciences of Saint Petersburg does not wish to lend itself to these ends and accept a few sets of these instruments to distribute to persons capable of taking on these observations.[140]

The Saint Petersburg Academy's response can only be described as incredulous: "The Secretary was charged with writing to the Abbé Mayer in Mannheim about this and demanding the ulterior details and conditions under which the Elector wishes to cede the instruments to the [Saint Petersburg] Academy." Mayer's return letter confirmed the situation in detail—there was no hidden agenda—and the offer "to associate yourselves in this way with the establishment of the Elector" was repeated. Knowledge that the project was real and underway precipitated this further development at the next meeting of the Petersburg Academy: "On the occasion of the Society for meteorological correspondence that the Palatine Elector has just founded, several Academicians were of the opinion to revive the old project of the late Professor Kraft to establish meteorological records in a dozen different places in the Empire of Russia."[141]

From that point, before the Mannheim Society had quite launched itself, the institutions in Mannheim and Saint Petersburg were closely linked in this common endeavor, and the prospect of the project's further extension into Russia remained an alluring possibility through the 1780s.[142] As it was, cases of instruments were forwarded to Saint Petersburg (a second was needed because the first arrived damaged), and observations were returned to Mannheim and published in the society's *Ephemerides*.[143] In addition to the Imperial Academy, the Petersburg Free Öconomical Society maintained a station for the Germans, and there was another in Moscow.[144]

In early 1781 the Mannheim Society began its own efforts to contact other scientific societies and to further realize its program. Latin letters of invitation were sent out to some thirty scientific institutions all over Europe.[145] Cases of instruments followed, and thirty-seven stations ended up reporting to Mannheim.[146] It was a vigorous program, for which the established system of scientific societies furnished the basis.

In addition to that of the Saint Petersburg Academy, another early success was scored with the Prussian Academy in Berlin.[147] The registers of the Berlin Academy record laconically that the academy "accepts the offers" of the Mannheim Society, and the secretary of the Mannheim Society, Hemmer, responded by expediting a case of instruments.[148] Again, the first case arrived damaged and another had to be sent.[149] As late as 1794, the Berlin Academy was still recording observations for Mannheim.[150] Elsewhere in Germany, the Mannheim Society found a strong ally in the Munich Academy of Sciences, which had fallen under the control of the Palatine Elector in 1763. As a coordinating institution along with the Mannheim Society, the academy in Munich separately collected observations from eighteen (then twenty-two) different German stations; it offered prizes for research in meteorology, and it published its own collection of meteorological *Ephemerides*.[151] The academies in Göttingen and Erfurt also participated in this project.[152]

These efforts by the Mannheim Society attracted a remarkable degree of support in Scandinavia. The Swedish Academy, already making regular weather observations, responded favorably to the German offer, and its newly appointed second secretary, Nicander, took up the special job of making the observations for Mannheim.[153] A concrete expansion of the program, similar to that envisioned for Russia, was put into effect in Sweden.[154] Similarly, the Danish Academy was sent instruments and forms for recording observations; Thomas Bugge took charge of the effort in Copenhagen.[155] In addition, as the Danish Academy and the Danish crown already supervised regular weather observations in Iceland and Greenland, these too were forwarded to the Mannheim Society and formed part of its larger corps of observations.[156] Finally, in that part of the world, the infant Norwegian academy at Trondheim likewise participated in the project with the Mannheim Society.[157]

In France, the undertaking generated some support, although not as much as in northern and central Europe. The academy in Dijon was the major French supporter. The Mannheim proposal was accepted by Maret of the Dijon Academy, but this time, after yet another broken case of instruments, Hemmer reasoned that the rigors of the post and customs were too much for the enterprise, and suggested that Maret arrange for their receipt in Mannheim.[158] Serviceable instruments arrived in Dijon in 1783, and observations were kept regularly thereafter, although at another point Hemmer asked that the Dijon Academy conform strictly to the model observations provided.[159] The Montpellier Society of Sciences was also approached, but in this case the results of the initiative are unknown.[160] The academies in Marseille, Clermont, and La Rochelle apparently participated.[161] The Paris Academy of Sciences was sent the original letter of early 1781, and Condorcet responded positively on the part of the academy; Cotte was highly praised in the first volumes of the *Ephemerides*, but nothing seems to have resulted from the connection to Paris.[162]

Elsewhere on the Continent, the Meteorological Society sent instruments to the academies in Brussels, Bologna, Turin, and Padua, and these institutions maintained stations for the Mannheim Society.[163] Several stations were set up in Switzerland.[164] Another station was operating in Prague, perhaps run by the science society there, and others were operating in Hungary and elsewhere in Germany.[165] Outside the Continent of Europe, the Mannheim Society attempted to enlist the Royal Society of London in the enterprise, but, ironically, Banks could find no one to perform the observations.[166] There is no evidence that the American Philosophical Society was on the list of institutions invited to participate or participating, but instruments were sent and a station established at Harvard College.[167] That was a long way from Mannheim. In addition, with what success it is unclear, the Mannheim Society attempted to establish a chain of stations in India.[168]

Finally, completing the circle and fulfilling its obligations to the network of participating institutions, the society printed twelve massive volumes of now impenetrable *Ephemerides* between 1781 and 1795 and distributed them among the learned societies in the normal fashion.[169]

The project sponsored by the Meteorological Society of Mannheim was one of the great cooperative endeavors engaged in by eighteenth century scientific societies. Although less spectacular than the transits of Venus efforts, it ranks along with them as a major accomplishment of the collective industry of those institutions. Different from those earlier astronomical undertakings, the Mannheim meteorological project shared with them a common purpose, a division of labor, and the essential institutional base provided by academies and societies of science.[170] Given the very real obstacle of merely getting the instruments to their destinations undamaged, one can empathize with the Mannheimer Fontaine when he wrote of the "grandure and beauty" of the plan.[171] It was a great idea to bring weather under the domain of reason, and if in the nature of things not a great deal was to be gained from it, the resulting network of institutional ties that spread across the learned world of the eighteenth century and that gave substance to this endeavor is, at least, strong testimony to the possibilities then imaginable when institutions cooperated in common scientific pursuits.

Triangulation

The project to determine accurately the difference in longitude between the observatories in Paris and Greenwich by the method of triangulation was initiated somewhat deviously by Cassini de Thury in

1783. In September or October of that year he sent a memorial to the English crown proposing a plan whereby a series of land-based triangles would be laid down between Paris and Calais and between Greenwich and Dover, and the two chains connected.[172] Cassini suggested that, if the English were ill-equipped to perform their half of the project, Frenchmen could be imported to do the job. This was a direct challenge, because the French and three generations of the Cassini family in particular had been working on a general map of France since the seventeenth century and were very experienced in the technique of triangulation.[173] Sir Joseph Banks discussed the proposal with George III and declared that persons could be found within the Royal Society "able and willing to undertake the work suggested in the memorial."[174] With this behind him, in May of 1784 Cassini approached the Paris Academy of Sciences, making it look as though the English had initiated the project:

M. de Thury said that he has received letters from England in which it was proposed to form triangles like those already established in France and to join these at Calais if the Academy wishes to extend a chain of triangles there with those of the English. This the Academy accepted and asked M. Cassini to write in order to coordinate the execution of this project which will give the respective position of the two capitals and the two observatories with an entire certainty.[175]

George III turned the entire project over to the Royal Society where it was approved in council in June of 1784.[176] Three thousand pounds were provided. Similarly, in France the Paris Academy of Sciences exercised the position as French overseer to the project.[177] Between 1784 and 1787, work continued on both sides of the Channel. In 1787, when the triangles were linked up, emissaries were sent to connect them, and once the project was completed, results were exchanged between the Paris Academy and the Royal Society.[178] General Roy, F.R.S., and leader of the English contingent, received the Copley medal of the Royal Society in 1785 for his efforts, and Ramsden, the man who made the English measuring instruments, was likewise honored in 1795.[179]

This was a not insignificant instance of inter-institutional cooperation. The observatories in Paris and Greenwich were the two major astronomical centers of the world, and the harmonizing of their observations was clearly an important undertaking. Using the Greenwich observations, the Board of Longitude published the *Nautical Almanach*, and from 1762 the Royal Society annually published its own set of Greenwich observations.[180] Similarly, the French astronomical calendar, the *Connaissance des Temps*, was put out under the auspices of the Paris Academy of Sciences, and all these publications were regularly exchanged as part of the sharing of materials between London and Paris.[181] With an accurate evaluation of the relative position of the two observatories, it became possible to coordinate their respective observations with consequent benefits to astronomy, geography, and navigation. With the observatories in Paris and Greenwich so conjoined through the cooperative labor of the Paris Academy and the Royal Society, international astronomy made an important step forward. So important was this link between England and France that in 1821 the Paris Academy requested of the Royal Society that the project be repeated in order to obtain an even greater accuracy in the findings.[182]

The Jurisprudence Prize

In 1785, twenty-seven years after the Lauragais prize, a most curious proposal was made by a certain and supposed Count de Windischgrätz, an influential Austrian with Belgian connections, for another prize to be judged communally by the Paris Academy, the Berlin Academy, and the Royal Society of London.[183] The count proposed a question that touched on both jurisprudence and mathematics. The object was to reduce the number of legal cases involving contracts by classifying the variables of contract law and, using tables and formulas from the competition, thenceforth to form contracts only of agreed-upon elements and combinations of elements. The taxonomy took a knowledge of law, identifying the combinations took a knowledge of probability and permutations, and it was required that candidates demonstrate their propositions with mathematical rigor.[184] Perhaps to avoid cheating and partiality, the mechanism of using the three learned societies to determine the winner was called upon again, but there was another probabilistic twist here, in that each judging institution (through its commissioners) would assign one of three grades to each entry, and the entry with the highest grade total would win.[185] The prize was worth three thousand *Risdales*.

[229]

Surprisingly, the proposal of this Austrian count to have his prize judged by the major scientific societies of France, Prussia, and England got further than the Lauragais plan of 1758. The proposal was received by the Paris Academy of Sciences and its commission was highly favorable in its report:

As it was exposed in the program, this question appeared important to us. Logic and the science of combinations are its subject as much as jurisprudence and politics, and the science of combinations can be justly regarded as a branch of mathematics. We believe that the Academy should not consider the subject of this prize as foreign to its concerns.[186]

The commissioners' report makes two further points regarding international relations among academies and societies as occasioned by this proposal. First, there should be no jealousy: "All the great academies of Europe for their own interests as for that of the sciences should aspire only to equality among themselves."[187] Second, the mechanism for determining winners "objectively" by a group of institutions was an appealing one.

This manner of procuring the result by the combined judgment of the three companies appeared to us ingenious. . . . This method of which this is the first example could be useful because there are many questions and this one seems to be among the number for which it would be good to procure the united wisdom of the most enlightened notions.[188]

Having each institution judge each entry independently had not been proposed before, but could the academy have forgotten about the earlier plan of one of its members, Lauragais?

The Count Windischgrätz contacted the Berlin Academy, indicating that its sister academy in France had already "accepted the office of Judge," and that, not having received a response from the Royal Society of London, he could do without its help. To stimulate the Berlin Academy a bit more, he offered to fund a separate prize of Berlin's choosing for fifty *louis d'or*. Another report was made, only this time by a lawyer and not by a mathematician, the privy counselor Anières. His report was considered important enough, "meriting a particular attention," that it was printed *in extensio* in the Berlin *Mémoires*.[189] It was damning: Impossible even in theory to reduce the complex relations involved in contracts to a fixed classification. In practice variations in local law and custom would render the project useless. There were problems of language and the meaning of words, and the method of judgment was impractical:

Each of these Academies or Societies, not understanding and not able to understand the terms of the art, of Jurisprudence, of routine, etc., used in these far away countries and for which the language is unfamiliar, will necessarily be unable to bring a solid judgment to the whole. . . . Hence I conclude to refuse the Count's offer or simply to send it to the king.[190]

That certainly ended the matter as far as the Berlin Academy was concerned. The Russian Academy of Sciences was contracted through the Viennese ambassador in Saint Petersburg, but its only response was to send the announcement to be printed in the public press.[191] In many respects this prize proposal was even less likely to be actually judged by the science societies than the Lauragais prize a quarter century earlier. The episode is interesting not so much because of its success or failure, but because of the international institutional framework in which the competition was conceived and considered.

Finally, Weights and Measures

The last eighteenth-century instance of a common effort (or plan for the same) on the part of the scientific societies was more a postscript to the chain of events already described than it was a serious episode in the common history of those institutions. Launched in France in 1791, under social (revolutionary) and scientific conditions very different from those of the 1760s or 1780s, the effort was initiated by the Paris Academy of Sciences for cooperation in the standardization of weights and measures, and it drew on elements of an older form of international cooperation.

The story is relatively simple. Seeking to preserve its position after the Revolution of 1789, the Paris Academy, through Talleyrand, proposed to the new National Assembly a standardization of French weights and measures.[192] At the time, there was a bewildering assortment of different weights and measures used in France, and reform of this chaotic system was high on the list of priorities of those holding political power.[193] After some debate, the proposal was accepted in 1790, and work continued under the academy's guidance until the academy was disbanded in 1793.[194]

The early stages of this project, as far as the academy was concerned, involved cooperation between it and the Royal Society in London. The English themselves had proposed a recasting of weights and measures, using an established pendulum length as the standard unit. The Talleyrand proposal incorporated this approach, "leaving," as Hahn says, "the technical details in the hands of the learned societies on both sides of the Channel."[195] For its part, the Paris Academy went back to the previous weights and measures exchange of the 1730s with the Royal Society as its starting point.[196] Soon, however, the Paris Academy shifted its researches to the establishment of a standard based on an arc of meridian, and nothing really came of English-French cooperation.[197]

Yet there remained a strong international flavor to the project of the Paris Academy because the goal was a truly universal measure. The Paris Academy in its post-revolutionary spirit in 1790 specifically invited foreigners to attend its meetings to discuss work on the new standards.[198] As Maurice Crosland notes, here was the beginning of something really new for the international organization of science, a proposal for what amounted to the first international scientific congress.[199] Later in 1798–99, a group of foreign experts did assemble in Paris under the aegis of the revivified academy to discuss the question of weights and measures. Nothing like this had ever been proposed or carried out before, and it is clear that, in the changing social and scientific contexts of the early 1790s, the endangered Paris Academy was forging new forms for international cooperation in science. Yet, almost as a postscript to this new approach, the academy could not forget old ways of doing things, and just two weeks after its first proposal of 1790 for an international congress:

It was decided that the Academy would write a circular letter to all the learned companies of Europe to notify them of the means it proposes to use to fix a unity of measure and to invite them for their part to cooperate in this. The Academy also invites those academicians [émigrés!] who would travel for their personal affairs to work together with the learned companies of the countries wherein they might find themselves on the adoption of the universal measure and on the operations necessary to achieve it.[200]

There is no evidence that this circular was ever sent, but the proposal is a strong indication that, late in the eighteenth century, international cooperation in science still meant concerting the forces of the scientific societies.

CHAPTER VII

Scientific Societies and the Making of the Scientist

I~N~ our introduction we posed "the problem of professionalization" for eighteenth-century science.[1] The idea, again, is that science becomes better assimilated in society in the seventeenth century, that something like an incipient "role" emerges for someone to pursue science, if not actually as a livelihood, then at least as a distinct engagement which other people would socially recognize as such. The "scientist," how- ever, is not seen until the nineteenth century. As is well known, William Whewell coins the term in 1840.[2] How, then, are we to think of the social development of the man of science in the eighteenth century? The age of the amateur? The scientist *malgré lui*?

While not sociologically rigorous, the answer from the point of view of this history at least is that the eighteenth century is a transitional stage, that the man of science—the scientist—as a socially defined type advanced great strides in the period, and that the scientific societies provided the primary institutional basis for this definition and ad- vance. It remains to consider the particular ways the scientific societies were effective in these regards.

First, the scientific societies clearly did provide an institutional basis for science and its leading practitioners: Haller, Buffon, Lagrange, Priestley, Franklin, Réaumur, Condorcet, d'Alembert, Linnaeus, Maupertuis, Lavoisier, Laplace, Clairaut, the Bernoullis, Euler, and others one might name. Not everyone we would recognize as making an important contribution to science in the eighteenth century was firmly connected to one of the scientific societies—the independent Bonnet in Geneva, for example. And again, many men held university positions as their primary or sole institutional appointments—Wolff in Halle and Marburg. But such men were often distinguished, nonresident members of the scientific societies—even the reclusive Cavendish naturally belonged to the Royal Society of London. Still and all, many more and many of the more important scientists of the eighteenth century were closely associated with the scientific societies and were often their leaders: Haller, president of the Göttingen society; Buffon, treasurer of the Paris Academy; Lagrange, founder of the Turin society, resident academician in Berlin and in Paris; Condorcet, permanent secretary of the Paris Academy; d'Alembert, permanent secretary of the *Académie française;* Franklin, president of the American Philosophical Society; Maupertuis, president of the Berlin Academy; Priestley, founding member of the Manchester Philosophical Society; Lavoisier, treasurer of the Paris Academy, and so on. Astronomy and mathematical physics are well represented in the academies, but the list of scientific society luminaries slights no discipline: Franklin, Nollet, Aepinus in electricity; Buffon, Linnaeus, Maupertuis, and later Lamarck and Cuvier in natural history and the life sciences; Lavoisier and entourage in chemistry, of course. Again, the scientific societies do not exhaust the set of institutions or circumstances in which men (and women) pursued their science in the eighteenth century, but they do constitute its core: the most powerful and prestigious institutions in which the best science was done.

The scientific societies provided a standard social niche and a recognized social role. Academies of science in particular institutionalized the role of academician. Concisely stated in the *Encyclopédie,* academicians are "members of an organization which carries the name Academy and which has for its object matters demanding study and application. . . . Science and wit are shared by the academician. . . . He strives toward and composes works for the advancement and perfection of a literature."[3] That academies and academicians occupied a distinct and formal place in late feudal corporate Europe is evident in the audience offered in 1776 to members of the Montpellier Society of Science by "Monsieur, frère du Roi." In a formal reception, best reported in French, "La Société a été reçue après La Cour des Comptes, Aydes et Finances, les Tresoriers de France et le Présidial, et avant les Universités [!]."[4] In a like vein, according to the 1772 letters patent of the Brussels Academy, the title of academician "will communicate to those decorated with it who are not already ennobled or of noble birth, by virtue of their admission to this company, the distinctions and prerogatives attached to the state of personal nobility."[5]

The records of the Saint Petersburg Academy provide an interesting instance of someone seeking recognition as a scientist and using the scientific societies as the legitimizing institution.

The Secretary read a petition from Mr. Eberhard Jean Schröter, inhabitant of Saint Petersburg, member of the Free Oeconomical Society of this town and correspondent of the Electoral Academy of Sciences at Munich and the Physical Society at Halle, who asks the Academy, in consideration of his application to the sciences and his knowledge of diverse parts of *la physique*, to be so kind as to accord him a certificate by virtue of which he will be acknowledged to belong to the class of learned men *(la classe des savants)* and rightfully entitled to the privileges attached to this state. The Conference acquiesced to his request, and the Secretary was charged with preparing such a certificate.[6]

The example is out of the ordinary, but in its extremes it suggests that the more established members of the international scientific community could rely on a defined social role backed up by the scientific societies.

Some men, in fact, were able to pursue actual scientific careers within the context afforded by the scientific societies. One would not say that scientific careers were feasible or actual in the Royal Society of London or other societies of science. The Royal Society (and the societal model) did not organize a body of men to a sufficient degree, nor did it pay them enough nor demand enough of their time to become the focus of much of a career. At the same time, one would not want to lose sight of the fact that a core group of Englishmen all through the eighteenth century were career scientists and Fellows of the Royal Society of London: Bradley, Halley, Sloane, Magellan, Maclaurin, Priestley, and so on.[7]

When it comes to the continental academies, the possibility that men pursued scientific careers as professional academicians was more real. The Paris Academy established a rigid hierarchical organization with clear lines of passage from the bottom to the top, paying positions. This form of organization, combined with the hegemony of the Paris Academy over other lucrative positions demanding scientific expertise really did allow academicians professional careers.[8] By late in the eighteenth century, in fact, the number of competent candidates exceeded the number of career positions available in the Paris Academy. The reform of the academy in 1785—regularizing supernumerary positions that had accreted to the academy, folding the slightly derogatory adjunct class into the associate class, and, reorganizing the disciplinary divisions of the academy—resulted.[9] The Paris Academy (as other Continental academies) was a restricted and elite organization with only a few "professional" positions, but, nevertheless, d'Alembert, Buffon, the Cassinis, Clairaut, Réaumur, Condorcet, Lalande, Laplace, Lavoisier, and the others constituted a non-trivial group which found homes and scientific careers in and around the Paris Academy.

There was less possibility in the French provinces of pursuing a career in science as an academician.[10] The structural or formal mechanisms were present in the organizational form of the academies, but the level of financial support and the role of provincial academies as civic institutions by and large precluded the possibility of serious careers in science in the French provinces. By the same token, because provincial academies offered some established positions as academicians, as institutional officers, and as teachers of public courses, some men on the provincial level (Guyton de Morveau, Maret, Chaptal, Bertholon, Nollet, de Ratte, Dubois de Fosseux) would seem to have lived as more or less full-time men of science.

In the Berlin Academy, the same kind of career possibilities as academician-scientist were open to men as in the Paris Academy, with the one important caveat that a person's success as an academician in Berlin depended on the good will of Frederick II. Those who received the approval of this monarch formed another well-established group of academicians of no mean standing: Maupertuis, Euler, Lagrange, Formey. With regard to Sweden, there is little doubt that there was a major community of men in that country, including Wargentin, Celsius, Linnaeus, Schelle, and others, who were able to make scientific careers within an institutional establishment that included the Swedish Academy of Science as its major part.[11]

Elsewhere on the Continent, men also found careers in academies, either because they were able to secure self-sustaining positions within institutions (e.g., Mayer in the Mannheim Academy) or because connections to other (pedagogical) institutions made such careers possible (e.g., Haller in the Göttingen Society). Conspicuous in this connection are the number of institutional officers, notably permanent secretaries, who were major academicians, who in a real sense kept the academies going, and who were active on the international plane in eighteenth-century science: Fontenelle and Condorcet are famous as secretaries of the Paris Academy, but then there are the less-well-known Zanotti of the Bolognese Academy, DeSaluces of the Turinese Academy, Lorgna of the *Società Italiana*, Lamey of the Mannheim Academy, Hemmer of the Mannheim Society, Kennedy of the Munich Academy, DesRoches of the Brussels Academy, Wargentin, Formey, J. A. Euler, and so on.

But nowhere were careers more explicitly or revealingly defined than within the Saint Petersburg Academy of Sciences.[12] In part, this situation derived from circumstances peculiar to Russia, and specifically the need to import and fully support academicians. But the ways and means by which the Saint Petersburg Academy created academic careers are indicative of standards and procedures applicable to other scientific societies. One could make a strong case for the career aspect of service in the Saint Petersburg Academy by discussing the series of outstanding scientific academicians that that institution, like the Paris Academy, turned out over the course of the eighteenth century. But the point can equally well be made by turning to the lesser example of the Swedish apothecary J. G. Georgi.

Georgi had served as an assistant to the academicians Pallas and Falk on two expeditions sponsored by the Saint Petersburg Academy. In 1774 he asked through an intermediary, the academician Orloff, if he could continue in the academy's services. Georgi produced testimonial letters from Linnaeus and Wallerius, both important members of the Swedish Academy, and the Russian Academy decided that Georgi should not be let go. In the meeting of the Archive-Conferences when all these details were reviewed:

Professor Pallas remarked that even though Mr. Georgi has not studied at some university, he should not be excluded on that account from the number of *savants* who can aspire to academic offices [*charges académiques*]. The greatest chemists of our century likewise have not attended universities; they have been apothecaries just as Mr. Georgi is. Just the same, the most celebrated academies take pride in counting them among their associates.[13]

Such openness to outside talent did elicit a negative reaction from some in the Saint Petersburg Academy, who raised objections to Georgi's being officially admitted into the academy on the grounds that he was a foreigner and that there were Russian students for whom places had to be reserved. But, based on precedent, there was an institutionalized mechanism whereby Georgi could advance in his career: "Even after he is naturalized the Apothecary Georgi must give proof of his capacity and present to this effect to the Conference a Latin memoir on a subject that relates to the science he wishes to take up as a profession [*dont il voudra faire profession*]."[14]

In early 1776, Georgi did submit the required Latin memoir and was elected as an adjunct of the Saint Petersburg Academy. In 1783 he was "advanced to the grade of Academician."[15] The example is not completely convincing, but it illustrates the possibilities of pursuing professional careers in Saint Petersburg and in academies in general.[16]

While it cannot be fully demonstrated here that scientific careers as academicians were widespread within the confines of even the major scientific academies of the eighteenth century, two further examples carry us a step closer to that goal and illustrate another extreme of career possibilities in science in and around the scientific societies. The briefest life histories of Leonard Euler and J. L. Lagrange indicate clearly that there was an established social, institutional, and professional system centering around the scientific societies "out there" in eighteenth-century society which was open to serious aspirants. Such a system offered the best opportunity to pursue science and live a life.

Both Euler and Lagrange began and ended overwhelmingly successful careers as scientists wholly within the institutional framework established by the academies.[17] Euler started out in 1726 in the Saint Petersburg Academy, after having attended university in his native Basel. Fourteen years later, in 1741, he was called to Berlin by Frederick II to head the mathematical class of the academy there. After more than two decades in Berlin, Euler returned to the Saint Petersburg Academy in 1766 where he reigned as its "dean" until he died in 1783. Lagrange began as an enterprising university student and organizing member of the Turinese Society of Sciences. Lagrange replaced Euler at the Berlin Academy in 1766, and later, in 1787, he was taken into the establishment of the Paris Academy of Sciences, changing his *associé étranger* status into *pensionnaire vétéran;* he survived the Revolution.

Interestingly enough, both Euler and Lagrange seem to have recognized the modernity and utility of their connections to the academies, and both rejected career alternatives in universities. Euler, dissatisfied in Berlin and looking to go elsewhere, reported in 1763 that he was offered a position in a Dutch university at an annual salary of 5,000 florins. His wife was won over to the proposition, "But for me," said the academician, "I feel a great repugnance for university life, and I will do all in my power to turn her away from this plan."[18] Similarly, Lagrange, also on the verge of leaving the Berlin Academy, in 1787 refused a position worth 5,000 lira annually as regius professor of mathematics at his own university of Turin, saying,

I have never liked mathematics in order to teach it, which many others can do, but uniquely for the pleasure of contributing to its advancement. The position offered me at Turin would have also been as little satisfying to me twenty years ago as it is at present . . . The Paris Academy holds a great deal of attraction for me as the first tribunal of Europe for the sciences.[19]

Compared to the ways the lives of major scientific talent unfolded in the seventeenth or nineteenth centuries, these bare outlines of the biographies of two of the most illustrious scientists of the eighteenth century indicate once again that for some men the scientific societies provided an important and attractive fixed basis for a life devoted to science.

Moreover, the key fact that both Euler and Lagrange migrated from one academy to another shows that the institutional framework for careers provided by the scientific societies was not limited to a single institution or nation, but rather that the international network of scientific societies created a common and collective institutional base that made possible scientific careers of truly international character and proportion: Euler, two stints in Saint Petersburg, one in Berlin; Lagrange, moving from Turin, to Berlin, to Paris; Maupertuis, going from Paris to Berlin; Delisle, from Paris, to Saint Petersburg, and back to Paris, and so on. Such career possibilities on the international level mark a major change from the seventeenth century and bring home once again how important scientific societies had become in organizing science on that level.

A related but hardly less influential factor—in addition to providing an institutional niche for a scientist—was the ability of scientific societies to control and impose professional standards on their individual members. The first and most obvious way scientific societies were able to impose standards was through their powers to elect members. Election procedures varied, but everywhere they demanded certain minimal requirements of candidates. Formal schooling and university degrees were not required, but the Latin memoir demanded of Georgi says a lot about standards. Everywhere, simple election had the effect of drawing a line between people an institution chose to admit and those it excluded.[20] In this way scientific societies were able to inject a new institutional mechanism in science that on the simplest level separated the masses from the smaller group of men who might achieve access to institutions, careers, and the international scientific community.

Once elected, men continued to be subject to the government of their home institutions in a variety of ways. Much of this further power of institutional control was, like rules for election, statutory, and such fine details typical of academic organizations concerning age, residence, and reading papers in turn were enforced to a high degree.[21] In addition, most academies exercised fairly stringent control over the actual scientific work of members through regular reports and judgments of committees and by controlling access to publication with the almost universal rule that the title of academician could not be used in a published work without prior approval by the institution in question.[22] In the extreme, the provision even extended to correspondents. One offender in the Dijon Academy was chastised by its president, de Ruffey, in 1764, as follows:

You have had a few memoirs published with the title of correspondent. This always compromises an academic body when a work is weak or reprehensible, which to a certain extent is your printed letter. In order not to again fall into these difficulties and in order for you to conform more exactly to our statutes, be careful when you wish to present a work to the public with the title of our correspondent to communicate beforehand the manuscript of the work to the Academy to obtain its approbation or at least its agreement.[23]

The best evidence that the scientific societies of the eighteenth century were in fact institutions of power and social control in science can be seen in the cases where they exercised their "muscle" against individuals who violated the accepted professional canons and norms. Throughout the world of scientific societies in the eighteenth century, people were ejected from institutions for immoral conduct, lack of participation, and as examples to others.[24] Another mundane case, while stopping short of outright ejection, shows the extent to which institutions could police and control the behavior of their members. The case occurred in the Montpellier Society of Sciences. The details speak for themselves.

20 August 1761. Mr. Peyre, having taken on the quality of Associate Chemist of the Royal Society of Sciences in a printed poster published by an operator-salesman of nostrums and other drugs who is presently in this town, it was decided that in the next assembly, which will be convoked extraordinarily for this purpose, the said Mr. Peyre will be admonished by the President of the session who will present to him the wrong he has done on this occasion in taking the title of Associate that he does not have, being simply an Adjunct, and in taking it under such circumstances. At the same time, he will be warned that if he does not conduct himself differently in the future, one will be forced to strike him from the catalogue of academicians.

27 August 1761. As a result of the deliberation taken in the last meeting of the 20th of this month, Mr. Brunn, Sub-Director, admonished Mr. Peyre and represented to him the gravity of the mistake he made by taking on the quality of Academician in a poster distributed throughout the city by charlatans and montebanks. Mr. Peyre responded that this quality had been given him without his knowledge and that he had expressed his great discontent. Mr. Brunn said to him that this disavowal in truth diminishes his mistake but does not excuse it entirely [because] in providing an authentic approbation of compounds prepared by charlatans he had patently forgotten what the dignity of the organization, of which he has the honor of being a member, demands of him. He promised to conduct himself more prudently and more decently in the future. In addition, the Company, using the right of inspection and discipline that it holds over each of its members, prohibits them from involving themselves in any manner in the future, either directly or indirectly in the preparation or distribution of drugs by charlatans or montebanks, reserving [the right] to pronounce judgment against those that contravene this deliberation.[25]

The view of Samuel Formey, encountered earlier, that academies were oracles of science with dictatorial powers finds a strong echo in this incident. Another instance of a similar character can be found in the decision of the Saint Petersburg Academy and its director, Razumovskii, to attempt to exclude Delisle from being accepted by other learned academies or societies after the latter had left Saint Petersburg under unpleasant circumstances. Wrote Razumovskii to Formey:

Sir, I must alert you that because of the bad conduct that M. Delisle maintains toward me, I have forbidden all the academicians and professors here from having any further literary commerce with him or communicating the smallest thing regarding the sciences. If I had the pleasure of learning that you had broken all contact with this ill-natured person who doubtless will pay the price of his impertinence, I will express my indebtedness on every occasion.[26]

The Royal Society of London also exercised similar powers. Its trial of Leibniz was the infamous case; its expulsion of David Riz, insignificant. But both illustrate the power of the institution. Yet another instance saw masked charges of fraud brought by the council of the Royal Society against John Canton and his experiments concerning the compressibility of water; this time, however, a special committee exonerated the accused.[27]

The most outstanding case involving the naked exercise of institutional power over an individual was the König affair in the Berlin Academy. In what really involved a larger dispute between the Leibnizians and the Newtonians of the Berlin Academy in the early 1750s, Samuel König was accused of forging a letter of Leibniz' which, if admitted as genuine, would have overthrown Maupertuis' priority to the principle of least action. This "affront to the scientific majesty of Maupertuis" resulted in the convocation of the academy in a full-fledged academic trial which, in the end, vindicated Maupertuis and destroyed König.[28] Euler, who was brought in to defend Maupertuis (and Newton), addressed himself in a letter to Merian to the basis of the academy's actions.

. . . other questions of this type are developed in the Judgment of the academy. They are all clearly those, the knowledge of which no legal tribunal could arrogate to itself. As they demand a deep knowledge of the sciences to which they appertain, I do not see to whom the right to judge them could be more befitting than to an Academy intended for the advancement of the sciences.[29]

Terence J. Johnson has explored theoretically the nature of professions and the process of professionalization. While not focused on science particularly, his remarks seem apt here to describe the place and function of the scientific societies relative to the professionalization of science in the eighteenth century.

In attempting to reconcile the inconsistent interpretations of the social role of the professions, the theory of professionalization has excluded the one element which was constant in earlier approaches: the attempt to understand professional occupations in terms of their power relations in society—their sources of power and authority and the ways in which they use them . . . Institutionalised forms of control of occupations are only to be fully understood historically through an analysis of the power of specific groups to control occupational activities. To achieve this understanding, we must make a clear distinction between the characteristics of an occupational activity (which may themselves change over time) and historically variant forms of the institutional control of such activities which are the product of definite social conditions . . . Professionalism, then, becomes redefined as a peculiar type of occupational control rather than an expression of the inherent nature of particular occupations. A profession is not, then, an occupation, but a means of controlling an occupation. Likewise, professionalization is a historically specific process which some occupations have undergone at a particular time, rather than a process which certain occupations may always be expected to undergo because of their "essential" qualities.[30]

In addition to each academy separately governing its own constitution and controlling local communities of scientists, scientific societies were able to influence the contemporary international scientific community by regulating their foreign members. One important development to be observed in this regard is that, as the eighteenth century progressed, the leading scientific societies became increasingly concerned about their foreign associates and tightened up on the standards to be applied to them.

This movement toward more strict control began in the Royal Society of London and the election reform instituted by Hans Sloane in 1727. After 1727, recommendations were required, and elections occurred only once a year. Later in the eighteenth century, the Royal Society further tightened up on its nonresident, foreign Fellows. In 1761 it instituted the rule that, in addition to previous recommendations, another three from Fellows on the "home list" were necessary for election as foreign Fellow, "Except the said Candidate shall be otherwise distinguished by his High Rank." Because of the "excessive number of foreign members," in 1765 the council of the society took a further step towards controlling them:

Resolved, that no foreigner be proposed for Election, that is not known to the learned World by some Publication or useful Invention, tending to the Improvement of natural knowledge, which may enable the Society to form a Judgement of his merit, and that till the number of foreign members be reduced to eighty, not more than two shall be admitted in one year.[31]

Note that at the next meeting of the council, it was "resolved unanimously that the Presidents of foreign Academies of Science . . . be excepted from the restrictions in the resolution."[32] In this reform, the Royal Society was reacting to the large number of foreign members, a total of one hundred sixty-nine in 1765, but the stress on more stringent demonstrations of a candidate's scientific capacity is noteworthy. In this, the 1765 reform by the Royal Society was in its intent more like similar reforms of Continental academies. Judging from the data presented by Lyons, this reform did have a significant limiting effect on the foreign fellowship of the Royal Society, and similar reforms in other institutions probably had the same effect.[33]

In the late 1740s and early 1750s, the Paris Academy also began a tradition of tightening up on the qualifications of its foreign correspondents. The position of correspondent of the Paris Academy was somewhat ad hoc until the middle of the eighteenth century. A member of the academy proposed a candidate from outside the institution with whom he wished to entertain a recognized correspondence, and if it approved, the academy elected the nominee its correspondent and assigned him to an academician, seemingly always the nominating individual.[34] The number of correspondents grew over time, and by 1747 the situation seemed to call for reform.[35] After off-and-on discussion, a new regulation was instituted in 1753.[36] Henceforth, after being proposed, a prospective correspondent had to wait for a period of at least a month, receive a favorable report of a committee set up to investigate his candidacy, and be elected by a two-thirds majority. In addition, correspondents had to correspond at least once every three years or they would be eliminated from the list of members. In 1762 the standards applicable to foreign correspondents were raised again, with the number of such men in association with the Paris Academy limited to one hundred and the requirement for communications raised to at least once a year. In 1767, a house cleaning of correspondents took place, with many dropped from the academy's list.[37]

Acting on the precedent set by its mother institution in Paris, the Royal Society of Sciences of Montpellier in 1769 likewise overhauled its rules and procedures for foreign correspondents: "It seems very important to consider this subject, even more since correspondencies having been accorded to a great number of persons, one has felt the necessity to regulate their form and laws."[38] The regulations adopted by the Montpellier Society were essentially the same as the 1753 rules of the Paris Academy, including the important provision that "correspondents will not be received except those that will give to the Royal Society an advantageous idea of their knowledge."[39] The Montpellier Society also provides an interesting indication of the growing concern of institutions concerning the use and misuse of academic titles. At the same time these new regulations for foreign correspondents were established:

It was decided, moreover, that correspondents will be alerted that their title can only be expressed by the French phrase, *Correspondant de la Société Royale des Sciences de Montpellier*, or by the Latin phrase which is its translation, *Regiae Scientiarum Societatis Monspeliensis Correspondons*. The words *Socius* and *Socius Extraneus* that some have employed, having always been a usage to distinguish the different orders that compose the body of the Society proper, it was decided, after the example of the Royal [Paris] Academy of Sciences, to use this more appropriate phrase, although [it is] hardly [good] Latin.[40]

In limiting the number of its correspondents to eighty in 1777, the Montpellier Society reasoned: "It seems necessary to fix the number of correspondents because, this number remaining unlimited and augmenting indefinitely, the position becomes less esteemed. The Royal Society [of Montpellier] is interested in giving to the state of Correspondent all the character of which it is susceptible, imitating on this important point the Academy of Sciences of the capital.[41]

A decade later, in another debate concerning foreign correspondents, the principle of exclusivity was reaffirmed in Montpellier: "If it is interesting for progress of the sciences to extend useful correspondencies, it is dangerous to lavish the tokens of esteem of a learned company and to weaken correspondencies by multiplying them."[42] What was needed, or so felt the Montpellier Society, was "a more regulated surveillance" and to this end the Society took the further step of limiting elections of correspondents to once a year.

In a similar effort to upgrade its correspondents, the Dijon Academy insisted on at least annual contributions from all its members. As Richard de Ruffey explained the new Dijon regulations in 1764:

The effect of the Academy's deliberation, which has cured several of its members of an indecent lethargy, proves its wisdom and necessity. In this the [Dijon] Academy has only followed the example of the Paris Academy of Sciences which has enacted the same regulation . . . When our Academy demands that its asociates produce one work each year, this is not a statute enacted by it alone, it is the fundamental law of all literary societies. An Academy is a society destined for the cultivation of the sciences and letters, and all members must be assiduous in fulfilling this goal and in working to be useful to it. In effect, an Academy composed of simple title holders who limit their vision to the sole honor of being inscribed on its roles is a paralyzed body, useless in the Republic of letters.[43]

Like the institutions in Dijon and Montpellier, it is not unlikely that other French provincial academies followed the lead of the Paris Academy of Sciences and increased restrictions on foreign members. Outside France, the Saint Petersburg Academy sought to supplement its foreign honoraries with "a class of foreign correspondents by means of which the Academy of Saint Petersburg, in imitation of that of Paris, can receive a considerable number of learned men from all countries and from which it will not fail to gain real advantages," but it was early agreed that "it is appropriate for the Academy to be circumspect in the choice of its correspondents and not to squander this title by according it without distinction to whoever demands to be honored by it." In 1784 a move was made within the Saint Petersburg Academy to lower the number of foreign associates, by then at fifty-four, "to make more precious the honor of this association."[44] Notably, in the new rules exceptions were allowed for "the superior genius or learned men of an extraordinary celebrity."[45]

Thus it would appear that the scientific societies not only regulated their own membership but exerted a shaping influence on the constitution of the international scientific community. Even as the number of persons pursuing science grew in the eighteenth century, the societies sought to make entry into their individual and collective ranks more select and difficult.

In yet another way the scientific societies helped define what it was to be a scientist in the eighteenth century. Just as the various scientific societies formed a hierarchy of institutions, so, too, did the import of learned society appointments vary. As Formey put it in 1775, "Admission to the academies, especially those that hold the first rank in the Empire of Sciences and Letters, is a prerogative as hoped for as it is to arrive at the highest civil or military honors."[46] Clearly, those who held the top posts within those first line institutions formed an elite group of men of higher status and prestige than those who were outsiders or who merely pretended to higher positions. An Euler or a Lagrange occupied a different position in science and the scientific community than did a Georgi or a lesser figure. At the top of the world scientific community stood a core of international scientific "superstars." The noteworthy point is that, as it turned out, a man's place and rank in the hierarchical community of the sciences in the eighteenth century was determined and symbolized by the number and quality of the scientific societies to which he belonged.

The tendency of men to seek and to accumulate memberships in institutions was widespread and even abused.[47] One J. A. Sallé, for example, sought admission to the Berlin Academy in the 1750s by humbling himself before Maupertuis during one of the latter's visits to Paris. Sallé wrote of this to Formey: "Eh! Who can blame my ambition in this regard. These Literary honors are the sole recompense worthy of a man of letters . . . [Election] will doubtless contribute toward augmenting my burgeoning reputation in our provinces."[48] The Russian Samoilowitz petitioned entry into the Berlin Academy over a course of years, saying at one point, again to Formey, "I cannot express how appreciative I would be if your illustrious academicians would wish to accord me the title of Associate of your celebrated Academy of Berlin. I am already an Associate of ten academies."[49] Another wrote to the Montpellier Society, saying, "I have the honor of belonging to a score of academic societies," and at the extreme Daniel Roche mentions the case of Titon du Tillet who belonged to over thirty institutions.[50] The fact that one Joseph Mandrillon could say of his election to the American Philosophical Society that, "If it does not enrich, at least it encourages and brings pleasure," implies that election to some societies did enrich in one way or another.[51] Even such notable figures in the international scientific community as Lalande and Formey were not above using all of their influence to extend the number of institutions in which they could claim membership.[52]

Another fact seems apparent in the accumulation of academic memberships and titles: once one had made it into one society, it became easier to gain entry into other institutions. Deparicieux, geometer in the Paris Academy, spoke of this in 1761 when he wrote to the Montpellier Society, the first academy to which he was elected back in the 1740s, saying, "The Royal Society of Sciences opened the door for me to the [Paris] Academy here, and the two together [opened the door] to the academies in Berlin and Stockholm."[53] More than simply "opening the door," being a member of one institution virtually became grounds for becoming a member of another. A few examples make this clear.

Regarding the election of one Batini, the records of the Montpellier Society report: "Doctor of Medicine of the University of Montpellier, member of the Royal Society of Medicine of Paris, of the *Accademia Curiosa Naturae* and that of Berlin. These Academic titles and a learned work . . . determined the choice and united the votes of the Society."[54] Of the Swede Melander, the Saint Petersburg Academy said: "M. Melander has distinguished himself in the Republic of Letters by astronomical works; he is already a member of several Academies, and he merits as a consequence that that of Saint Petersburg honor him with the same distinction without his even asking for it."[55] In appointing Louis Godin professor of mathematics, the University of Lima in Peru noted that he "needs no other recommendation to give a full idea of his learning and his great experience in mathematical sciences than to be known as one of the worthy members of [the] illustrious and royal Academy [of Paris]."[56] Finally in this regard, wrote Condorcet of the candidacy of Malvezzi as correspondent of the Paris Academy:

M. d'Alembert and I were charged with informing the Academy if the Count Malvezzi is qualified to be admitted to the ranks of its correspondents. Here is what we think in this respect: Count Malvezzi is a member of the Institute of Bologna, one of the most celebrated learned societies of Europe. This consideration alone seems to us sufficient to obligate the Academy to accord him the title for which he is asking.[57]

In back of this general pattern of men multiplying their learned society memberships was the larger utility that such multiple associations held. The same Samoilowitz just encountered also badgered the Montpellier Society as part of his campaign to increase the number of his memberships. At one point, he crudely asked its secretary that "if this illustrious Company accords me the title of member, be so good as to inform me of it as soon as possible so that I can use it at the head of my works which I am about to publish."[58] Anyone familiar with the published works of men of science in the eighteenth century will recognize the common practice of authors to list their learned society titles after their names. Indeed, learned society titles constituted *the* means of establishng a man's credentials in the eighteenth century. The "Tableau de l'Académie Royale des Sciences," a list of the members of the Paris Academy prepared by the Abbé Rozier in 1774, makes this clear.[59] Two representative entries read:

Mr. de la Lande, royal reader in mathematics; royal censor; member of the Royal Society of London, the Imperial Academy of Petersburg, the Institute of Bologna, the Royal Academy of Sciences of Sweden, the Royal Society of Göttingen, the academies of Rome, Florence, Cortona, Mantua, Haarlem, the Academy of Arts established in England, and the Royal Marine Academy.

Mr. Duhamel de Monceau, member of the Royal Society of London, the Imperial Academy of Saint Petersburg, the Institute of Bologna, the Academy of Palermo; honorary of the Edinburgh Academy and the Marine Academy.

Rozier, a lesser figure to be sure, styled himself on the title page of this work as follows:

Chevalier of the Church in Lyon; member of the Royal Academy of Sciences, Beaux-Arts and Belles-Lettres of Lyon; member of academies in Villefranche, Dijon, Marseille; member of the Imperial Society of *Physique* and Botany of Florence; correspondant of the Society of Arts in London; member of economic societies in Berne, Auric, Lyon, Limoges, Orléans, etc.; and former Director of the Royal Veterinary School of Lyon.

Rozier's titles, not all of which are academic or purely scientific, accurately reflect his background and minor position in the eighteenth-century Republic of Sciences and Letters.[60] By comparison, the difference between his set of titles and those of Lalande or Duhamel du Monceau indicate that the latter, more prestigious and important men of science obtained a measure of their career status and place in the international scientific community from their membership in the more important academies and societies of science of the day.

Such titles and the accumulation of titles served the same function in the eighteenth century as do modern university degrees today. Just as we associate someone's background and status with what degrees he holds and where he got them (John/Jane Doe, B.A., M.A., Ph.D.), so, in the eighteenth century, did scientific society titles (Membre de l'Académie de . . ., etc.) certify a man's position in the more limited confraternity of scientists and the learned of that day.[61]

The problem of the professionalization of science in the eighteenth century still eludes us. Yet it seems safe to say that the scientific societies were central to the definition of a new, eighteenth-century kind of scientist. They were the most important forums for the best science. Scientifically they were largely self-governing, and they exercised considerable power and control over the scientific community. They provided status and socially recognized roles; they provided recognition and certification. They offered employment and access to employment possibilities. They gave science a new set of leading institutions, and they organized and activated science on all levels. They defined local, national, and international communities of scientists. At the highest level they embodied and supervised the elite in the world of science. If the ranking men who made up these institutions were not scientists and committed professionals, then the words have no meaning.

One can add an ironic footnote. For particular scientific societies the great men of science, the "Nobel prize winners" of the day, were rare and to be treasured. Scientists of high standing were valuable pieces of property who reflected and added to the prestige of institutions. This rather crass view is evident in the following proposal made in the Saint Petersburg Academy, shortly after the death of Euler.

Professor Kraft proposed on this occasion that, as the number of effective academicians has suffered a considerable diminution in the last years, it would be wished that the Academy think about making a new acquisition, in engaging in its services a man of recognized celebrity . . . The other academicians agreed to the utility of such an acquisition.[62]

Along these lines, in 1787, when it became known in the Berlin Academy that Lagrange was leaving, strenuous efforts were made to keep the great scientist there. A memorandum reports:

It seems advantageous and even necessary for the Academy to conserve this *savant* of the first order . . . It is a question of humoring a bit the hypochondriac state and amour-propre of this great geometer, without whom our academy would have no celebrated mathematician and would only find one with great difficulty in Germany or the rest of Europe.[63]

The new Naples Academy, which "has a great need of a genius," also tried to recruit Lagrange.[64] Such was the status and desirability of Lagrange and other top scientists. Lagrange went to the Paris Academy, as we know.

The End of an Era

THIS story reaches a natural end with the French Revolution. In France, all evidence indicates that the vitality of the many French academies of Paris and the provinces had already declined by 1789.[1] The Paris Academy and other provincial academies made overtures to new local and national authorities after 1789, but certainly no new academies were founded in France, and it had to have been the increasingly pressing political situation after 1789 that drew men's attention, rather than science.[2] From a European-wide perspective, the whole system of scientific societies continued: exchanges, correspondence, visitors, elections, publications, prizes, public meetings, and so on. But a watershed in the by-then long history of these institutions had been reached. It seems safe to say that, as far as concerns eighteenth-century scientific societies considered as a group, the period 1789–1815 is a long hiatus. As far as organized science and its history are concerned, on the other side of that watershed, the scientific societies, as they emerged after 1815, were no longer the vanguard institutions they had once been.

If, in the day-to-day lives of the scientific societies and the men who constituted them, the year 1789 did not mark the sudden onset of a new age, 1793 certainly did. On the 8th of August 1793 the Jacobin Convention voted its famous decree proscribing the academies of France: "All academies and literary societies patented or endowed by the nation are eliminated."[3] At a blow, the entire complex of academies that had grown up in France over the previous century and a half and that had made France the heartland of scientific societies was completely undone. The Paris Academy of Sciences, it is true, was revived in 1795 as the First Class of the Institut de France, and it was certainly France's leading scientific institution for years to come. French provincial academies, on the other hand, reappeared after 1815 as a rule, and they were more decidedly local institutions than before.[4]

From 1792 until 1815, with brief moments of respite, France and Europe were almost constantly at war. European war may have brought science closer to the war machines of nations, but it did little for the scientific societies.[5] Where actual hostilities took place, disruption or closure were the results. The Belgian Academy of Science, for example, was disrupted in 1793 and closed completely between 1794 and 1816.[6] In 1806 Napoleon broke into the Berlin Academy of Sciences to carry off equipment and collections.[7] The Turinese Academy was closed between 1792 and 1801; the Bolognese Academy was disrupted by military campaigns in 1796, 1799, and 1800, and closed completely in 1804.[8] Doubtless other academies and other societies were similarly affected.[9]

As a sidelight to the generally disruptive effects of the Napoleonic Wars on scientific societies in the heart of Europe, one needs to note the French tendency to reorganize science and scientific societies, once they had passed through. Louis Napoleon, for example, appointed himself permanent president of the Dutch Society of Sciences and created the Royal Institute of the Batavian Republic, after the Paris example.[10] Napoleon recast the Turinese Academy as the national Academy of the Cisalpine Republic, with himself as president, and created an Institute to compete with the Bolognese Academy.[11] The *L'Institut de l'Egypte* (1798–1801), created by Napoleon at the time of his early Egyptian campaign, is another example of this organizing tendency of imperial France.[12] But these developments are of little long-term consequence. They are more interesting for what they say about the vision of the Grand Empire and what might have been, had a stable peace been achieved in the period.

But this transition in the histories of scientific societies and organized science involves more than the immediate effect of war and revolution after 1789. For a variety of reasons, something of the earlier spirit seems to have gone out of the major academies and societies of science in the last decades of the eighteenth century.

In Sweden, for example, a real decline in the scientific vitality of that country had already set in by 1772, when Gustav III reasserted the full power of the Swedish monarchy and ended the "Era of Freedom."[13] National interests turned away from science and toward belles lettres, poetry, and romanticism in general. The Swedish Academy of Sciences was not immune from effects, and after the death of Wargentin in 1783, it became almost wholly preoccupied with questions of trade, agriculture, and "practical" science in general. It was not until a generation later, when Berzelius became president of the academy in 1818 and instituted the reforms of 1821, that the Swedish Academy again emerged as a strong institution of science and a leader on the world scene.

Similarly, by the mid-1780s, the Berlin Academy had declined seriously from its former glory of the 1750s and was a stagnant institution. That Frederick-William II chose to support the *Gesellschaft Naturforschender Freunde* after the death of Frederick the Great in 1786 is an interesting sign of the times.[14] Past the deaths of Frederick-William II and Formey in 1797, the Berlin Academy underwent a vacillating period of reform wherein the German faction, then the French faction, then the German faction again alternately held the upper hand within the academy and attempted to reshape the institution according to its own lights.[15] When the period of reform ended in Prussia in 1812, the *Akademie der Wissenschaften* (no longer the *Académie*) was completely recast as a separate research institute aligned with the new University of Berlin. It was no longer an academy in any eighteenth-century sense, but a small part of a larger university-based and research-oriented Prussian scientific establishment.

In Russia, the death of Catherine the Great in 1796 ended the strictly ancien régime period in the history of the Saint Petersburg Academy. During the reign of Paul I (1796–1801), "The Russian Academy virtually went out of existence."[16] Even under the more benevolent rule of Alexander I (1801–1825), the academy fared hardly better. It was given institutional control over patents and inventions in 1803, but in reaction to events in France, the academy was subjected to new degrees of bureaucratic pressure and attacks by nationalistic, theological, and anti-science forces in Russia. Largely cut off from the West, for a while the academy in Saint Petersburg was little more than a glorified censorship bureau for Western books. Russian science in the academy regained an equilibrium after 1815, but it was not until the 1860s and the advent of Alexander II that the academy finally established itself as an accepted and fully Russian scientific institution.

The Royal Society of London presents no convenient parallel to match the dramatic changes affecting scientific societies on the Continent in the period at and after the French Revolution. Yet, the Royal Society, too, was in transition. Its president, Sir Joseph Banks, was a major force in revitalizing the society in the late 1770s, but until his death in 1820, his continued stranglehold on the Royal Society ensured an institutional decline which mirrored his own.[17] The Royal Society in 1820 was not a serious center of science, and it took twenty years of reform before the Royal Society threw off its social club aura and transformed itself into a society of professional scientists.[18]

Indicative of the individual and collective circumstances of scientific societies in the early nineteenth century is the fact that academies and societies did not undertake another joint project until 1827. Beginning then and continuing through 1848, an international network of observing stations was set up to record variations in the earth's magnetic field.[19] The exercise was run by Humboldt (of the Congress of German Naturalists), and it was taken up by the Saint Petersburg Academy in 1829 and with government support by the Royal Society of London in 1839. The result was a coordinated international network of stations producing two million observations.[20] This geomagnetic survey was in the grand tradition of eighteenth-century common endeavors, but the four decades that separate this project and the end of Meteorological Society of Mannheim signify something of the divide between the eighteenth and nineteenth centuries as far as cooperation among scientific societies is concerned.[21]

Specialization was a major factor which had consequences for organized science and scientific societies in the nineteenth century. The exponential growth of scientific knowledge over the course of the eighteenth century and the growing number of practitioners of science meant that the effectiveness of the umbrella academy or society devoted to all the sciences was bound to be short-term. At some point science was going to break up into disciplines. Actually, the creation of academies and societies exclusively for science has to be seen as a form of specialization already. Within some academies which were wholly scientific in their orientation, a high degree of specialization existed already. In the Paris Academy of Sciences, this specialization was institutionalized in disciplinary sections for members. Certainly for many of the disciplines of science, particularly the mathematical ones, but even including the more empirical Baconian sciences, the age of the generalist was over in the eighteenth century. Buffon, though originally a mathematician, really had very little in common with Euler, or vice versa.

Furthermore, specialized societies did not suddenly materialize at the end of the eighteenth century. They had been there all along. In some senses, one wants to single out the emergence of specialized medical societies as a feature of the specialization of science in the eighteenth century. In a backhanded way the appearance of an academy like the *Société royale de médecine* in Paris suggests a more narrow notion of science and the possibility that specialized societies could appear which aped the form and function of the general academy of eighteenth-century vintage. But further, single-discipline scientific societies already existed in the eighteenth century in the Meteorological Society of Mannheim, the *Accademia Bontanica* of Florence, and the cosmographical societies in Nuremberg and Uppsala.[22]

Thus, the appearance of the Linnean Society in London in 1788 did not mark something fundamentally new. It was like any other (private) grouping of scientific men, only especially concerned with botany and natural history.[23] By the same token, with the Linnean Society specialized scientific societies increasingly were less the institutional oddity and more the norm for organized science. In England, the Geological Society of London followed in 1807, the Zoological Society of London in 1826, the Royal Astronomical Society in 1831, the Chemical Society of London in 1841, and so on.[24] The trend continued elsewhere throughout the nineteenth century, and continues today.[25] Specialized societies increasingly became the focus of practicing communities of scientists. They were the places that engaged the energies of (particularly younger) practitioners, places that first received the results of their researches.

Related to the appearance of specialized scientific societies and their effect on the general academy or society of science were notable developments in the scientific press. We have spent much time investigating the state of the scientific press in the eighteenth century, the ways in which the publications of the scientific societies were the primary means of communicating original science, the deficiencies that arose in the latter part of the century, and alternatives that essentially built on the pattern of scientific communication and publishing established by the scientific societies. But Rozier's Journal was a jerry-built affair, and as the system of eighteenth-century scientific societies declined, the publication of specialized scientific journals arose with a vengeance.[26] The famous *Chemische Journal* of Lorenz Crell (from 1778 with various titles), was followed by the *Botanical Magazine* (1787; later *Curtis's Botanical Magazine*), the likewise famous *Annales de Chimie* (from 1789, as the publication monopoly of the Paris Academy of Science slipped away), the *Annalen der Physik* (1790), and so on.[27] The large and late volumes of the academies did not suddenly become obsolete, and their nineteenth-century *Comptes-rendus* and specialized series did much to make up for the inroads created by specialized, independent journals. Just the same, as contemporary scientific practice indicates, specialized journals, periodicals, and proceedings are the main sources where scientists get information and present their findings.

Career patterns in science changed considerably as the nineteenth century unfolded. We suggested that eighteenth-century careers in science centered around the scientific societies, state employment, and universities. In the nineteenth century, new institutions began to employ scientists, and the state and status of the old scientific societies likewise changed. In France, new and revivified centers such as the *Ecole polytechnique*, the *Bureau des Longitudes*, the *Muséum d'Histoire naturelle*, and the *Société d'Arcueil* increasingly became the places where scientific work was done.[28] Although the full effects of this denaturing were not felt in the Paris Academy until 1830, it increasingly became a pantheon of persons rewarded ex post facto for their scientific achievements attained at earlier stages of their professional careers.[29] In Germany, the emergent university system became a greatly revitalized network which employed academic scientists and enforced work from them.[30] Liebig's establishing his lab at Geissen in 1826 is the most potent symbol. In England, the creation of the Royal Institution is an example of new institutional niches in which science could then be pursued.[31] Later in the nineteenth century, of course, scientists found gainful professional employment in capitalist industry, which doubtless made their specialized societies more important to them than national academies and societies of science.[32]

In the second and third decades of the nineteenth century the organi-

[258]

zational and institutional character of science was complicated by the appearance of professional societies. The variously titled *Deutsche Naturforscher Versammlung* began in 1822 as a union of German scientists with a professional interest in the pursuit of knowledge.[33] This example was taken up in England among provincials and the leaders of scientific reform and resulted in the creation of the British Association for the Advancement of Science in 1831.[34] With peripatetic annual meetings attracting a large attendance, reports, publications, and support for research, the British Association quickly became an important new institution for rallying science and men of science. The institutional example has been taken up almost universally.

Most of the scientific academies and societies studied here still exist, and their numbers have been multiplied many times over as scientific culture has spread from its eighteenth-century limits. Some, as we suggested at the outset, are honorary organizations, some are ceremonial consultants fit for blue-ribbon panels, some are great bureaucratic entities, some are backwater associations of local notables. With closer contacts among science, government, and industry and with the growth of specialized and professionalized societies since the eighteenth century, scientific societies of the kind studied here lost the once preeminent place they held for organized and institutionalized science.

Official Scientific Societies:
1660–1793

APPENDIX 1 lists alphabetically by town scientific societies officially recognized in the period 1660–1793 by some public authority. Private societies are included if they give rise to an official successor; otherwise they are listed in appendix 2. Correct official names of institutions, along with contemporary variants and the names of unofficial predecessors, are given. Societies are characterized as to type: academies, societies, "Renaissance" academies. Political sources of official recognition, date of foundation, and years of activity are indicated for each institution. Where known, sources of material support, degree of orientation toward science, size and organization of membership, extra institutional facilities (beyond the near universal possession of a "house"), publication record, and prize and other activities are specified for individual institutions. Sources of this information are likewise indicated.

Agen, France.
    Société des sciences, arts et belles lettres.
      Official academy: 1776–1793.
      (Cercle de lecture, antecedent.)
      Recognition: French king.
      > 50% science.
      30–35 ordinary members.
      No prizes; no publications.
      Source: Roche.

Amiens, France.
Académie des sciences, belles-lettres et arts.
Private *Société littéraire:* (1702 . . .) 1745–1750.
Official academy: 1750–1793.
Recognition: French king; provincial governor, patron.
Support: crown, city.
> 60% science.
30–35 orginary members. (10 core at meetings.)
Library, botanical garden, prizes.
Public courses: natural history (1758); chemistry (1787).
Sources: Roche; Leleu; Bouillier; Almanach royal.

Angers, France.
Académie royale d'Angers.
Official academy: 1685–1793.
Lapses: 1700–1745.
Revived 1747 as *Académie royale des sciences et belles-lettres.*
Recognition: French king.
Support: crown, city, private.
30–35% science. (More after 1747.)
30–35 ordinary members.
Prizes; public courses.
Sources: Roche; Boreau; Roche, "Histoire"; Tisserand.

Arras, France.
Académie royale des belles-lettres.
Authorized *Société littéraire:* 1737–1773.
(Most active: 1737–1745; supported by dues.)
Official academy: 1773–1793.
Recognition: French crown.
Support: provincial estates.
40% science; 30–35 ordinary members.
Prizes 100% science and arts.
Numerous publications: 1739–1791.
Sources: Roche; Berthe; Lefèvre-Pontalis; Bouillier; Almanach
royal.

Auxerre, France.
Académie des sciences, arts et belles-lettres.
Authorized as the *Société des sciences, arts et belles lettres:* 1749–1769.

Official academy: 1769–1772 [sic].
Recognition: French crown.
50% science.
No prizes; 30–35 ordinary members; no publications.
Sources: Roche; Chaillou de Barres.

Barcelona, Spain.
Real Academia de Ciencias Naturales y Artes.
    Official *Real Conferencia Física*: 1764–1770.
    Official academy: 1770–present. (Lapses 1808–1814.)
    Recognition: Spanish crown (Carlos III).
    Support: crown, Catalan estates.
    100% science; no publications.
    60 resident members.
    Observatory, library, museum, natural history cabinet.
    Sources: Murua y Valerdi; Diccionario Enciclopedio Hispano-
        Americano (1887).

Batavia, East Indies.
Bataviaasch Gnootschap van Kunsten en Wetenschappen.
    Official society: 1778–1795 . . .
    Recognition: Governor General of East Indies; Dutch East India
        Company.
    Support: dues, private support.
    100% science and economy.
    30 Batavian members; 200 European.
    Verhandelingen, 21 vols. (1779–1847); vols. 1–4 (1781–1786), re-
        printed in Amsterdam; German tr., vol. 1, Leipzig, 1782.
    36 prize contests: 1778–1792.
    Close contact with Dutch societies.
    Sources: Bataviaasch Gnootschap, *Gedenkboek;* Scudder.

Berlin, Prussia.
Societas Regia Scientiarum.
    Official academy: 1700–1744.
    Recognition/Support: Prussian crown.
    4 classes; 50% science.
    Monopolies: Almanac, silk manufacture.
    8 vols. *Miscellanea Berolinensia*, 1710–1746.
    Pensions; no prizes; observatory, anatomical theater.
    Succeeded by . . .

Académie royale des sciences et belles-lettres de Prusse.
    Official academy: 1744 . . .
    (Succeeds *Societas Scientiarum.*)
    Recognition/Support: Prussian crown (Frederick the Great).
    50% science.
    55 vols. (assorted) *Mémoires,* 1745–1804.
    Samuel Formey, secretary.
    32 prizes: 1745–1788.
    Sources: Harnack; Bartholmèss; Scudder.

Besançon, France.
Académie des sciences, belles-lettres et arts.
    Private *Société littéraire:* 1748–1752.
    Official academy: 1752–1793 (inactive after 1789).
    Recognition: French crown.
    Support: private, town, patron.
    (+/−) 40% science/arts.
    40 ordinary members (10 core).
    143 members 1748–1793.
    19 *Séances publique,* 1754–1783.
    Library; prizes. Académiciens-Nés category.
    Sources: Roche; Pingaud; Cousin.

Béziers, France.
Académie royale des sciences et belles lettres.
    Authorized academy (*Société des sciences et belles-lettres*): 1723–1766.
    Official academy: 1766–1793.
    60% science; *Recueil,* 1 vol. (1736).
    Observatory; library; no prizes; less than 20 ordinary members.
    Sources: Roche; Bouillier; Almanach royal; Taton, *Enseignement.*

Bologna, Italy.
Accademia delle Scienze dell'Istituto.
(Scientiarum et artium institutum bonoiense atque academia.)
    Official academy: 1714–1804.
    Bologna Senate, papal recognition support.
    100% science.
    36 core members; pensions; library.
    Publications: 8 vols. (in 10) *Commentarii,* 1731–1791.
        German tr., Brandenburg, 2 vols., 1781–82.
    Sources: Rosen; Maylender; Scudder.

Bordeaux, France.
Académie royale des sciences et belles-lettres.
    Official academy: 1712–1793.
    Recognition: French crown.
    Support: estates, town, private benefaction.
    80–90% science. Less than 20 ordinary members.
    Public library (40,000 vols.); botanical garden; cabinets; prizes.
    Courses: natural history, experimental physics.
    6 vols. natural history of Guyenne, 1715–1739.
    6 vols. *Recueil des Prix*, 1715–1735; continues from 1740.
    Sources: Barrière, Roche, Vivie; Bouillier.

Boston, Massachusetts.
    American Academy of Arts and Sciences.
    Official society, 1780 to Present.
    Recognition: State of Massachusetts.
    Publications: *Memoirs* (1785, 1793).
    Sources: Bates; Hindle; Stearns; Stone.

Bourg-en-Bresse, France.
Académie des sciences, belles-lettres et arts.
    Private society (*Société d'émulation*): 1755–1757.
    Official academy: 1783–1793.
    60 + % science/arts; prizes; no publications.
    Dozen core members.
    Sources: Roche; Buche.

Brest, France.
Académie royale de Marine.
    Official academy: 1752–1793. (Reformed, upgraded 1763.)
    Recognition/Support: French crown.
    100% science; astronomy = high; library; no prizes; weekly meet-
        ings.
    Employed instrument maker. 30–35 ordinary members.
    Mémoires, 1 vol., 1773.
    Sources: Roche; Charliat.

Brussels, Austrian Netherlands.
Académie royale et impériale des sciences et belles-lettres.
    Private société littéraire, 1769–1772.

Official academy: 1772–1794; continued, 19th century.
Recognition/Support: Austrian crown.
50% science; *Mémoires*, 5 vols., 1777–1788; German tr., 1 vol., Leipzig, 1783.
Cabinets; pensions.
Sources: Quetelet; Académie royale, Brussels; Scudder.

Caen, France.
Académie des arts et belles-lettres.
Official academy: 1705–1793.
(Incorporates previous *Académie royale de belles-lettres*, 1652–1705; lapses 1715–1731.)
Recognition: French crown.
40% science; prizes = 96% science.
5 vols. (belles lettres) *Mémoires*, 1754–1760.
Public courses.
Sources: Roche; Roche, "Histoire"; Almanach royal; Brown; Tisserand.

Cap Français, Saint-Domingue, Hispaniola.
Cercle des philadelphes.
Private academy: 1784–1789.
Approved, funded by local authorities.
Société royale des sciences et des arts du Cap Français.
Official academy: 1789–1792.
20 resident members; 125 total 1784–92; dues.
Library; botanical garden; cabinets; prizes; jetons.
1 vol., *Mémoires*, 1788.
In contact with Paris Academy of Sciences.
Sources: Maurel; Hindle; Scudder.

Châlons-sur-Marne, France.
Académie des sciences, arts et belles-lettres.
Private society: 1750–1753.
Authorized *Société littéraire:* 1753–1775.
Official academy: 1775–1793.
Recognition: French crown; town support.
Less than 20 ordinary members.
60% science; public courses; no publications.
Library.
Sources: Roche; Génique; Tisserand.

Cherbourg, France.
  Société Académique.
  Private society: 1755–1773 with local authorization.
  Official academy: 1773–1793.
  Inactive: 1757–1767; 1783–1793.
  60% science; connections to Marine Ministry.
  Less than 20 ordinary members.
  Publications, 1755–1793.
  Sources: Roche; Noel; Scudder.

Clermont-Ferrand, France.
  Société royale des sciences, arts et belles-lettres.
  (A.k.a.: Académie, Académie royale.)
  Authorized *Société littéraire:* 1759–1780.
  Official academy: 1780–1793.
  50% science; library; botanical garden; no prizes.
  30–35 ordinary members; no publications.
  Sources: Roche; Cohendy.

Copenhagen, Denmark.
  Det Kongelige Danske Vidensakbernes Selskab.
  Official academy: 1742 to Present.
  Recognition/Support: Danish crown.
  50% Science. 30–40 ordinary members; 10–15 foreign.
  125 prize contests, 1767–1788 (24 winners). Jetons.
  17 vols. various *Skrifter* and *Samling:* 1743–1799.
  23 vols. *Dansk Historisk Almanach,* 1760–1782.
  Dictionary of the Danish language.
  Considerable geographical, cartological work.
  Sources: Lomholt; Kronick.

Dijon, France.
  Académie des sciences, arts et belles-lettres.
  Private, endowed Pouffier Foundation: 1725–1740.
  Official academy: 1740–1793.
  Recognition: French crown; support from Estates.
  70% science; 53 prizes, 1741–1793; pensions.
  Courses: chemistry, botany, materia medica.
  Observatory; library; botanical garden; cabinets.
  10 vols. *Mémoires,* 1769–1785; + misc. publications.

[267]

20 ordinary members; 341 members 1740–1793.
Sources: Roche; Tisserand; Misland.

Dublin, Ireland.
Royal Irish Academy.
Official society: 1785 . . .
Recognition: British monarchy.
1/3 science; library, *Transactions* (23 vols., 1787–1859).
Sources: Bonfield; Irish *Transactions* and *Index*.

Edinburgh, Scotland.
Philosophical Society of Edinburgh.
(A.k.a.: Society for Improving Arts and Sciences.)
(Preceded by Society for the Improvement of Medical Knowledge,
1731–1737.)
Private society: 1737–1745; 1750–1772.
45 members; science and medicine; 3 vols. *Essays and Observations*
(1754, 1756, 1771; French tr., 1 vol., Paris, 1759).
Is loosely succeeded by:
Royal Society of Edinburgh.
Official society: 1783 . . .
Recognition: British crown.
50% science; literary and philosophical sections; 163 members in
1783.
Published *Transactions*, 27 vols., 1788–1876.
German tr., 1 vol., Göttingen, 1789.
Sources: Emerson; Shapin; Hoppen; Forbes; Kendall; Scudder.

Erfurt, Germany.
Akademie gemeinütziger Wissenschaften.
(Academia electoralis moguntina scientiarum utilium.)
Official academy: 1754 to 19th c.
Recognition: Mainz elector.
Utilitarian interests.
Prizes; library; observatory; laboratory; botanical garden; anatom-
ical theater; 14 vols. *Acta*, 1757–1807; 2 vols. *Commentationes*,
1780.
Sources: Hammermayer; Hubrig; Kraus; Hufbauer; Scudder.

Florence, Italy.
Academia Botanica.
Official academy: 1739–1783.
Recognition/Support: Austrian crown (Francis I).
100% science (botany and natural history); 50 ordinary members; 100 in toto 1739–1783.
Incorporated into *Reale Accademia Georgofili* in 1783.
Source: Maylender.
Reale accademia dei Georgofili.
Private society: 1753–1767.
Official academy: 1767 . . .(?).
Recognition/Support: Tuscan authorities.
8 vols. *Atti*, 1791–1817.
Mostly economic and agricultural interests.
Sources: Maylender; Cochrane; Giuliani.

Göteborg, Sweden.
Kungl. Vetenskaps-och Vitterhets Samhället.
Private society: 1773–1778.
Official society: 1778 . . .
Recognition: Swedish crown.
25% science + local history/belles lettres; prizes.
100 or so members.
Handlingar, 4 vols., 1778–1788.
Sources: Beckman; Nachmanson; Almhult.

Göttingen, Germany.
Königliche Societät der Wissenschaften.
Official society of science: 1752 . .
Recognized by Hanoverian (and English) crown.
100% science; 12–15 ordinary members; 56 prizes 1753–1782.
Almanac monopoly; Haller = president.
Publishes *Göttingische anzeigen von gelehrten Sachen*; 117 vols., 1753–1801; 12 vols. *Comentarii* and *Novi Commentarii*, 1751–1778; 16 vols. *Commentationes*, 1779–1808.
Sources: Joachim; Heilbron; Kraus.

Grenoble, France.
Académie Delphinale.
Private literary society: 1772–1780.

Authorized *Société littéraire:* 1780–1789.
Official academy: 1789–1793.
Recognition/Support: French crown.
50% or so science.
Prizes; library; *Mémoires* (3 vols. 1787–1789); public courses.
40 members like *Académie française.*
Sources: Roche; Gautier; Almanach royal; Tisserand.

Haarlem, Netherlands.
Hollandsche Maatschappij der Wetenschappen.
Official society: 1752–1804 . . .
Recognition: *Staten* of Holland and West Friesland.
100% science/arts.
Cabinet; prizes (86 competitions, 40 winners, 1752–1800).
Publications: *Verhandelingen*, 30 vols., 1754–1793. German tr., 1
vol., Frankfurt-a.-m., 1758; 2 vols., Leipzig, 1775–76.
Economic branch from 1777.
Sources: Heilbron; Bierens de Haan; Forbes; Scudder.

Hesse-Hamburg, Germany.
*Société patriotique de Hesse-Hombourg pour l'encouragement des connais-
sances et des moeurs.*
Official academy: 1775–1781.
Recognition/Support: Landgraff of Hesse-Hamburg.
Mostly economic, utilitarian interests, but closely connected to
scientific societies.
Sources: Hammermayer; Hubrig; Voss.

Leipzig, Saxony, Germany.
Fürstlich Jablonowskische Gesellschaft.
Official society: 1768 . . .
Recognition/Support: Polish elector.
Sources: Voss; Hammermayer; Alex von Harnack; Staszewski.

Lisbon, Portugal.
Academia real das ciências de Lisboa.
Private academy: 1779–1783.
Official academy: 1783 . . .
Recognition: Portuguese crown.
3 classes; 2/3 science; pensions; press; prizes; lottery proceeds.

Strong economic, agricultural interests.
Memorias; 10 vols. *Ephemerides Nauticas,* 1789–1796.
Library, museum, laboratory; jetons.
Sources: Ribeiro; Hammermayer; Ferrâo.

London, England.
Royal Society of London for the Promotion of Natural Knowledge.
Private society: 1660; official society: 1662 to present.
Recognition/Patronage: English crown.
Dues; private and public support.
100% science, natural philosophy.
300 members in 1741; 545 in 1800.
Philosophical Transactions from 1665, 81 vols., 1665–1791; Latin, German, French, Italian and Dutch tr., totalling 39 vols., 1672–1760.
Copley medal; expeditions.
Sources: Lyons; Stimson; Heilbron; Scudder.

Lund, Sweden.
Kungl. Fysiografiska Sällskapet.
Private society: 1771–1778.
Official Society: 1778 to present.
Recognition: Swedish crown.
100% science (natural history & economy).
Publishes: 2 vols. *Handlingar* (1776, 1786); *Magazin* (3 vols., 1781–1786).
100 ordinary members, 1772–1789;
50 foreign members, 1773–1789.
Library.
Source: Gertz.

Lyons, France.
Académie royale des sciences, belles-lettres et arts.
Private society: 1700–1724.
Official society (*Société royale des sciences et belles-lettres*): 1724–1752.
Official academy (*Académie royale des sciences et belles-lettres*): 1752–1758.
Official academy (merged with *Académie des beaux-arts*): 1758–1793.
Recognition: French crown.
Support from town, king (600).

40 ordinary members.
60% + science; 163 prizes = 96% science;
Library; cabinet; jetons.
Mémoires couronnés (1788).
Sources: Trénard; Roche; Dumas; Bouillier; Almanach royal;
Lyons Academy, Centenaire.

Mannheim, Germany.
Academia Electoralis Scientiarum et Elegantiorum Literarum Theodoro-Palatina.
(A.k.a.: *Kurpfälzische Akademie.* . . .)
Official academy: 1763–1795.
Recognition: Palatine elector.
50% science.
Pensions; library; observatory; collections; press.
Andreas Lamey, secretary.
Published *Mannheimer Zeitung* and 11 vols. *Historia*, 1766–1794.
Sources; A. Kraus; Böhm; Walter; Cassidy.
Societas Meteorologicae Palatinae.
Official academy (Palatine elector): 1780–1795. Offshoot of Mannheim *Academia.*
100% science (meteorology).
Ephemerides, 12 vols., 1781/1783 to 1792/1795.

Mantua, Lombardy, Italy.
Accademia Virgiliana.
Private, "Renaissance" academy: 1686–1752.
Official "Renaissance" academy: 1752–1767. Became:
Reale Accademia di Scienze, Lettere ed Arti.
Official academy: 1768–1797.
Recognized/Supported: Austrian crown (Maria-Theresa, Joseph II).
4 classes; 50% science; library; anatomical theater.
Copyright; censorship.
1 vol. *Memorie*, 1795.
Sources: Maylender; Scudder.

Marseille, France.
Académie des belles-lettres, sciences et arts.
Official academy: 1726–1793.

Known as *Académie des belles-lettres:* 1726–1766.
Reformed: 1766.
Recognition: French crown.
Support: royal and town.
Less than 20 ordinary members.
40% range science, esp. late.
Prizes (+/− 100); 45 vols. *Recueils,* mostly belles lettres, 1727–1786.
Library; ran royal naval observatory for Ministry.
15 scientific "sociétaires"; public courses.
Sources: Roche; Lautard; Dassy; Academie, Deuz Siècles; Tisserand; Bouillier; Lefèvre-Pontalis.

Messina, Sicily, Italy.
Accademia Peloritana dei Pericolanti.
Official "Renaissance" academy, 1728 to 19th c.
Recognition: Messina Senate; Spanish Viceroy.
Confederation with *Accademia dei Dissonanti,* Modena.
20% science.
Source: Maylender.

Metz, France.
Société royale des sciences et arts.
Official academy (French crown): 1757–1793.
Private support, benefaction.
80% science range; prizes; ran musée.
Less than 20 ordinary members.
Sources: Roche; Fleur.

Middelburg (Vlissingen), Netherlands.
Zeeuwsch genootschap der Wetenschappen.
Official society: 1768 . . .
Recognition: Zeeland states.
15 vols. *Verhandelingen,* 1769–1792.
Sources: Singels; R. J. Forbes.

Montauban, France.
Académie des belles-lettres.
Official academy (French crown): 1741–1793.
(Preliminary private group 1730s).

Supported by city.
30–35 ordinary members.
20% science range; prizes (60% science).
2 vols. *Recueil*, 1745–1747.
Sources: Roche; Ligou; Almanach royal.

Montpellier, France.
Société royale des sciences.
Official academy (French crown): 1706–1793.
Town, provincial support.
100% science; 20 ordinary members.
Observatory; library; laboratories; botanical garden.
Courses in chemistry and *physique*.
2 vols. *Mémoires*, 1766, 1768; 35 *Séances publiques*, 1774–1788.
Published a natural history.
Sources: Roche; Kindelberger; Castelnau; Le Grand.

Munich, Bavaria, Germany.
Churbayerische Akademie der Wissenschaften.
Official academy: 1759 to 19th century.
Recognition: Bavarian elector.
50% science (historical and philosophical classes).
Pensions; paper, almanac and calendar monopolies; map tax; prizes.
Observatory; library; cabinets; J. P. Kennedy, secretary.
10 vols. *Abhandlungen*, 1763–1776.
7 vols. *Neue Abhandlungen*, 1778–1797.
9 vols. meteorological *Ephemerides*, 1781–1789.
Sources: Hammermayer; Heilbron; Hufbauer; Kraus.

Nancy, Lorraine, France.
Société des sciences et belles-lettres: 1750–1793.
(A.k.a.: *Académie de Stanislas.*)
Official academy: 1750–1793.
Recognition/Support: Stanislas (Leszczynski) of Lorraine; then, from 1766, French crown.
40–60% science; library; 20 ordinary members.
Prizes; university professors (Pont-à-Mousson) automatically members.
4 vols. *Mémoires*, 1754–1759.
Sources: Roche; Roche, "Histoire"; Bouillier; Almanach royal.

Naples, Italy.
Reale Accademia delle Scienze e Belle-Lettere: 1778 . . .
Official academy: Ferdinand IV de Bourbon.
4 classes: 50% science; 10 pensioned academicians.
1 vol. *Atti,* 1787.
Source: Maylender:

Nîmes, France.
Académie royale de Nîmes.
Official academy (French crown): 1682–1700; 1752–1793.
Town, royal support.
26 ordinary members.
40-50% science, esp. after 1752 revival.
Library; 2 vols. of *Recueil;* prizes; taught Hebrew/Greek.
Sources: Roche; Terrin; Simon.

Olmütz, Moravia, Austrian Empire.
Societas Eruditorum Incognitorum.
Private society: 1746.
Official society: 1747–1751.
Recognition/Support: Maria Theresa, Austrian court.
Press privilege: *Monatlich Auszüge,* 1747–1748;
    von Petrasch = founder.
Sources: Vávra; Voss; Hammermayer.

Orléans, France.
Académie royale des sciences, arts et belles-lettres.
(A.k.a.: *Société royale de physique, d'histoire naturelle et des arts.*)
Ministerial authorization: 1781.
Royal authorization (French crown): 1784.
Official academy (Letters-patent): 1786–1793.
Preceding private groups in Orléans: 1725–1741.
Supported by city; crown; dues.
36 ordinary members.
90% science range; prizes; public courses: chemistry, botany, med-
    icine.
Cabinets; laboratory; botanical garden; no publications.
Sources: Roche; Fouchon; Guillon.

Padua, Italy.
Accademia di Scienze, lettere ed Arti.
  Official academy: 1779 . . . (1797?).
  Recognition/Support: Venetian Senate.
  4 classes (including speculative philosophy); 50% science; 36 ordinary members, 26 pensions; connected to university—pensions to professors.
  3 vols. *Saggi*, 1786–1794.
  Sources: Moschetti; Maylender; Franzoja; Cesarotti.

Paris, France.
Académie royale des sciences.
  Official academy: 1666–1793.
  Recognition/Support: French crown.
  100% science.
  45 ordinary members; 20 pensions; prizes (67 awarded); expeditions.
  92 vols. *Mémoires*, 1699–1793.
  9 vols. *Recueil des . . . prix*, 1721–1772.
  7 vols. *Machines . . . approvées*, 1735–1777.
  24 vols. *Description des arts et métiers*, 1761–1782.
  11 vols. *Savants Etrangers*, 1750–1786.
  Sources: Hahn; Jaeggli; Maury; Maindron; Scudder.

Pau, France.
Académie royale des sciences et beaux-arts.
  Official academy (French crown): 1718/1725–1793.
  Supported by provincial estates.
  Prizes (63% science); library; 15 ordinary members; no publications.
  Sources: Roche; Desplat; Almanach royal.

Philadelphia, Pennsylvania.
  American Philosophical Society.
  Private society: 1743–1747.
  American Society for Promoting and Propagating Useful Knowledge.
  Private society: 1758–1768.
  American Philosophical Society.
  Private society: 1767–1768.

American Philosophical Society, Held at Philadelphia, for Promoting Useful Knowledge.
Official society: 1768 to present.
Chartered by Pennsylvania Assembly (1780).
3 vols. *Transactions,*, 1771–1793; Magellan prize fund.
Sources: Hindle; Stearns; Bates.

Prague, Bohemia, Austro-Hungary.
Privatgesellschaft in Böhmen: 1769–1785.
Böhmische Gesellschaft der Wissenschaften: 1785–1791.
Königlich Böhmische Gesellschaft der Wissenschaften: 1791 . . .
Recognition/Support: Austrian crown.
100% science; 13 vols. (various) *Abhandlungen*, 1775–1798.
Sources: A. Kraus; Teich; Kalousek.

La Rochelle, France.
Académie royale des belles-lettres.
Official academy: 1732–1793.
Recognition: French crown.
+/− 30% science, esp. late.
Prizes; library; experimental physics course.
3 vols. *Recueil* (belles lettres), 1747–1763.
Sources: Roche; Torlais; La Rochelle Bibliothèque catalogue.

Rotterdam, Netherlands.
Bataafsch Genootschap der Proefonderwindelijke Wijsbegeerte.
Official society: 1769 . . .
Recognition: Staten of Holland.
100% science/arts; 12 vols. *Verhandelingen*, 1774–1798.
Sources: Kuenen; Singels.

Rouen, France.
Académie royale des sciences, belles lettres et arts.
Private reunions: 1736–1744.
Authorized academy: 1744–1756.
Official academy (French crown): 1744–1793.
Supported by city.
60–70% science/arts; library; botanical garden; public courses: medicine, hydrography; 20 ordinary members; no publications.
Sources: Roche; Scudder.

Rovereto, Italy.
Imperiale Regia Accademia degli Agiati.
  Official society: 1752 to 20th century.
  Recognition/Support: Austrian crown. Maria-Theresa = patron.
  Mostly letters and agriculture; some science; 3 vol. memoirs.
  500+ members in 1775; center for corresponding agricultural
    societies. Small prizes.
  Sources: Maylender; Imperiale Reale Accademia, *Memorie.*

Saint Petersburg, Russia.
Academia Scientiarum Imperialis Petropolitanae.
(Académie Impériale des Sciences de Saint
Pétersbourg.)
  Official academy (Russian crown): 1724 to 1917.
  40 ordinary members.
  100% science; pensioned academicians; prizes; full complement of
    departments; almanac monopoly; expeditions.
  14 vols. *Commentarii,* 1726–1746.
  21 vols. *Novi Commentarii,* 1747–1778.
  17 vols. *Acta* and *Nova Acta Academiae,* 1777–1803.
  Sources: Vucinich; Heilbron; Boss.

Siena, Italy.
Reale Accademia delle scienze di Siena (detta de' Fisiocritici).
  Private "Renaissance" academy: 1690 . . .
  Official academy (Tuscany): 1759 . . .
  100% science/arts; 128 members; 6 vols. *Atti,* 1761–1794.
  University professors automatic members.
  Sources: Maylender.

Stockholm, Sweden.
Kungl. Vetenskapsakademie.
  Official academy: 1739 to present.
  Recognition/Support: Swedish crown.
  100 members; 100% science/arts; observatory.
  Prizes; expeditions; almanac privilege; botanical garden.
  77 vols. *Handlingar* and *Nya Handlingar,* 1739–1812.
    German tr., 53 vols., Hamburg, 1749–1794 + misc. Danish,
    Dutch, and Latin tr. = 14 vols.
  Sources: Lindroth; Heilbron; Dahlgren; Scudder.

Toulouse, France.
Académie royale des sciences, inscriptions et belles-lettres.
Authorized *Société des sciences,* 1729–1746.
Official academy (French crown): 1746–1793.
Supported by city; provincial estates.
30–35 ordinary members.
75% science range; prizes; 4 vols. *Mémoires,* 1782–1790.
Library; botanical garden; observatory; *Catalogue des plantes,* 1782.
Public courses; sponsored musée.
Sources: Roche; Bouillier; Almanach royal.

Trondheim, Norway.
Det Kongelige Norske Videnskabers Selskab.
Private society (*Det Trondhjemske Selskab*): 1760–1766.
Official society from 1767.
Recognized/Supported by Danish court.
Science, history, and philosophy sections; library; prizes.
32 members, 1768.
5 vols. *Skrifter* 1761–1774; 2 vols. *Ney Samling,* 1784–1788. (Printed
   in Copenhagen and translated into Dutch.)
Sources: Midbøe; Lindroth; Scudder.

Turin, Lombardy, Italy.
Societas Privata Taurinensis: 1757–1759.
Private society; 100% science; 1 vol. *Miscellanae.*
Société royale des sciences: 1759–1783.
Official society; recognition: Kingdom of Sardinia; 4 vols. of
   Mélanges.
Académie royale des sciences: 1783–1792.
Official academy; 20 core members; 6 vols. of *Mémoires,*
   1784–1800.
Pensions.
Source: Accademia delle scienze, *Il primo secolo.*

Uppsala, Sweden.
Societatis Regiae Scientiarum.
(A.k.a.: *Kungl. Vetenskapssamhället; Kungl. Vetenskaps-Societeten.*)
Began as private society (*Collegium Curiosorum*): 1710–1728.
Official society: 1728. . .
Recognition/Support: Swedish crown.

24 members; 9 vols. *Acta*, 1720–1751; 5 vols. *Nova Acta*, 1773–1792.
Sources: Bring; von Sydow; Frängsmyr; Karlberg.

Utrecht, Netherlands.
Provinciaal Utrechtsh genootschap van Kunsten en Wetenschappen.
    Private society *(Konstgenootschap):* 1773–1777.
    Official society: 1777 to 1791; 1800 . . .
    Recognition/Support: House of Orange; Staten of Utrecht; town.
    Partly science/arts; prizes; 10 vols. *Verhandelingen*, 1781–1821.
    Sources: Singels; Forbes.

Valence, France.
Société Académique et Patriotique.
    Private society: 1784–1786.
    Official academy (French crown): 1786–1793.
    80% science/arts range; prizes; 20 ordinary members.
    Sources: Roche; Colonjon.

Variable locale:
*Sacri Romani Imperii Academia Caesareo-Leopoldina-Carolina Naturae
    Curiosorum.*
    Private "Renaissance" academy: 1652–1677.
    Official "Renaissance" academy: 1677 to 20th century.
    Recognition/support: Holy Roman Emperor.
    100% science and medicine.
    46 vols. *Miscellanea Curiosa* (a.k.a.: *Acta*), 1670–1791.
    Sources: Aberhalden; Reichenbach; Meding; Ornstein.

Notable Private and Semi-Private Scientific Societies: 1660–1793

Appendix 2 lists by town notable scientific societies of the period 1660–1793 without official charters. Because of the more ephemeral nature of private societies, appendix 2 is perhaps not as complete as appendix 1, although most experts would agree that the list represents an essentially comprehensive survey of such institutions. The private status of most of the societies listed is unquestionable, but there are judgment calls on the taxonomic borderline between private and official societies. The various *Naturforschenden* societies in Berlin, Halle, and Zurich seem to have been all but official societies, and they and others have been classified as semi-private here, with the "authorizing" authority indicated. Cases in Italy of societies recognized privately by dukes, counts, and the like also seem to belong more appropriately here in appendix 2.

Insofar as possible with more informal organizations, societies in appendix 2 have been classified as academies, societies, "Renaissance" academies, and clubs. As before, I have striven to provide the correct name for each society, dates of activity, and other relevant and available information previously described for societies in appendix 1.

Alexandria, Virginia.
    Alexandrian Society for the Promotion of Useful Knowledge.
        Private society: 1790. . . (?).
        Source: Hindle.

Altdorf, Germany.
    Collegium Curiosum sive Experimentale.

Private "Renaissance" academy: 1672–1695.
Cimento imitator: experimental science.
20 original members; 2 vols. *Experimenti tentamina* (1676, 1683).
Christopher Sturm, prime mover.
Source: Ornstein.

Arezzo, Italy.
Accademia Aretina.
Private "Renaissance" academy: 1787. . .
Grand Duke Leopold (Tuscany) = patron.
< 20% science.
Source: Maylender.

Bergamo, Lombardi, Italy.
Accademia dei Naturalisti.
Private society: 1782. . . (?).
Doctors, surgeons, professions = members.
Source: Maylender.

Berlin, Prussia, Germany.
Nouvelle Société littéraire.
Private Society: 1742–1747.
Gesellschaft Naturforschender Freunde.
(A.k.a.: *Privategesellschaft Naturforschender Freunde.*)
Semi-private society: 1773 to present.
Authorized by Prussian crown.
Support: dues.
Library; natural history cabinet; prizes.
37 members, 1773, 300 members, 1773–1793 (5 women).
4 vols. *Beschäftigungen*, 1775–1779.
6 vols. *Schriften*, 1780–1785.
5 vols. *Beobachtungen*, 1787–1794.
Sources: Becker; Harnack; Bolton.

Birmingham, England.
"Lunar Society of Birmingham."
Private club: 1766–1791.
No rules, publications, minutes; 6-10 members.
Sources: Schofield; Bolton.

Bologna, Italy.
  Accademia della Traccia.
    Private "Renaissance" academy: 1665–1667.
  Accademia Fisico-mathematica.
    Private "Renaissance" academy: 1667. . .(?).
    (Same as *Filosofia Naturale*, 1677?).
  Accademia dell'Arcidiacono.
    Private "Renaissance" academy: 1687.
    Felice Marsigli = founder.
    Experimental society.
  Accademia degli Inquieti.
    Private "Renaissance" academy: 1690–1710s.
  Accademia dei Concordi.
    Private "Renaissance" society: 1777–1795.
    100% science and medicine.
    Sources: Rosen; Maylender.

Boston, Massachusetts.
  Boston Philosophical Society.
    Private society: 1683–1688.
    Sources: Bates; Hindle; Stearns; Beall.

Bratislava (Pressburg), Austro-Hungary.
  Gesellschaft der Freunde der Wissenschaften.
    Private society: 1761. . . (1763?).
    Source: Hammermayer.

Breman, Germany.
  Physikalische Gesellschaft.
    Private society: 1790s.
    Source: Hammermayer.

Caen, France.
  Académie de Physique.
    Authorized academy: 1666–1675.
    100% experimental science.
    Source: H. Brown.

Cosenza, Calabria, Italy.
  Accademia dei Pescatori Cratilidi.
    Authorized marquisate "Renaissance" academy: 1756–1784.
    50% science (?); became agricultural society, 1784.
    Source: Maylender.

Cunea, Piedmont, Italy.
"Physical Society."
Private society: 1760's (?).
1 vol. memoirs.
Source: Becker.

Danzig, Poland.
Naturforschende Gesellschaft.
(A.k.a.: *Societas Physicae experimentalis.*)
Private society: 1743. . .(1790s).
20 ordinary members; dues; support from Leipzig elector,
Jablonowski.
100% science; 3 vols. *Versuche und Abhandlungen*, 1747–1756;
1 vol. *Neue Sammlung*, 1778; library; cabinets; observatory; prizes.
Sources: Voss; Hammermayer; Staszewski; Lindroth; Schumann.

Derby, England.
Literary and Philosophical Society.
Private society, 1783. . .
Library; monthly meetings.
Source: Robinson; Hume.

Dublin, Ireland.
Dublin Philosophical Society.
Private society: 1683–1687; 1692–1698; 1707–1708.
33 members (1684); 49 (1693). Dues.
Source: Hoppen; Weld.
Physico-Historical Society.
Private society: 1739–1778.
1 vol. *Observations*, 1739.
Source: Irish *Transactions;* Royal Society Archives.

Edinburgh, Scotland.
Philosophical Society.
Private society: 1705. . . (1710?).
Source: Weld; Hoppen.

Florence, Italy.
Accademia del Cimento.
Private ducal "Renaissance" academy: 1657–1667.

Grand Duke Ferdinand and Prince Leopold = patrons.
100% science; 12 members; published *Saggi* (1667).
Source: Middleton; Ornstein; Maylender.

Görlitz, Germany.
Oberlausitzischen Gesellschaft der Wissenschaften.
Private society: 1779 into 20th c.
Science and local history; 52 members, 1783; library;
cabinets; takes over provincial newsletter *(Lausitzisches Magazin);*
25 vols., 1768–1792.
Source: Jecht.

Haarlem, Netherlands.
Teylers Tweede Genootschap (Second Society).
Private society: 1778. . .
Supported by Teyler's Foundation *(Stiching).*
6 members; mainly science; prizes.
34 vols. *Verhandelingen,* 1781–1857.
Museum; library; cabinets (minerals; electrical).
Sources: Heilbron; Bruijn.

Halle, Germany.
Gesellschaft der Naturforschenden Freunde.
(A.k.a.: *Naturforschende Gesellschaft.*)
Semi-private society: 1779 . . .
Authorized by Prussian court.
10 original members; 115 elected 1780–1789.
Library; cabinets; *Abhandlungen,* 1783.
Sources: Voss; Hammermayer; Schmieder.

Hartford and New Haven, Connecticut.
Connecticut Society of Arts and Sciences.
Private society: 1786 . . . (1790?).
(Revived 1799 as an official society.)
Sources: Bates; Hindle.

Jena, Germany.
Societas Physica.
(A.k.a.: *Naturforschende Gesellschaft.*)
Private society: 1793 . . . (?).

[285]

1 vol. *Schriften*, 1802.
Source: Hammermayer; Scudder.

Lausanne, Switzerland.
Société des sciences physiques.
   Private society; 1783 . . . (1790s?).
   3 vols. *Mémoires*, 1783–1788.
   Source: *Mémoires;* Scudder.

Lexington, Kentucky
   Kentucky Society for Promoting Useful Knowledge.
   Private society: 1787 . . . (1790).
   Sources: Hindle; Gray.

London, England.
   Temple Coffee House Botany Club.
   Private club: 1689–1720.
   Botanical Society.
   Private society: 1721–1726.
   (Emerges from Coffee House Club?)
   Source: Stearns.
   Linnean Society.
   Private society: 1788 . . .
   100% botany and natural history.
   Transactions from 1791.
   Source: Gage.

Mainz, Germany.
   Medizinische und Physikalische Gesellschaft.
   Private society: 1793–1804. . . (?).
   Source: Hammermayer.

Manchester, England.
   Literary and Philosophical Society.
   Private society: 1781 . . .
   40 founding members; rules; weekly meetings; library.
   Published 5 vols. *Manchester Memoires*, 1785–1805.
   Sources: Sheenan; Nicholson; Fleure; Thackray; Shapin and
      Thackray; Barnes.

Milan, Italy.
Accademia Clelia de'Vigilanti.
Private "Renaissance" academy: 1700 . . . (?).
50% science; a press; Donna Clelia = patron.
Source: Maylender.

Modena, Italy.
Accademia Ducale dei Dissonanti di Modena.
Private "Renaissance" academy: 1680–1752.
(Merges with *Accademia Peloritana*, Messina, 1729; gets Spanish support.)
Authorized ducal academy: 1752–1817.
Francesco III d'Este = patron.
Science orientation.
Became *Reale Accademia Modenese di Scienze, Lettere ed Arti.*
Source: Maylender.
Accademia Rangoniana.
Private "Renaissance" academy: 1783–1796. Marquis
Rangone = patron.
100% science; dues; prizes.
Source: Maylender.

Naples, Italy.
Accademia degli Investiganti.
Private "Renaissance" academy: 1650–1668. (Lapses 1656–1662.)
100% science; experimental physics.
Source: Maylender.
Accademia del Cimento.
Private "Renaissance" academy: 1680 . . . (?).
100% science; after Florentine academy.
Source: Maylender.
Accademia Galianiana.
Private "Renaissance" academy: (± 1700).
100% science.
Source: Maylender.

Newcastle-Upon-Tyne, England.
Literary and Philosophical Society.
Private society: 1793. . .

[287]

Dues; monthly meetings; library.
Source: Watson.

New York, New York.
New York Society for Promoting Useful Knowledge.
Private society: 1784. . . (1790?).
Source: Hindle.

Nuremberg, Germany.
Society of Physicians.
Private society: 1730. . .(1750s?).
Christopher Trew = prime mover.
15 vols. *Commercium litterarium Noribergense*, 1731–1745.
Source: Royal Society of London archives; Scudder.
Cosmographical Society.
Private society: 1747. . .(1760s?).
Cartography; connected with Nuremberg globe factory;
Published memoirs.
Sources: Forbes; Andreef.

Oxford, England.
Oxford Philosophical Society.
(A.k.a.: Oxford Experimental Club.)
Private society: 1651–1690. (Lapses 1660–1683.)
100% science; 36 members in 1684; no publications; dues;
Director of experiments position.
Sources: Hoppen; Gunther; Weld.

Palermo, Sicily, Italy.
Accademia Palermitana (del Buon Gusto).
Private academy: 1752 to present.
Patrons = notables; 60 members; maybe 20% science.
2 vols. *Saggi di Dissertazioni*, 1755, 1791.
Sources: Maylender; Sampolo; Scudder.

Palma, Italy.
Accademia Boreliana.
Private "Renaissance" academy: 1673. . .1720(?).
Founded by Borelli; *Cimento*-like experimental physics.
Source: Maylender.

Peterborough, England.
  Literary and Philosophical Society.
    Private society: 1730–1809.
    10 members at meetings.
    Sources: Weld; H. J. J. Winter.

Pistoia, Italy
  Accademia Enciclopedica.
    Ducal "Renaissance" academy: 1763. . .(1793?).
    Recognition: Grand Duke of Tuscany.
    Mainly science; laboratory; botanical garden.
    Source: Maylender.

Reggio d'Emilla, Italy.
  Accademia degli Ipocandriaci.
    Private "Renaissance" academy: 1746–1784 (with lapses).
    1/3 science; 469 members (1757); university setting;
    Spallanzani a member.
    Became *Reale Accademia Reggiana di Scienze, Lettere ed Arti* (1819).
    Source: Maylender.

Richmond, Virginia
  Académie des sciences et beaux-arts des États-Unis de l'Amérique.
    Private academy and university: 1786–1789.
    Sources: Hindle; Bates; Kronick.

Rio de Janeiro, Brazil.
  Academia Scientifica (Cientifica) *do Rio de Janeiro.*
    Private academy: 1772. . .(1775?).
    100% science; support from local viceroy.
    Wargentin, Bergius = members; shipped botanical specimens.
    Sources: Ribeiro; Hammermayer; Varnhagen.

Rome, Italy.
  Accademia di Fisico-mathematica.
    Private "Renaissance" academy: 1677–1698.
    100% science; supported by Queen Christina of Sweden.
    Sources: Middleton, "Rome"; Maylender.

Salerno, Italy.
Accademia degli Irrequiti.
Private "Renaissance" academy: 1709–1719.
Devoted to mathematics.
Source: Maylender.

Spalding, England.
Philosophical Society. (A.k.a.: Gentlemen's Society.)
Private society: 1710–1758 . . .(1810 to present).
Library; museum; Newton = honorary member.
Sources: Weld; H. J. J. Winter.

Trenton, New Jersey.
Trenton Society for Improvement in Useful Knowledge.
Private society: 1781. . .(1783?).
Sources: McCormick; Hindle.

Turin, Lombardy, Italy.
Accademia Filopatria.
Private "Renaissance" academy: 1782. . .(1792?).
Count Balbo = patron.
History and natural history; 3 vols. *Ozi Litterari*, 1787–1791.
Sources: Maylender; Accademia delle scienze, *Il primo secolo*.

Uppsala, Sweden.
Cosmographiska sällskapet.
Private society: 1758–1824.
Royal support: Swedish crown.
Associated with globe factory.
Sources: Bratt; Lindroth, *Historia*.

Venice, Italy.
Accademia Corrara.
Private "Renaissance" academy: 1673 . . . (1678?).
Private patron; especially astronomical.

Accademia Geografico-storia-fisica.
Private science society: 1680–1714.

Accademia dei Planomaci.
  Private "Renaissance" academy: 1740–1772.
  Patron = Doge.
  50% science; published journal, *Le Novelle Letterarie.*
  Source: Maylender.

Verona, Italy.
  Accademia degli Aletofili.
    Private "Renaissance" academy: 1768–1774.
    (Revival of 17th-century predecessor.)
    Semi-private "Renaissance" ducal academy: 1774–1780.
    Authorization: Venetian Senate.
    +/− 70% science; merges into *Società Italiana delle scienze.*
    Source: Maylender.
  Società Italiana delle Scienze.
    Private society: 1782–1796.
    Anton Marie Lorgna = founder.
    100% science/arts; 7 vols. *Memorie,* 1782–1796.
    Revived 19th century.
    Sources: Maylender; Pace; Accademia di Verona, *Memorie.*

Williamsburg, Virginia.
  Virginian Society for the Promotion of Useful Knowlede.
    Private society: 1772–1774.
    Held meetings; awarded one medal.
    Strong economic orientation.
    Source: Hindle; Gray.

Zurich, Switzerland.
  Naturforschende Gesellschaft. (A.k.a.: *Physikalische Gesellschaft.*)
    Semi-private society: 1746 to present.
    Support: dues; city of Zurich.
    100% science/arts; library; cabinets; gardens; expeditions(!);
    economic commission; public lottery; prizes (50 to 1804).
    100 members; 3 vols. *Abhandlungen,* 1761–1766.
    Sources: Hansen; Rübel.

Notes

Abbreviations and Conventions

Manuscript Sources:

AD-CD = Archives départementales de la Côte-d'Or, Dijon. Registers and miscellaneous records of the Dijon Academy.

AD-H = Archives départementales de la Hérault, Montpellier. Registers and miscellaneous records of the Montpellier Society.

AdW = Akademie der Wissenschaften der DDR, Berlin (DDR), Zentrales Akademie-Archiv. Manuscript collections of the Berlin Academy; references are to the MS catalogue, "Der älteren Akten der Preussischen Akademie der Wissenschaften: 1700–1811."

APS = American Philosophical Society Library Archives, Philadelphia.

BI = Bibliothèque de l'Institut, Institut de France, Paris.

BM-D = Bibliothèque municipale, Dijon. Principally cited are the papers of Richard de Ruffey.

BM-M = Bibliothèque municipale, Montpellier. Miscellaneous papers of the Montpellier Society.

DS-Formey = Deutsche Staatsbibliothek, Berlin (DDR); Formey correspondence; cited by box, folder, page no. in folder.

KVA = Kungl. Vetenskapsakademie Library, Stockholm. KVA-Bergius = Bergius correspondence; cited by volume from 15 vol. MS collection, "Bergianska Brefsamlingen." KVA-Wargentin = Wargentin correspondence; KVA-Wilckes = Wilckes correspondence; cited by letter and date.

PA = Archives, Académie des Sciences, Institut de France, Paris. MS minutes (Procès-verbaux) cited PA-PV, by vol., year, and page.

RS = Royal Society of London, Library. Separate collections cited: RS-CM = Council Minutes; RS-JBC = Journal Book Copy; RS-LBC = Letter Book Copy. All cited by vol. and page numbers.

Printed sources are cited by short title only and are listed in the bibliography.

Quoted material, from whatever source, if originally in English, is given in the original in the text, and the source cited in the notes. Material originally in a language other than English is given in translation in the text, and unless otherwise indicated, the translations

[293]

are mine. If the non-English original of the translation is taken from a published source, that source alone is indicated in the notes. All translations and quotations from unpublished manuscript sources are from French originals; in these cases the original is also given in the notes along with the source. Original spelling, punctuation, and markings have been preserved, *without* special indication of original usage or errors.

Preface

1. In Conradi see pp. 424–26 for further nineteenth-century bibliography. Important bibliography is also in Russo, *Éléments de bibliographie*, pp. 50–58.

2. It should be pointed out that from the eighteenth century academies and societies have been accurately depicted in what one might call an encyclopedia tradition. See the various entries in the *Encyclopédie* of Diderot and d'Alembert and the *Supplément* to the *Encyclopédie;* articles "Academies," by F[rancis] S[torr] and N[icholas] H[ans], *Encyclopedia Britannica,* 11th ed. (1910) and 14th ed. (1967), respectively; see also Oberhummer, "Die Akademien der Wissenschaften," for a presentation standard by the 1950s.

3. In Roddier see esp. pp. 45, 53.

4. Unfortunately Kopelevich's book, which surveys the major scientific societies treated here, is available only in Russian; see review by Griffiths.

5. Ornstein died accidentally in 1915, two years after the publication of her Columbia doctoral dissertation. Her work has stood as it was written through editions and reprints in 1928, 1938, 1963, and 1975.

Introduction

1. See Hartung, "Science as an Institution."

2. On organized science in antiquity see Lloyd, *Early Greek Science,* esp. chs. 7 and 9; and *Greek Science after Aristotle,* chs. 1, 6, 7, and 10; Ben-David, *The Scientist's Role in Society,* ch. 3. For discussion of scientific societies before the 17th century—see Kronick, *History,* pp. 118–119; Conradi, "Learned Societies"; Thornton and Tully, *Scientific Books,* ch. 7.

3. This is a standard view in the literature. Hahn, *Anatomy,* p. 2, remarks that "the organizational revolution must now be considered an essential component of the Scientific Revolution rather than its by-product." See also Hahn, *Anatomy,* pp. 1–14; Ornstein, *Rôle,* "Conclusion"; Westfall, *The Construction of Modern Science,* ch. 6; A. R. Hall, *The Scientific Revolution,* ch. 7. See also, here, chapter 2.

4. This, too, is a standard view. See Hahn, *Anatomy,* pp. 304–18; Crosland, *The Society of Arcueil,* esp. chs. 3 and 4; Fox and Weisz, eds., *The Organization of Science and Technology in France,* esp. "Introduction"; Morrell and Thackray, *Gentlemen of Science,* pp. 12–16; Berman, *Social Change and Scientific Organization,* pp. xx–xxi and 187–90; Mendelsohn, "The Emergence of Science as a Profession," pp. 27, 47. Still serviceable and to the point is Merz, *A History of European Thought,* esp. vol. 1, chs. 1–3.

5. It is important in what follows not to lose sight of universities as institutions of continuing importance for organized science. By no means did the university simply disappear between the 17th and 19th centuries. Leyden, Edinburgh, Göttingen (1732), Halle, Jena and many others were active centers which provided placement for many

scientists. As we will see, several universities were also closely allied with academies and societies of science. See Heilbron, *Electricity*, passim and pp. 128–30; Hufbauer, *The Formation of the German Chemical Community*, p. 46 and esp. appendix 2, pp. 225–69. Through their studies and emphasis on 18th-century universities, these authors provide a useful corrective to any tendency to overplay the importance of 18th-century scientific societies.

6. For a sense of the development of these scientific "institutions," see Taton, *Enseignement;* see also Gillispie, *Science and Polity.*

7. Kronick, *History*, p. 78; see also, McClellan, "The Scientific Press," pp. 429–30. The unaffiliated *Journal des Sçavans*, the first real scientific periodical, did antedate the *Philosophical Transactions* of the Royal Society by a month or two in 1665.

8. See note 4 and Williams, "Science, Education and the French Revolution"; "Science, Education and Napoleon I"; Sharlin, *The Convergent Century*, ch. 1; Fayet, *La Révolution française.*

9. See, for example, Furet, "La 'Librairie' du royaume de France au 18e siècle," in Furet, ed., *Livre et société*, pp. 3–32; Darnton, *Mesmerism and the End of the Enlightenment in France*, esp. ch. 1; Gay, *The Enlightenment*, esp. vol. 2, chs. 3 and 4. See also Gillispie, *The Montgolfier Brothers;* Darnton, *The Business of Enlightenment;* W. and A. Durant, *The Age of Voltaire*, ch. 16; Gillispie, *Science and Polity*, ch. 4.

10. See particularly Ben-David's *The Scientist's Role in Society*, esp. chs. 5 and 6; see also his "Scientific Role," and "The Profession of Science and Its Powers;" see also T. S. Kuhn, "Scientific Growth."

11. Mendelsohn, "Emergence," esp. pp. 4–7. For Mendelsohn it was the increasingly middle-class and lower-middle-class social origins of men of science, the coming of industrialization, and especially the liberating effects of the French Revolution that brought about the metamorphosis of the 18th-century amateur into the 19th-century professional scientist.

12. Elliott, *The Sociology of the Professions*, esp. "Preface"; Johnson, *Professions and Power;* see also Ben-David, "The State of Sociological Theory," pp. 448–58; Gaston, "Sociology of Science and Technology," esp. pp. 473–88; Geison, *Professions*, pp. 1–4.

13. See, here, chapter 7. The academicians and institutional officers of the major scientific academies on the continent of Europe come to mind as the "professional scientists" of the 18th century, but one should not rule out *de facto* circles around the Royal Society of London, for example.

14. Butterfield, *The Origins of Modern Science*, pp. 203–21. The primary obstacles were the phlogiston theory and the lack of clear conceptual and instrumental means of distinguishing gases and especially their weight.

15. Bernal, *Science in History*, 2:508–18, esp. The period between 1690 and 1760 in particular was "a time for the digestion of and reflection on the enormous scientific advance of the seventeenth [century]." This time looked forward to another surge of progress "comparable to the seventeenth century in its scientific importance." In that interim science was "fashionable and revolutionary," a time when a "bored aristocracy and a thwarted bourgeoisie" botanized in a perverse effort to return to nature. Scientifically the period was barren, "a dip . . . pause in the curve of scientific progress." See also Lilley, "Social Aspects of the History of Science."

16. In addition to previously cited works by Gillespie, Hahn, Heilbron, Hufbauer, and Roche, which certainly do not underplay the 18th century, see Rousseau and Porter, eds., *The Ferment of Knowledge*, passim and "Introduction." Unfortunately this volume omits

any systematic discussion of 18th-century scientific societies (see p. 6). The review of Rousseau and Porter by Cantor, "The Eighteenth Century Problem," makes many of these same points (e.g., p. 55). See also Truesdell, "A Program toward Rediscovering the Rational Mechanics of the Age of Reason," in Truesdell, *Essays*, pp. 84–137. In this connection one might signal the historiographical speciality of science and literature; e.g., Nicholson, *Science and Imagination*.

17. Kuhn's "The History of Science," with its key discussion of the Merton thesis, is to be consulted in this connection also, and, of course, Kuhn's position derives fundamentally from his *Structure of Scientific Revolutions*.

18. See Kuhn, "Mathematical versus Experimental Traditions," p. 52. Kuhn focuses almost exclusively on the physical sciences in the 18th and 19th centuries, with the consequence that the scientific societies seem less important for some Baconian sciences than craft traditions or industrial connections. While this may be true, it seems probable that with regard to other Baconian enterprises such as botany, natural history, geology, and meteorology, the societies were of major significance in organizing large-scale research and in operating as clearing houses.

19. Several noted histories of 18th-century science, because they focus on schools and traditions, in retrospect fit the kind of model Kuhn suggests. See Cohen, *Franklin and Newton*; Schofield, *Mechanism and Materialism*; and Thackray, *Atoms and Powers*. All three of these bipolar works consciously seek to transcend single-discipline studies and encyclopedic treatments and to remedy what Schofield (p. v.) calls "a failure of understanding" with regard to large views of 18th-century science.

1. The Age of Scientific Societies: A Taxonomy

1. See appendix 1. In some instances (e.g., the Berlin *Societas Scientiarum* and the *Académie Royale de Prusse*) institutions succeeding one another have been counted as two societies, while in other instances (e.g., in Philadelphia, Turin) antecedent institutions and rechartered societies are counted as a single institution. Such judgment calls lead to a certain vagueness in the overall numbers.

2. Specific traits of the societies of concern here rarely receive discussion; see remarks concerning earlier types of societies in Maylender, *Storia*, 1:355; 4:199; Ben-David, "The Scientific Role."

3. Hahn, *Anatomy*, p. 105; Bouillier, *Institut*, pp. 320–22.

4. Based on data taken from appendixes. See also figure 4.

5. Ribeiro, *Historia dos estabelecimentos scientificos*, listing vol. 1; Hammermayer, "Akademiebewegung," p. 18. Considering science and nonscience societies together, the phenomenon of the learned societies was even more global than the map reveals. There were nonscience Portuguese societies in Brazil from the 1720s; the Spanish founded economic societies throughout their empire (e.g., Manilla, 1780); and the English established the Asiatic Society of Calcutta in 1784. See Shafer, *Economic Societies in the Spanish World*, pt. 2; Hammermayer, "Akademiebewegung," p. 18.

6. Vienna did possess a noteworthy astronomical observatory, and Madrid was well endowed with other humanistic academies. Several other scientific societies received direct and indirect Austrian and Spanish support.

7. On the Republic of Letters see Gay, *The Enlightenment*, 2:56–83; Becker, *The Heavenly City of the Eighteenth-Century Philosophers*, pp. 33–35; Polanyi, "The Republic of Science."

8. RS-JBC, 21:413–14. The occasion was the presentation of the Copley medal to Benjamin Franklin. Hufbauer, *The German Chemical Community*, pp. 2, 147, makes the valuable point that, as far as science is concerned, the Republic of Letters was multidisciplinary, and not specialized in its orientation.

9. "Les Académies etant les diverses colonies de la république des Lettres. . . ." BM-Ruffey, MS. 1627, p. 79. The date is 1763.

10. "Discours sur la necessité d'admettre des Étrangers dans les sociétés littéraires," Berlin Academy *Mémoires* (1746), 2:427–28. Note that, according to contemporary usage, the term *société littéraire* very much included the scientific societies.

11. "Considérations sur ce qu'on peut regarder aujourd'hui comme le but principal des Académies," Berlin Academy *Mémoires* (1767/1769), 23:380.

12. See especially Roche, *Siècle*, ch. 2 on this point; see also, Goubert, *The Ancien Régime*, ch. 9.

13. On the social and professional background of society members, consult Roche, *Siècle*, esp. table 43; McClellan, "The Académie Royale des Sciences," pp. 561–66; Lyons, *The Royal Society*, appendix 2B. See comments by Darnton, "The Rise of the Writer," p. 28. A handful of women were members and participants in an equal number of provincial academies; see Bouillier, *Institut*, p. 85; Maylender, *Storia*, e.g., at *Accademia Clelia*; Jaeggli, "Recueil," p. 410.

14. For more on the general social and economic background to the 18th century, see Goubert, *The Ancien Régime*, esp. chs. 2–6; Anderson, *Europe in the Eighteenth Century*, ch. 4; Braudel, *Capitalism and Material Life*, esp. chs. 2–3.

15. Roche, *Siècle*, 1:31, speaks of a "national debate on the usefulness of academies," and he cites tracts by Roupmel de Chenilly, "Réflections sur l'utilité des compagnies littéraires" (1753) and one Gourcy, "Est-il à propos de multiplier des académies?" (1769).

16. See Delandine, *Couronnes académiques*, "Montauban" entry. The academies at Nancy and Pau proposed two, like topics ("On the progress of the sciences and arts since the establishment of academies" and "On the utility of an academy") in 1751. The *Académie française* itself proposed but did not award for 1755 the question, "To what point is it desirable to multiply literary societies?" See Delandine, *Couronnes*, under these academies and Roche, *Siècle*, 1:31.

17. Société royale de Montpellier, *Mémoires* (1766), 1:iii.

18. "Ne Voyons nous pas Plusieurs des Provinces du Royaume décorées de Compagnies Littéraires qui animée du même esprit s'empressent de Concourir à ses travaux par leur recherches[?] Que disje[?] L'europe entiere pres qu'entièrement remplie d'Acad^es célébres formées par les soins des plus illustres sçavants et sous la protection des plus grands Princes[.] L'Acad^e de LInstitut de bologne etablie par M Le Comte marsigli, nôtre illustre Confrère, L'Acad^e Imperiale de Petersbourg formée par Pierre legrand au quel nous pouvons donner icy le même titre de Confrère dont il s'honoroit protégée sans interruption par ses successeurs et à present par cette Impératrice si digne deluy succéder . . . Comme elle; L'Acad^e Royale deberlin établie autrefois par le célébre Leibnitz depuis réformée et protégée par cet aléxandre du Nord qui sçait également Conquerir, gouverner, et éclairer des états, Les deux acad^es de suède florissantes sous les auspices d'un Jeune monarque que ses Vertus ont déjà rendu l'admiration de L'europe et L'amour de ses peuples, celles de Naples[,] celle de florence et un grand nombre d'autres que Les bornes qui me sont prescrites ne me permettent pas même de Nommer[,] ne doivent-elles pas presque toutes leur éxistence à La Noble émulation et à cette Masse de Lumières qu'ont répandu dans L'europe les travaux de L'Acad^e et pour tout dire aussi ceux de cette autre Illustre Compagnie que L'angleterre se Glorifie d'avoir produit et que nous avons toûjours

regardée comme nôtre sœur sans rivalité et sans que les Guerres les plus vives qui se sont élevées entre les deux Nations ayent Jamais pû altérer L'espéce de fraternité qui nous unis[?] L'amérique même en a receu des Rayons[.] Boston et Philadelphie voyent au milieu de la guerre la plus animée fleurir dans leur sein deux Acad[es] déjà célébres par les soins de cet homme unique que la multiplicité de ses talents et L'usage qu'il en a sçu faire ont rendu à Lafois Le solon Le Brutus et Le Platon de ses Compatriotes." PA. Dossier de Fouchy. The occasion was a presentation before the library committee of the Paris Academy.

19. Several sources confirm this latter range. See Roche, *Siècle*, 1:359 and graphique 16, which charts attendance for ten French provincial academies at this level. Published minutes of the Saint Petersburg Academy, and the American Philosophical Society likewise list 10 to 20 members attending on average. See Imperial Academy, Saint Petersburg, *Protokoli;* Eduard Winter, *Die Registres der Berliner Akademie;* American Philosophical Society, "Early Proceedings." The unpublished minutes of the Paris Academy of Sciences indicate an average attendance of 30. The Royal Society of London received upwards of 20 guests at any particular meeting, which indicates that larger and less regulated societies, like the Royal Society of London, had more members in attendance at meetings than academies.

20. This sense of the calendrical and the ritual is emphasized by Roche, *Siècle*, 1:355ff. and graphique 18.

21. Concerning learned society proceedings and their place in the scientific press of the 18th century, see Kronick, *History*, chs. 5 and 6 and table 1; see also Tully and Thornton, *Scientific Books*, esp. chs. 4 and 5.

22. Although there is yet no comprehensive study of the prize contests sponsored by 18th-century societies, considerable inroads have been made. Roche, *Siècle*, 1:324–55 and graphiques 10–15, elaborates a thorough study of the prize contests of the French provincial academies, by far the most numerous. Barrière, *Académie de Bordeaux*, pp. 115–39, and Tisserand, *Académie de Dijon*, pp. 535–94, report fully on prizes offered by the academies in Bordeaux and Dijon as well as other academies. Delandine, *Couronnes académiques* lists most 18th-century prize contests in France. Other histories of individual institutions likewise detail other prize series. Prizes of the Paris Academy of Sciences are inventoried in Maindron, *Fondations*, and winners signaled in Jaeggli, "Recueil." See also Nordin-Pettersson, "Prisfrågor," for prizes of the Swedish academy of sciences. The most closely studied single scientific contest is Kopelevich, "The Petersburg Astronomy contest." See also Tisserand, *Les Concurrents de J. J. Rousseau*, and Bouchard, *L'Académie de Dijon* for studies of the most famous society contest of the 18th century, that of the Dijon Academy for 1754, "What is the source of inequality among men and is it authorized by natural law," won by Rousseau.

23. A corollary of the way in which such prizes were administered is that entries were theoretically anonymous and keys to the identities of losing authors were supposed to be destroyed. Such practice makes historical research on prize contests especially difficult; see Roche, *Siècle*, 1:327–28 and Gillispie and Youschkevitch, *Lazare Carnot*, esp. pp. vii–viii, where Gillispie and Youschkevitch happily rediscovered a manuscript of Lazare Carnot's so "lost" in the archives of the Berlin Academy.

24. See Maindron, *Fondation*, pp. 17–22ff.; Harnack, *Geschichte* (1901), pp. 301–18; Tisserand, *Académie de Dijon*, pp. 535–99; Barrière, *Bordeaux*, pp. 129–37. Delandine, *Couronnes*, lists over one thousand society prize contests in the 18th century. Roche, *Siècle*, 1:342–43, puts the figure at over two thousand, 60 percent of which were questions

concerning pure science and estimates total money paid out in France at "several tens of thousands of livres" for the last decades of the century alone. See, here, appendixes.

25. "*Theoria magnetis:* Ce sujet si epineux avoit déjà été proposé sans succès pour les annuées 1742 et 44 de sorte que pour cette fois le prix fut triple. M. de Maupertuis qui se trouvoit à Bâle dans l'intervalle de tems entre 1744 et 1746 nous exhorta fortement à nous mettre sur les rangs disant que 7500# en valoient bien la peine et que suivant les loix de l'acad. il ne pouvoit pas être renvoyé; alors mon frère m'avoua qu'il avoit même jetté quelques idées sur le papier, mais qu'il en étoit si peu satisfait, que cela lui avoit fait totalement abandonner son dessein; il m'offrit en même tems comme en badinant, de me communiquer ses idées, si je voulois les mettre en œuvre et partager le gateau avec lui en cas de succès: je le pris au mot et il en resulta la pièce dont il est ici question . . . le succès surpassa mes espérances, car elle partagea le prix triple avec celles de MM. L. Euler et du Four Ecuyer Correspondant de l'Acad R^le des Sciences. Elle a été imprimée sous le nom des deux frères." PA, Dossier [Jean] Bernoulli II. These details were sent in a letter (April 2, 1782) from Jean Bernoulli for an éloge for the late Daniel Bernoulli. See also Roche, *Siècle*, 1:129ff.; Barrière, *Bordeaux*, pp. 64–67. Of all society prize contests, questions concerning the natural sciences reached a peak after 1750 and dropped off in the last decades of the 18th century; see Roche, *Siècle*, 1:344–54 and graphique 13; Barrière, *Bordeaux*, pp. 133–35. Hansen, "Scientific Fellowship," p. 200, reports that Euler won 30,000# in 12 contests of the Paris Academy and that three Bernoullis (Daniel, Jean I and II) won more than that in 16 contests of the Paris Academy.

26. For this reason the scientific societies are virtually invisible in Taton, *Enseignement*. Contemporaries did distinguish universities and colleges (whose role was teaching) from learned societies (whose role was to promote research). The point is made by many commentators, including Albrecht von Haller, Christian Wolff, and the authors of the *Encyclopédie;* see Roche, *Siècle*, 1:151–53; Sonntag, "Albrecht von Haller," pp. 380–81; Hammermayer, "Akademiebewegung," p. 6; Voss, "Die Akademien," p. 45; and *Encyclopédie*, article "Académie," 1:52; Hufbauer, *The German Chemical Community*, p. 45; Heilbron, *Electricity*, pp. 128–31, 134–37.

27. Heilbron, *Electricity*, pp. 134–66 and table 2.5. See also Hufbauer, *The German Chemical Community*, appendix 2.

28. The case of the Dutch microscopist van Leeuwenhoek, who was without institutional affiliation in Leyden, exemplifies the point, in that he was known to the scientific community largely through work that appeared in the *Philosophical Transactions* of the Royal Society in London; Leeuwenhoek submitted 375 papers to the Royal Society; see Weld, *History*, 1:243–45; Ornstein, *Rôle*, pp. 129–30; Kronick, *History*, p. 80.

29. See Purver, *Concept*, pt. 2, for an extreme presentation of the Royal Society case. Note, however, that Louis and Peter founded societies after first receiving advice and counsel from Huygens and Perrault via Colbert and from Leibniz. See Bertrand, *L'Académie*, pp. 4–10; Vucinich, *Science in Russian Culture*, pp. 45–48.

30. Roche, *Siècle*, ch. 1, esp. pp. 33–38; Bouillier, *Institut*, pp. 7ff.

31. Accademia delle scienze di Torino, *Il primo secolo*, pp. 3–5.

32. Leibniz' role as propagandist for societies has been extensively documented. See Ornstein, *Rôle*, pp. 182–97 W. Schneiders, "Gottesreich und gelehrte Gesellschaft"; Harnack, *Geschichte* (1901), pp. 21–55.

33. Sonntag, "Albrecht von Haller," pp. 379–80ff.; Joachim, "Anfänge." Given these patterns to the prehistories of the scientific societies, assigning precise dates of official incorporation for the beginnings of institutions is somewhat misleading. The Paris

Academy, for example, began informally on royal authority in 1666; it received an official charter in 1699, which was not legally registered until 1713. (See Aucoc, *Lois*, pp. lxxxiv–xcii.) When one wants to mark the official beginning of the Paris Academy is a matter of taste, and the dates assigned to the foundation of institutions in the appendixes should be considered with this point in mind.

34. See, for example, budget for the academy for 1770 printed in Imperial Academy, *Protokoli*, 2:776–77.

35. Kopelvich, *Vozniknovenie*, would seem also to see the Saint Petersburg Academy as the archtypical government-funded academy; see Griffiths, "Review," p. 155.

36. See Hahn, *Anatomy*, pp. 81–82; McClellan, "The Académie Royale des Sciences," p. 547n.; Harnack, *Geschichte* (1901), pp. 260ff. Rappaport, "Liberties," explores the history of election and appointments in the Paris Academy, finding, overall, little meddling.

37. Lyons, *The Royal Society*, p. 127.

38. The numbers equal average number of members per category. For the supporting details, see McClellan, "The Académie Royale des Sciences," esp. pp. 543–53; Hahn, *Anatomy*, ch. 3; Aucoc, *Lois*; Institut de France, Académie des Sciences, *Index Biographique*; Chapin, "Les Associés Libres."

39. This division of the sciences made in the Paris Academy and other scientific academies was based on a distinction between what were considered the mathematical and deductive sciences and those thought of as qualitative, material, and inductive. See Hahn, *Anatomy*, pp. 53–54, 98–99; McClellan, "The Scientific Press," pp. 437–38.

40. Originally 30,000 livres were allocated for pensions within the academy. This figure was augmented by another 6,000 lv. in 1716, the total distributed among the three pensionnaires in each section in amounts of 3,000 lv., 1,800 lv., and 1,200 lv., depending on seniority. The secretary and treasurer of the Paris Academy each received 3,000 lv. per annum. By the end of the 18th century total funds for salaries had risen to 50,000 lv. In addition, 12,000 lv. was made available for institutional expenses, some of which went as gratifications to junior members of the academy for special services rendered. See Bertrand, *Académie*, pp. 85–107; Hahn, *Anatomy*, p. 79; Hahn, "Scientific Research," pp. 506–508; Heilbron, *Electricity*, p. 117; Kopelvich, *Vozniknovenie*, p. 108; Chapin, "The Academy of Sciences," pp. 385–86; PA-PV (1750), 69:409–11.

41. Heilbron, *Electricity*, pp. 118–28, presents the most thoroughly researched information concerning the payment of academicians in other academies (Berlin, Saint Petersburg, Göttingen, Bologna, Turin, Munich, Stockholm, and Haarlem). Roche, *Siècle*, 1:114–23, in addition describes the budgetary situations of a number of the French provincial academies, including payment of small pensions: 50 lv. for twelve pensionnaires at Dijon; 100 lv. per academician at Montpellier; 400 lv. at Metz divided among all academicians; 250 lv. annually at Lyon. Payment of academicians (in addition to considerable other state support) at Mannheim is indicated by Friedrich, *Geschichte Mannheims*, pp. 599–600, and Böhm, "Die Kurpfälzische Akademie der Wissenschaft," pp. 118–19. Similarly Kraus, *Vernunft und Geschichte*, pp. 252–54 and 296–97, suggests that academicians at Erfurt and Prague also received direct financial support. Quetelet, "Premier Siècle," p. 4, makes evident the tremendous range of support for academicians in Brussels, including the payment of pensions. Some sort of remuneration would thus seem a common feature of academies of science. Although pensions paid to academicians were rarely sufficient to provide a living, Heilbron's statistics on the salary levels of university professors (*Electricity*, pp. 152–57) indicate that they did hardly better, if as well.

[300]

42. The rights and obligations of the Parisian academicians were spelled out in the royal ordinance of 1699, reprinted in Aucoc, *Lois*, pp. lxxiv–xcii.

43. McClellan, "The Académie Royale des Sciences," pp. 547–49.

44. McClellan, "The Académie Royale des Sciences," p. 546, where the figure 713 is reported; there were three omissions in that study, and the number should be 716. Statistics on other scientific societies are hard to come by. Roche's numbers for the provincial academies put their average membership at slightly less than the figures for the Paris Academy; *Siècle*, 1:105–08 and table 18. Data from Lyons, *The Royal Society*, appendix 3, and Royal Society of London, *Record*, appendix 5, portray a total membership of approximately 1,900 in the parallel period, 1699–1793, and an average membership of 325. The Royal Society of London was doubtless the largest 18th-century scientific society; *Teylers Tweede Genootschap* of Haarlem, with six members at any one time, was the smallest. Various membership lists for different academies and societies are available, but, pending further study, one can say only that other academies and societies probably maintained about the same relative average and total memberships. See further, appendixes. Roche (*Siècle*, table 18) counts 6,400 or so members of French academies in the period, 1680–1789, and I would estimate a universal scientific society population in the 18th century at perhaps 10,000.

45. This theme is developed at length by Hahn, *Anatomy*, esp. ch. 3. Note his remarks, p. 75; "[The Academy satisfied] the scientific demands of the age while functioning within the framework of contemporary society. In many instances these two demands reinforced each other. The over-all organization of the scientific community along elitist lines, for example, was readily accepted by a society that cherished corporative privileges and respected proper authority."

46. This analysis of the organizational makeup of the Royal Society is derived from Lyons, *The Royal Society*, and Stimson, *Scientists and Amateurs*. The numbers, taken from Lyons, equal average numbers of Fellows per category.

47. Derived from Lyons, *The Royal Society*, appendix 3A.

48. See Lyons, *The Royal Society*, appendix 2A.

49. Lyons, *The Royal Society*, appendix 2C.

50. Lyons, *The Royal Society*, appendixes 3A and B. For a long period foreign membership of the Royal Society was complicated by a semi-official category of "colonial" members. These were treated as regular, English F.R.S., who had not officially signed the register book and paid their dues. With the election reform of 1766 and the American Revolution, the category of "colonial" Fellow gradually disappeared; see Stimson, *Scientists and Amateurs*, pp. 151–60; here, chapter 2.

51. Bierens de Haan, *De Hollandsche Maatschappij*, p. 274; Heilbron, *Electricity*, p. 128.

52. See Roche, *Siècle*, 1:45.

53. For these and similar Germanic academies (e.g., the Danish *Videnskabernes* academy) the distinction between sciences and nonscience sections or internal divisions is less appropriate than elsewhere, given the union of other "sciences" under the term "Wissenschaften." See Kraus, *Vernunft*, and Kraus, "Die Bedeutung der deutschen Akademien des 18. Jahrhunderts." Hammermayer, "Akademiebewegung," p. 2, remarks that "Geisteswissenschaften" are not so far removed from "Naturwissenschaften" in academies as one might think. For a French comparison, see Daniel Roche, "L'Histoire dans les activités des académies provinciales."

54. Roche, *Siècle*, 1:143.

55. Roche, *Siècle*, 1:15–72, esp. pp. 17–18. Roche provides the most extended analysis of the process of receiving official recognition and its significance. He at once emphasizes the solidfying effect of integrating a private body into the corporate structure of France, and at the same time he underscores tendencies toward an affable sociability, even libertinage, in unregulated private groups.

56. Hahn, *Anatomy*, pp. 72–73, discusses these extra privileges for the case of the Paris Academy; see also Bouillier, *Institut*.

57. Turgot's unsuccessful effort to rationalize and subdue the Parisian guilds in the 1770s or the efforts of Catherine the Great of Russia in 1767–68 to reform the Russian nobility and legal system are instances of this rationalizing drive on the part of *ancien régime* governments; see Gillispie, *Science and Polity*, ch. 1, and esp. pp. 21–31; Parker, "French Administrators;" Rappaport, "Government Patronage."

58. Hammermayer, "Die Benediktiner und die Akademiebewegung," pp. 65–92, pinpoints 18th-century reunions of Benedictines devoted to historical and ecclesiastical study. He similarly notes three nonscience academies (*Storia ecclesiastica*, 1741; *Storia di liturgia*, 1748; *Storia Romana e antichità*, 1754) founded in Rome by Pope Benedict XIV and the foundation in 1747 of an *Academia liturgica pontifica* in Lisbon under papal sponsorship; see "Akademiebewegung," pp. 4–5 and notes 17 and 20, and Ribeiro, *Historia*, 1:259–66. All things considered, these developments were very much peripheral to the mainstream of learned and scientific societies in the 18th century. Benedict XIV in particular was more like Frederick II of Prussia or other enlightened monarchs in his founding and supporting of academies. More truly feudal and reactionary elements were not associated with the societies.

59. Hahn, *Anatomy*, pp. 65–66ff.; Parker, "French Administrators," p. 92, mentions other academies appraising inventions.

60. BM-M, MS. 52 ("Correspondance de la Société Royale des Sciences de Montpellier"), #104. The answer was 6:24:43 p.m.

61. See Hankins, *Jean d'Alembert*, pp. 132ff.

62. Berlin Academy, *Mémoires* (1753), 9:511–21.

63. *Encyclopédie* of Diderot and d'Alembert, article "Académie," signed "o" (d'Alembert). See also, Christian Bartholmèss, *Histoire philosophique de l'Académie de Prusse*, 1:162n; McClellan, "International Organization," p. 289.

64. See Formey, "Considerations sur ce qu'on peut regarder aujourd'hui comme le but principal des Académies," Berlin Academy *Mémoires* (1767), 23:370–80 and "Second discours sur . . . ," Berlin Academy *Mémoires* (1768), 24:358–66; see also Bartholmèss, 1:162n.; Hahn, *Anatomy*, p. 101. Formey's addresses appear in the *Supplément* to the *Encyclopédie* in the article "Académies, avantages des."

65. Formey, "Second Discours," Berlin Academy *Mémoires* (1768), 24:364.

66. Formey, "Second Discours," Berlin Academy *Mémoires* (1768), 24:366. See also Lomonosov, "Dissertation sur les devoirs des journalistes," (1754), in Kunik, ed., *Sbronik materialof*, 2:519–30; Cesarotti, "Riflessioni sopra i doveri Accademici" (1786).

67. Hahn, *Anatomy*, pp. 63–64; Marguet, "La Connaissance des Temps."

68. See Lyons, *The Royal Society*, p. 139; Manuel, *A Portrait of Isaac Newton*, ch. 15; A. R. Hall, *Philosophers at War*.

69. See Gillispie, *Science and Polity*, pp. 261–90ff; Darnton, *Mesmerism*, pp. 48, 62, and 83.

70. Harnack, *Geschichte* (1901), pp. 250–57.

71. RS-CM, 7:33 (April 10, 1783), and pp. 5, 8, 12, 30. For expulsions from the Royal Society in the 17th century, see Hunter, "The Social Basis," pp. 54–57.

72. See Hahn, *Anatomy*, ch. 5, esp. pp. 116–18, 135–58; Gillispie, *Science and Polity*, ch. 4. See also Gillispie, "The *Encyclopédie* and the Jacobin Philosophy of Science."

73. Weld, *History*, 2:464–65.

74. Edward Forbes, "Address by Principal Forbes," p. 17. This 19th-century view is not restricted to English commentators; witness Bouillier, *Institut* (1879), p. 303, where he writes: "La Société royale de Londres est d'ailleurs une société particulière, qui s'est fondée, soutenue, développée, par le zèle et l'initiative des simples particuliers, amis des sciences, et non une institution publique qui relève de l'État et qui ne puisse subsister sans lui. Tel est le caractère propre qui distingue cette Société, non pas seulement de l'Institut de France, mais des autres grandes académies de l'Europe." See also similar comments by Ornstein, *Rôle*, pp. 91ff., and A. R. Hall, "Introduction," *Essays of Natural Experiments*, p. viii, where the latter remarks that "poverty was the price of freedom" for the Royal Society.

75. Privileges of the Royal Society are spelled out in its various charters; these are reprinted and translated (from the Latin) in Royal Society of London, *Record*, pp. 215–86. See also Lyons *The Royal Society*, appendix 1; Thompson, *History of the Royal Society*, appendix 1 and appendix 2, "Patent Granting Chelsea to the Royal Society, together with some additional privileges and powers." See also Purver, *Concept and Creation*, p. 136.

76. Books receiving the imprimatur of the Royal Society are listed in Royal Society of London, *Book Catalogue*, 3:541ff.; for the example of Hales, see RS-CM, 3:143 (February 28, 1733).

77. See Weld, *History*, 1:470. A request to send an agent to France to judge a machine was refused in an ordinary meeting of the Royal Society in 1729, "Since it is not Consistent with their Custom and practice to pass their Judgement on Works or Inventions which are neither directly laid before them nor Can be brought under the Inspection or Examination of a Committee appointed by them." RS-JBC, 15:376–77 (November 20, 1729).

78. Chelsea Patent (n. 75).

79. Lyons, *The Royal Society*, pp. 183–87; Weld, *History*, 2:27–31; Stimson, *Scientists and Amateurs*, p. 169; here, chapter 6.

80. Stimson, *Scientists and Amateurs*, pp. 176–77.

81. Weld, *History*, pp. 508–09; RS-JBC, 34:62, where the phrase "at the request of Government" occurs; see also RS-CM, 4:308–18.

82. A literally explosive affair before it was over—the magazine at Purfleet blew up—the episode developed strange political overtones, pitting Franklin and the Royal Society against the king. The Society sided on scientific grounds with the rebellious Franklin and chose pointed rods; the King took loyalists and blunted ones. See Weld, *History*, 1:95–102; Stimson, *Scientists and Amateurs*, pp. 169–72; Lyons, *The Royal Society*, pp. 192–93.

83. Weld, *History*, 1:400; Lyons, *The Royal Society*, pp. 184–85; Ronan, *Astronomers Royal*, pp. 45–55.

84. See Lyons, *The Royal Society*, pp. 145, 189, 201; see also RS-CM, 4:308–18; RS-JBC, 27:1, 554; 28: 235, 381, etc.

85. A special committee was set up to handle the gifts from the Hudson's Bay Company. See RS-JBC, 27:140, 197–201, 443–48, 523, etc. The Hudson's Bay Committee reported to the full Society in 1772, "From the great use not only to the study of natural

History, but also perhaps to commerce and manufactures, from what hath been presented to the Royal Society by the Hudson's Bay Company, we cannot but wish that Application was made from the Royal Society to the Directors of the East India, Turkey, Russia and African companies, for the same sorts of collections to be transmitted annually" RS-JBC, 27:199. See also Weld, *History*, 2:123–25.

86. See Lyons, *The Royal Society*, pp. 187–90; Weld, *History*, 1:231.

87. Weld, *History*, 2:43; Lyons, *The Royal Society*, pp. 77–82ff. See also Jacob, *The Newtonians and the English Revolution*, and the discussion of the work of Margaret and J. R. Jacob in Steven Shapin, "Social Uses of Science," Rousseau and Porter, eds., *Ferment of Knowledge*, esp. pp. 98–105.

88. Lyons, *The Royal Society*, appendix 2A, puts the figure at around 10 percent. This does not include rich but non-noble Fellows.

89. Weld, *History*, p. 146; see also RS-CM, 4:304; 5:145.

90. See Royal Society of London, 2nd Charter, in Lyons, *The Royal Society*, appendix 1, p. 339. The Charter likewise exhorts justices, mayors, aldermen, sheriffs, baliffs, constables, and other officers and ministers to aid and support the work of the Society.

91. This is not to say, again, that there were not real and important differences between academies and societies but only that the obviously nationalistic bias of the traditional view of the Royal Society and societies like it obscures some of the important ways in which the society as an institution was integrated into, rather than independent from, its social and governmental context. On other societies, see Hindle, *Pursuit*, pp. 140, 170, 263–67; Shapin, "Property, Patronage, and the Politics of Science."

An interesting contrast in this regard is supplied by the Saint Petersburg Academy of Sciences. In 1779 an adjunct of the academy, Oserezhovsky, notified the academy of his intention to leave it, "Upon which the academicians observed unanimously that, as an Academy of Sciences is by its nature a company of volunteers, aiding each other by their reciprocal lights and working without constraint for the progress of the sciences, one could in no way forcefully oppose the wish of one who asked to leave it." *Protokoli*, vol. 3, December 9, 1779.

92. Roche, *Siècle*, 1:117–25.

93. Hindle, *Pursuit*, pp. 142–45ff.

94. Desplat, *Académie royale de Pau*.

95. "Discours de Reception de M. Bitaubé," Berlin Academy *Mémoires* (1766), 22:525.

96. Kindelberger, "The Société royale," table 2.

97. Thompson, *History of the Royal Society*, passim.

98. See McClellan, "The Académie Royale," pp. 561, 564.

99. See Gelfand, *Professionalizing Modern Medicine*, esp. ch. 4; Hannaway, "Medicine, Public Welfare and the State." Learned medical societies would seem largely uncharted, historiographically; see Garrison, "The Medical and Scientific Periodicals of the Seventeenth and Eighteenth Centuries"; Kronick, "The Fielding H. Garrison List."

100. On the economic, patriotic, arts, and agricultural societies of the 18th century, see Hubrig, *Die patriotischen Gesellschaften*; Shafer, *The Economic Societies in the Spanish World*; Hahn, "The Application of Science to Society"; Vierhaus, ed. *Deutsche patriotische und gemeinnützige Gesellschaften*; Justin, *Les Sociétés Royales d'Agriculture*; Hammermayer, "Akademiebewegung," pp. 14–23 (and notes); Roche, *Siècle*, 1:61–63; J. Brown, "The Publication and Distribution of the *Trudy*"; Prescott, "The Russian Free Economic Society."

101. These distinctions are developed by Hubrig, *Patriotischen Gesellschaften*, pp. 36–39.

102. Shafer, *Economic Societies*, p. 345.

103. See Justin, *Sociétés;* Roche, *Siècle,* 1:55, 280–84.

104. See Wood, *A History of the Royal Society of Arts;* Hudson and Luckhurst, *The Royal Society of Arts;* Allan, *William Shipley.*

105. Kraus, *Vernunft,* pp. 251–61.

106. See Voss, "Die Societe [sic] Patriotique de Hesse-Hambourg."

107. Tisserand, *Académie de Dijon,* pp. 267, 330–33. On the turning of regular provincial societies toward economic and utilitarian concerns, see also Desplat, *Pau,* pp. 95–106; Barrière, *Bordeaux,* pp. 69, 129–32; Cousin, *Beçanson,* pp. 335, 344; Roche, *Siècle,* 1:55ff.

108. Imperial Academy, *Protokoli,* vol. 2, October 4, 1770; vol. 3, August 17, 1771, September 9, 1771.

2. *Origins: Scientific Societies in the Seventeenth Century*

1. This is the standard view of the role of universities in the Scientific Revolution. See Ornstein, *Rôle,* ch. 8; Ben-David, *Role,* ch. 4; Crombie, *Medieval and Early Modern Science,* 2:14–119. In the universities, science teaching was relegated to the inferior *quadrivium* (arithmetic, music, geometry, astronomy) of the Bachelor's degree, a portion of study (natural philosophy) for the Master's degree, and aspects of medical instruction. In all, science enjoyed a subordinate position to law and theology in antique universities. For a more refined case study, see Schmitt, "Science in the Italian Universities."

2. See M. Boas (Hall), *The Scientific Renaissance,* pp. 18–19, 22–28ff.; Crombie, *Medieval,* 2:103–06; Schmitt, "Italian Universities," pp. 49–50; Smith, *Science and Society,* pp. 45–46.

3. See Hans, "Academies"; Maylender, *La Storia delle Accademie in Italia.* On the phenomenon of these "Renaissance" academies, after Maylender's encyclopedic volumes, see Buck, "Die humanistischen Akademien"; Cochrane, *Tradition and Enlightenment in the Tuscan Academies,* pp. 1–34; A. R. Hall, "Introduction," *Essays of Natural Experiments,* pp. vii–xvi; Drake, "Introduction," *Discoveries and Opinions* pp. 7–11; Brown, *Scientific Organizations,* pp. 1–3; Rosen, "The Academy of Sciences," ch. 1; Roche, *Siècle,* "Conclusion."

4. See Maylender, *Storia,* 4:294–315 and 327–37.

5. See Maylender, *Storia,* 1:125–30; Middleton, *The Experimentors,* p. 6. For further discussions on the origins and character of "Renaissance" academies, see n. 3 and Maylender, *Storia,* 1:377–93; 2:84–93; 3:350–62; 4:199.

6. The advent of printing was important for science in the 15th and 16th centuries, and the association of "Renaissance" academies with a press was close. See Ornstein, *Rôle,* pp. 75–76; Eisenstein, *The Printing Revolution;* Stillwell, *The Awakening Interest in Science.*

7. More should be said, perhaps, about the *Accademia della Crusca,* a ducal language academy which published a dictionary and initiated the tradition of national language academies, such as the *Académie française;* see Maylender, *Storia,* 2:122–46; Cochrane, *Tradition,* pp. 50–52; Biagi, *L'Accademia de Belle Arti,* pp. 11–14. The *Accademia del Designo* supervised civic design and architecture.

8. The term is Cochrane's, *Tradition,* p. 6; see also Drake, *Galileo Studies,* p. 80; *Encyclopédie,* article "Académie."

9. In Maylender six different academies were known as *dei Fantastici,* thirteen as *degli Incogniti;* see Maylender, *Storia,* entries for these associations and 4:24 for the continuation of this tradition into the 18th century.

10. Maylender, *Storia*, 3:162 and 165. Another *Immobili* appeared later in Venice in 1618, possibly with anti-Galilean overtones; see Maylender, *Storia*, 3:163–64. Note added in press: Offering good insight into the phenomenon of "Renaissance" academies and sixteenth-century science, Eamon and Paheau, "The Accademia Segreta," pinpoints a new "Renaissance" science academy which existed at Naples between 1542 and 1547.

11. For these details concerning the *Accademia Secretorum Naturae*, see Drake, *Galileo Studies*, p. 80; Maylender, *Storia*, 4:150–51; Middleton, *The Experimentors*, pp. 6–7; Ornstein, *Rôle*, pp. 73–74.

12. Apparently two other short-lived associations of the 16th century dealt in part with natural philosophy, the *Accademia degli Occulti* (Brescia, 1563) and the *Accademia dei Secreti* (Venice, approximately 1570); see Maylender, *Storia*, 4:87–91 and 5:148–49.

13. For further background to the *Accademia dei Lincei*, see Maylender, *Storia*, 3:430–503; Carutti, *Breve Storia*; Drake, *Galileo Studies*, pp. 79–94; Ornstein, *Rôle*, pp. 74–76.

14. Drake, *Galileo Studies*, p. 81, translates portions of the initial declaration of principles of the *Accademia*.

15. Notable volumes include Galileo's *Letters on Sunspots* (1613) and the *Assayer* (1623), and a natural history of Mexico; see Drake, *Discoveries and Opinions*, pp. 76–78, 85ff.; *Galileo Studies*, pp. 91–93; Ornstein, *Rôle*, p. 76.

16. Ornstein, *Rôle*, p. 75.

17. The fullest study of the *Accademia del Cimento* is the previously cited one by Middleton. See also Ornstein, *Rôle*, pp. 76–90. Middleton (*The Experimentors*, p. 9) notes that another science academy the *Accademia degli Investiganti* existed in Naples for a short period between 1650 and 1656; see also Maylender, *Storia*, 3:367–69.

18. The *Saggi* were translated into English by Richard Waller in 1684 as *Essays of Natural Experiments*. Middleton provides a new translation of the *Saggi* and complete bibliographical information, *The Experimentors*, pp. 347–54.

19. Rosen, "The Academy of Sciences," p. 2; Middleton, "Science in Rome"; Maylender, *Storia*, passim. See also 18th-century "Renaissance" academies listed in appendixes.

20. This account of the origins of the Royal Society glosses over two debated questions. The significant one concerns the role of Puritanism in the growth of science in 17th-century England and the foundation of the Royal Society. As an entrée to the literature, see Hill, *Intellectual Origins of the English Revolution*; Merton, *Science, Technology, and Society*; Kearney, "Puritanism & Science"; Rabb, "Religion and the Rise of Modern Science"; A. R. Hall, "Merton Revisited." The other, more antiquarian question concerns groups preceding the Royal Society, their nature, and connection to the institution of the 1660s; for a sense of this issue, see Purver, *Concept and Creation*, passim, and respectively A. R. Hall, ed., "Introduction," *Essays of Natural Experiments*, and M. B. Hall, "Sources for the Early History of the Royal Society," in Birch, *History*, 1:xiv–xvi, esp. For more on the prehistory of the Royal Society, see McKie, "Origins"; Ornstein, *Rôle*, ch. 4.

21. See Johnson, "Gresham College."

22. See T. M. Brown, "The College of Physicians." Professorships in geometry and astronomy were endowed at Oxford in 1619, natural philosophy at Oxford in 1621; the Lucasian professorship (later Newton's professorship) was established at Cambridge in 1663; see Ornstein, *Rôle*, pp. 236–37, 248.

23. See Gunther, "Introduction," *Early Science in Oxford*, 4:1–16; see also, Weld, *History*, 1:33–34; McKie, "Origins," pp. 25–26.

24. The 1660 organization was called "The Society for the Promotion of Physico-Mathematicall Experimental Learning"; see McKie, "Origins," p. 36.

25. The name was changed in the second charter to its modern name. See Lyons, *The Royal Society*, pp. 36–37. The charters are reprinted and translated in The Royal Society, *Record*; see also Lyons, appendix 1, and Thompson, *History of the Royal Society*, appendix 1. A sense of the close connections between the early Royal Society and the English court can be had in surveying the tercentenary volume edited by Harold Hartley.

26. For the prehistory and foundation of the Paris Academy of Sciences, see Hahn, *Anatomy*, pp. 4–6; Bertrand, *L'Académie*; Brown, *Scientific Organizations*, chs. 1, 2, 4, and 6; Taton, *Origines*; Bigourdan, *Sociétés savantes*; Ornstein, *Rôle*, ch. 5. Gauja, *L'Académie des sciences*.

27. See note 26 and Dainville, "Foyers du culture." Brown, *Scientific Organizations*, p. 3, calls Peiresc (1580–1637) "the chief link between the academies of Italy and those of Paris." Peiresc was based in Aix.

28. Consider, for example, Mersenne's *La Verité des sciences contra les sceptiques* (1627); see Thorndike, *History*, 7:436–444; Ornstein, *Rôle*, pp. 140–41; and Brown, *Scientific Organizations*, p. 38, for discussions of this and other of Mersenne's and Gassendi's work. In this connection see also Kearney, *Science and Change*, chs. 1 and 2.

29. See *Correspondance du P. Marin Mersenne, Religieux Minime*, 12 vols. See also Lenoble, *Mersenne*, esp. ch. 15.

30. Private correspondence had drawbacks in that letters were seen by only a few people and did not facilitate an easy exchange of ideas among diverse groups; see Thornton and Tully, *Scientific Books*, pp. 41, 256; Kronick, *History*, pp. 53–61; McKie, "The Scientific Periodical," p. 122; Ornstein, *Rôle*, pp. 198–99, lists Mersenne, Peiresc, John Collins, and Wallis as other notable correspondents; Kronick, *History*, pp. 57–58, adds Theodore Haak and Samuel Hartlib to those whose correspondence was especially systematic and voluminous; see also, Brown, *Scientific Organizations*, p. 33.

31. See Solomon, *Public Welfare, Science and Propaganda*; also Bigourdan, *Sociétés savantes*, pp. 5–10; Brown, *Scientific Organizations*, pp. 17–31; Kronick, *History*, p. 66.

32. See Brown, *Scientific Organizations*, esp. chs. 6 and 7 for details concerning these "academies."

33. The coming of Leonardo da Vinci to the Paris court of Francis I in 1516 marks this tradition. See Frémy, *L'Académie des derniers Valois*.

34. See Hahn, *Anatomy*, pp. 9–10, 13, 46–47; Bertrand, *L'Académie*, pp. 1–2. It is to be noted again that the *Académie française* was thus the first of the 18th-century type of state society; the Paris Academy and the Royal Society, the first of the scientific societies. For more on the *Académie française*, see Stackelberg, "Die Akademie Française"; Robertson, *A History of the French Academy*; Gaxotte, *L'Académie française*.

35. The *Académie royale des Inscriptions et Belles Lettres* (originally *des Inscriptions et médailles*) was founded in 1663; *L'Académie royale de Peinture et de Sculpture*, established in 1648, received renewed support under Colbert; and the *Académie française* continued in existence; after the *Académie des sciences*, the *Académie royale d'architecture* was created in 1671 under the same Colbertean thrust; see Hahn, *Anatomy*, pp. 11–16, 46–47.

36. Andreas Kraus has pointed to this characteristic of corporate status as the decisive quality of 18th-century societies; see *Vernunft und Geschichte*, p. 207. Recall that the Paris Academy did not achieve official incorporation until after 1699. To say that the Paris Academy was not an official, state institution before then, however, is to be excessively legalistic and overlooks the extent to which the Paris Academy was very different from the

Accademia del Cimento from its beginnings; on this point see Hahn, Anatomy, ch. 1, esp. pp. 4–5.

37. This was certainly the opinion of Samuel Formey, permanent secretary of the Berlin Academy, who noted in 1775, "The Royal Society of London seems to have been the first attempt and the first model, for I count as nothing all these little Italian societies." Berlin Academy, Mémoires (1775), "Histoire" section, p. 17.

38. Middleton, The Experimentors, p. 346.

39. On Bacon, see Farrington, Francis Bacon; Bowen, Francis Bacon; Ornstein, Rôle, pp. 39–44; Purver, Concept and Creation, ch. 2; Thorndike, Magic and Experimental Science, vol. 7, ch. 4.

40. See note 39; Ornstein, Rôle, reprints Bacon's "House of Salomon" as an appendix, pp. 264–70.

41. Bacon, quoted in Ornstein, Rôle, p. 270.

42. Ornstein, Rôle, p. 44, calls Bacon "the veritable apostle of the learned societies."

43. Middleton, The Experimentors, pp. 56 and 331–33.

44. Middleton, The Experimentors, pp. 3, 263, 332–33.

45. Sprat, History, pp. 35, 115. For more on Sprat and his History, see Stimson, Scientists and Amateurs, pp. 70–76; Purver, Concept and Creation, pp. 9–19ff. See also Lyons, History, pp. 41–42, for a similar statement by Robert Hooke: "The business and design of the Royal Society is. . .To examine all systems, theories, principles, hypotheses, elements, histories, and experiments of things naturall, mathematicall and mechanical, invented, recorded or practised by any considerable authors ancient or modern. In order to the compiling of a complete system of solid philosophy for explicating all phenomena produced by nature or art, and recording a rationall account of the causes of things."

46. See Birch, History, 1:15ff.; Dorn, "The Art of Building," esp. ch. 6; Weld, History, 1:175 and note; M. B. Hall, "Science in the Early Royal Society"; Hoppen, "Nature of the Early Royal Society." See also Royal Society of London, Record, pp. 37–38.

47. The Royal Society was founded in 1660 by twelve men. One of their first acts was to compose a list of forty-one potential members; see McKie, "Origins," pp. 33–35. Lyons, History, appendix 3, lists the average number of Fellows for each five-year period; the figures for 1666–70 are 203; for 1671–75, 215. On the Society's finances in its formative period, see Lyons, History, pp. 77–84; Lyons, "The Society's Finances"; Stimson, Scientist and Amateurs, pp. 64–65.

48. The relative weight, so to speak, of "scientists" versus amateurs is an important theme of both Lyons, The Royal Society, and Stimson, Scientists and Amateurs; see Lyons, pp. 72–115 and appendixes 2A and 2C; Stimson, pp. 95–97.

49. See Stimson, Scientists and Amateurs, pp. 106–07; Ornstein, Rôle, pp. 128–29. Royal Society of London, Record, pp. 37–38, lists some two dozen works that were sanctioned by the Society through the 17th century. Newton's Principia, of course, figures among that number, but hardly as a Baconian work.

50. For discussions of the theme of Baconianism in the early Paris Academy, see Hahn, Anatomy, pp. 19–28; Bertrand, Académie, pp. 10–28; Maury, Les Académies d'autrefois, pp. 15ff.; Ornstein, Rôle, pp. 174–59. For a close-up view of the workings of the early Paris Academy in one area see Stroup, "Wilhelm Homberg."

51. Hahn, Anatomy, pp. 13–18; Bertrand, L'Académie, pp. 2–3, 5. Twenty-two men (sixteen academicians and six students) were originally associated with the Paris Academy; a total of 56 men made up its entire membership during the period 1666–1699; see Ornstein, Rôle, pp. 147–56 (esp. 145n. and 147n.); McClellan, "The Académie Royale des Sciences," pp. 543–44.

52. See note 50. In a memorandum approved by Colbert, Christian Huygens set out the aims of the new academy as he saw them: "The principal and most useful occupation of this Assembly should be, in my view, to work on a natural history more or less according to the plan of Bacon. This history consists of experiences and observations and is the unique way of arriving at a knowledge of causes of all that one sees in nature. . . . One must, following diverse routes, of which I have just named a few, distinguish chapters in this history and amass to it all observations and experiences which pertain to each particular. . . . The utility of such a history extends to the whole human race and to all the centuries to come. . . . Comprehensive surveys are always a solid foundation upon which to base a natural philosophy." Colbert's marginal notations read "Good" ("Bon") throughout. Memorandum is reprinted in Bertrand, *Académie*, pp. 8–10, and translated in part in Ornstein, *Rôle*, p. 149.

53. That is not to say that all the work of the Paris Academy in the 17th century took place behind closed doors. The beginnings of the General Map of Frnce, undertaken by Picard, and expeditions to Tycho's Uranibourg and to Cayenne brought academicians out into the wider world, but not into any greater institutional contact with others.

54. The academy did exert a negative foreign effect, as it were, by draining off major scientific talent from other countries, e.g., Huygens, Roemer, and J. D. Cassini. Foreigners were made associate members of the academy if they did not give up their native citizenship; see Ornstein, *Rôle*, p. 156; McClellan, "The Académie Royale des Sciences," p. 543.

55. See in this regard Turnbull, "Early Scottish Relations"; Hoppen, "The Royal Society and Ireland I and II"; Stearns, "Colonial Fellows of the Royal Society," "Fellows of the Royal Society"; Andrade and Martin, "The Royal Society and its foreign relations [sic]."

56. Lyons begins his classification of foreign members at 1680–85, congruent with the election reform of 1682 in the Royal Society. (See Lyons, *The Royal Society*, appendix 3A; Weld, *History*, 1:279–80.) Yet this data does not agree with that presented by Lyons in appendix 2A, which notes foreign membership beginning in the 1670s. Stimson adds to the confusion by implying that there existed two categories, one for English colonials, another for foreigners. (Stimson, *Scientists and Amateurs*, pp. 148–50.) Foreign membership was limited in the 18th century, so at some point foreign members were recognized separately. What probably happened was that non-resident Fellows were distinguished in 1665 (a known fact), foreign Fellows were also elected along the line (Cassini, for example, in 1672), and these were later (either in 1682 or subsequently) classed as a special category of members.

57. See Hahn, *Anatomy*, p. 17; Ornstein, *Rôle*, p. 153. J. B. Du Hamel (1623–1706) was the nominal secretary of the Paris Academy until its reform in 1699. Du Hamel's minutes of the meetings of the Paris Academy were published in Latin in 1698 and in 1733 in French as *Histoire de l'Académie Royale des Sciences, 1666–1699*. Du Hamel himself, however, apparently did not consider the position to be full time, writing at one point, "I seldom go to the King's Library." See Du Hamel letter to Henry Oldenburg, Paris, June 28, 1670, in *The Correspondence of Henry Oldenburg*, Hall and Hall, eds., vol. 7, letter #1471, and the Hall's note 4. All translations from Oldenburg's correspondence are by the Halls.

58. The Royal Society actually had two secretarial positions. John Wilkins and Oldenburg were the first two elected (in 1663); see Weld, *History*, vol. 2, appendix 9, "Secretaries of the Royal Society"; Lyons, *The Royal Society*, pp. 50, 67–68.

59. See A. R. Hall and M. B. Hall, eds., *The Correspondence of Henry Oldenburg*, 10 vols., passim. These volumes and the important introductions and notes by the Halls provide full

and detailed information on Oldenburg and his network of correspondence. Marie Boas Hall presents an overview of that material in "Oldenburg and the Art of Scientific Communication"; see also Stimson, *Scientists and Amateurs*, pp. 65ff.; Lyons, *The Royal Society*, pp. 55ff.; Ornstein, *Rôle*, pp. 123ff.

60. Quoted variously in Ornstein, *Rôle*, p. 123; Lyons, *The Royal Society*, p. 46; Kronick, *History*, pp. 57–58, and Weld, *History*, 1:135n. Originally Oldenburg's position was unsalaried, but after this petition, he received forty pounds per annum. For details concerning Oldenburg's remuneration, see the Halls in Oldenburg, *Correspondence*, 5:xxv.

61. For discussions of the historical rise of scientific journals and their significance, see Brown, *Scientific Organizations*, ch. 9; Ornstein, *Rôle*, ch. 7; Thornton and Tully, *Scientific Books*, ch. 8; Kronick, *History*, chs. 4 and 5; Barnes, "The Editing of Early Learned Journals;" McKie, "The Scientific Periodical from 1665 to 1798," pp. 122–23.

62. Thus, Thornton and Tully, *Scientific Books*, p. 236: "it was not until scientists had grouped themselves into societies that we encounter the early journals"; Ornstein, *Rôle*, p. 209, "The scientific journal must be thought of, therefore, as an invaluable instrument which the seventeenth century created partly in the service of, partly cooperating with, the learned societies."

63. See Weld, *History*, 1:177–82; Kronick, *History*, pp. 135–37.

64. See Hahn, *Anatomy*, pp. 17, 27; Ornstein, *Rôle*, p. 200.

65. Other noteworthy 17th-century periodicals which included news and reports of scientific interest are the *Giornale de litterati* (Rome, 1668–97), the *Miscellanea Curiosa Medico-Physica* (1670–1712), the *Acta Eruditorum* (Leipzig, 1682–1782), and Bayle's *Nouvelles de la République des Lettres* (Amsterdam, 1684–1718). Kronick, *History*, p. 78, counts thirty-five scientific periodicals appearing in the 17th century. Garrison, "Medical and Scientific Periodicals," p. 338, counts 107 medical and scientific periodicals for the same period.

66. See Kronick, *History*, table 1, p. 78. For the importance of learned society proceedings in the scientific press, see Kronick, *History*, p. 112; McKie, "The Scientific Periodical," pp. 125–26, and "Scientific Societies," p. 138; Thornton and Tully, *Scientific Books*, pp. 262–65; McClellan, "The Scientific Press," pp. 427–31.

67. "Because news ages so fast," as Denis de Sallo, first editor of the *Journal des Sçavans*, put it; see Ornstein, *Rôle*, p. 199; McKie, "The Scientific Periodical," p. 123.

68. See, for example, Hahn, *Anatomy*, p. 27. In a notable instance of this use of learned journals, Oldenburg in early 1672 sent news of Newton's reflecting telescope to Huygens with the request that for priority purposes an insertion be made in the *Journal des Sçavans*. (See Christian Huygens, *Œuvres complètes*, 7:129–31 and note; Isaac Newton, *The Correspondence of Isaac Newton*, Turnbull, ed., 1, 76 and note.) As it turned out, Huygens was called upon to defend Newton's originality later in 1672. (See Huygens, *Œuvres*, 7:186.) Huygens himself was still attached to the older mode of using coded ciphers, but in one encounter with the Royal Society, according to the Halls, "the Fellows expressed the opinion that it was better to establish priority by publication." See Oldenburg, *Correspondence*, 4:xxii.

69. See E. Barnes, "International Exchange," for a review of the dissemination of information in the 1680s through the medium of learned journals. See also Kronick, *History*, p. 79; Ornstein, *Rôle*, p. 199; McKie, "The Scientific Periodical," p. 122.

70. "A Letter of Mr. Isaac Newton . . . Containing his New Theory of Light and Colours," *Philosophical Transactions* for February 19, 1672); see also Weld, *History*, 1:237.

71. See Barnes, "International Exchange." Oldenburg's correspondence network was a natural channel for distributing the *Philosophical Transactions*, and through it the Royal Society journal became known in France, Germany, Italy, and elsewhere. The Parisian

academicians were extremely interested in the *Phil. Trans.*—they translated it privately in their confines. (See Halls' comments in Oldenburg, *Correspondence*, 4:xxiii; Brown, *Scientific Organizations*, pp. 203–05). Moreover, the fact that both the *Philosophical Transactions* and the *Journal des Sçavans* were sporadically reprinted and translated (into Latin) in Amsterdam and in Germany is testimony to how much international demand for these journals outstripped supply. (On this point see the Halls' comments in Oldenburg, *Correspondence*, 7:66n; Ornstein, *Rôle*, p. 128; Thornton and Tully, *Scientific Books*, p. 240.) All private 17th- and 18th-century journals regularly reprinted and in essence shared materials; see Kronick, *History*, pp. 101ff.; Ornstein, *Rôle*, pp. 127, 201n., 202–03; Thornton and Tully, *Scientific Books*, pp. 240–41; the Halls in Oldenburg, *Correspondence*, 8:xx.

72. Oldenburg to Carcavy, January 2, 1667/8, Oldenburg, *Correspondence*, vol. 4, letter #744; see also Halls' notes to this letter.

73. Oldenburg to Gallois, July 18, 1668, in Oldenburg, *Correspondence*, vol. 4, letter #925.

74. See Harnack, *Geschichte* (1901), p. 14.

75. See Harnack, *Geschichte* (1901), pp. 20–27, for further details concerning these plans.

76. Ornstein, *Rôle*, pp. 175–76.

77. For more on this notable institution see Winau, "Zur Frühgeschichte der Akademia Curiosorum"; Reichenbach and Uschmann, *Beiträge zur Geschichte*; Meding, *Académie Impériale*, pp. 1–28; Aberhalden, *Berichte über den Verlauf*, pp. 1–2; Ornstein, *Rôle*, pp. 169–75; Kronick, *History*, pp. 81–83.

78. Several 18th-century style societies established in Italy (Messina, 1728; Modena, 1752; Mantua, 1767; Padua, 1779), were either reformed versions of preexisting "Renaissance" academies or saw their origins in "Renaissance" predecessors. See Maylender, *Storia*, under these entries, and below, chapter 4.

79. Roche, *Siècle*, 1:18–31.

80. This provincial group received tacit recognition from Colbert and Paris and had some of its operating costs supported by the central government. The loss of its principal member, Huet, brought the Caen academy to a close, however, and indicated the frail level of social support for scientific activities in the French provinces in the second half of the 17th century. See H. Brown, "L'Académie de Physique," and *Scientific Organizations*, pp. 216–27.

81. For more on the Oxford Philosophical Society, see Gunther, *Early Science in Oxford*, vol. 4, *The Philosophical Society*, and vol. 12, *Dr. Plot and the Correspondence of the Philosophical Society of Oxford*; see also Weld, *History*, 1:34–35; Ornstein, *Rôle*, pp. 99n.–100n. It seems something of a question whether the organization that was definitely alive after 1683 was the same philosophical society as that of 1650; see McKie, "Origins," p. 37n.; Purver, *Concept and Creation*, p. 126n.

82. Weld, *History*, 1:34.

83. See Hoppen, *The Common Scientist*, passim, and his "Royal Society and Ireland I and II."

84. See Hoppen, *The Common Scientist*, pp. 89, 114–15ff. Earlier contact between the Royal Society and the Paris Academy has been mentioned, but it was not until the middle of the 18th century that these two institutions cooperated formally. See here chapter 5.

85. Weld, *History*, 1:302n.

86. See Hoppen, *The Common Scientist*, pp. 139–40, and "The Royal Society and Ireland II," appendix, which lists articles of Dublin F.R.S. published in the *Philosophical Transactions*.

87. See Hoppen, *The Common Scientist*, pp. 89ff., and "Royal Society and Ireland I," pp. 127–28, "Royal Society and Ireland II," pp. 79–80. See also Gunther, *Early Science*, 12:vii, where, in a contemporary pamphlet, the three societies in London, Oxford, and Dublin are referred to as "corresponding societies."

88. Bates, *Scientific Societies*, p. 3; Hindle, *Pursuit*, p. 60; Stearns, "Colonial," p. 84.

89. Weld, *History*, 1:305–06; Hoppen, *The Common Scientist*, p. 210; A. R. Hall, *Cambridge Philosophical Society*, p. 4.

90. Weld, *History*, 1:232; Hoppen, *The Common Scientist*, pp. 210–21.

91. Hoppen, *The Common Scientist*, p. 211; Weld, *History*, l:383.

92. Birch, *History*, 1:34–35; Weld, *History*, 1:166–69; Brown, *Scientific Organizations*, pp. 91–116. There was also a failing effort to establish contact between the Royal Society and the *Accademia del Cimento* before the latter closed in 1667; see Middleton, *The Experimentors*, pp. 282–96; Hall, ed., "Introduction," *Essays of Natural Experiments*, p. x.

93. See Oldenburg, *Correspondence*, passim and 3:xvi.

94. Oldenburg, *Correspondence*, vol. 4, letter #774.

95. Hahn, *Anatomy*, pp. 16–17.

96. Oldenburg, *Correspondence*, vol. 4, letter #899.

97. Oldenburg, *Correspondence*, vol. 4, letter #925.

98. See Halls' comments in this regard in Oldenburg, *Correspondence*, 7:xxv.

99. See Oldenburg, *Correspondence*, vols. 7 and 9, passim, A. R. and M. B. Hall, "Les Liens publics," pp. 52–53. Du Hamel also at one point wrote Oldenburg in a more traditional mode to introduce "three gentlemen of France [who] will perhaps need your influence to attend the Royal Society." Oldenburg, *Correspondence*, vol. 9, letter #2218. See also Brown, *Scientific Organizations*, p. 205.

100. M. B. Hall, "Oldenburg," p. 290.

101. See Weld, *History*, vol. 2, appendix 9, and Lyons, *The Royal Society*, p. 51.

102. Quoted in Weld, *History*, 1:267–68.

103. Stimson, *Scientists and Amateurs*, p. 67.

104. Brown, *Scientific Organizations*, p. 258.

105. For more detail to this standard story, see again Lyons, *The Royal Society*, chs. 2 and 3; Stimson, *Scientists and Amateurs*, chs. 5 and 6. Both authors divide the 17th-century history of the Royal Society into the two periods 1660–70 and 1670–1700. See also Ornstein, *Rôle*, ch. 4; Hunter, "The Social Basis," pp. 23–32.

106. See Stimson, *Scientists and Amateurs*, pp. 86–94, for a discussion of the satirical attacks on the "virtuosi" of the Royal Society, a tradition of public humor at the Society's expense that climaxed with Swift's *Gulliver's Travels* and the Grand Academy of Lagado. The Dublin Philosophical Society was likewise the object of similar ridicule; see Hoppen, *The Common Scientist*, ch. 6. Regarding finances, Lyons, *The Royal Society*, p. 79, notes that by 1676 dues to the Society were in arrears by 2,000 pounds. After early enthusiasm, membership in the Royal Society had fallen to 115 by the period 1691–95, below the initial 145 Fellows of the period 1663–65; attendance at Council meetings had also declined to record levels; see Lyons, appendices 2C, 3A, and 3B.

107. Weld, *History*, 1:70.

108. See Stimson, *Scientists and Amateurs*, p. 149; Weld, *History*, 1:279–80; Lyons, *The Royal Society*, pp. 80 and 94. The tradition of increasing council power continued through Newton's presidency and until 1730; at that time, in a legal brief, the council asserted powers it clearly did not have, and its assertions were rejected by the attorney-general; see Lyons, *The Royal Society*, pp. 151–53.

109. Discounting Lord Brouncker, whose tenure as president paralleled Oldenburg's as secretary, the average tenure of P.R.S. until Newton was 2.6 years; see Weld, *History*, appendix 7, "Presidents of the Royal Society"; see also Stimson, *Scientists and Amateurs*, pp. 99–101.

110. See Hahn, *Anatomy*, ch. 1; Bertrand, *Académie*, pp. 1–47; Ornstein, *Rôle*, ch. 5, for further details concerning the development of the Paris Academy through the end of the 17th century. Salomon-Bayet, *Institution de la science*, sheds a valuable light on some of the internal workings in science in the academy in this period and through the 18th century.

111. Hahn, *Anatomy*, pp. 21–22, points out how much early advisory procedures were unsystematic.

112. Hahn, *Anatomy*, pp. 19ff., plays down outside political influences on the development of the academy in the 17th century, in contrast to the traditional view. Compare Ornstein, *Rôle*, pp. 156–57; Bertrand, *Académie*, pp. 39–41; Maury, *Les Académies d'autrefois*, pp. 37–40.

113. Reprinted in Aucoc, *Lois*, pp. lxxxiv–xcii.

114. Article 20, Letters Patent of 1699; Aucoc, *Lois*, p. lxxvii.

115. Article 30, Letters Patent of 1699; Aucoc, *Lois*, p. lxxix.

116. Article 31, Letters Patent of 1699; Aucoc, *Lois*, p. lxxix.

117. Articles 5, 27, and 28, Letters Patent of 1699; Aucoc, *Lois*, pp. lxxv, lxxviii.

118. In this connection see Paul, *Science and Immortality*.

119. See Kronick, *History*, ch. 6.

120. See Maindron, *Fondations*, esp. pp. 1–22, for these details.

121. The Meslay prize was the first in the Paris Academy. A literary prize had been established earlier in the *Académie française*, likewise private, and Meslay wished to carry over this custom to the Academy of Sciences. Science prize questions were first proposed in the Bordeaux Academy in 1716; see Maindron, *Fondations*, p. 3.

3. The Scientific Society Movement to 1750

1. Figure 4 is derived from appendixes 1 and 2. The category of private societies is added on top of official societies, giving the total number of institutions over time. Private societies are not to be understood as constituting a separate graph. On the exponential growth of science since the 17th century, see Price, *Little Science, Big Science*, esp. pp. 1–32.

2. This summary of the early Berlin Academy is taken from Harnack, *Geschichte* (1901), pp. 1–228; see also Bartholmèss, *Histoire philosophique*, 1:1–204; Pasquier, *Leonard Euler*, pp. 34–50.

3. Harnack, *Geschichte* (1901), p. 176.

4. Harnack, *Geschichte* (1901), p. 187; Calinger, "Frederick the Great and the Berlin Academy of Sciences."

5. Quoted in Harnack, *Geschichte* (1901), p. 198. Euler apparently was thinking of Daniel Bernoulli.

6. See Berlin Academy, *Histoire et mémoires de l'Académie* (1745), 1:viii.

7. Harnack, *Geschichte* (1901), p. 231. See also AdW, MS. I:IV:31, 32. "Registres de l'Académie, 1746–1766, 1766–1786." Extracts from these registers for the period 1746–66 have been published by Eduard Winter.

8. Harnack, *Geschichte* (1901), p. 222. See also AdW, MS. I:IV:3, "Registre Général de l'Académie royale des Sciences et Belles-Lettres, contenant tout ce qui concerne les affaires publiques," pp. 509–11.

9. See H. Brown, "Maupertuis philosophe."

10. On Formey, see W. Krauss, "Ein Akademie Sekretär vor 200 Jahren: Samuel Formey," in his *Studien*, pp. 53–62.

11. See Vucinich, *Science in Russian Culture*, chs. 1 and 2; Vucinich, Review of *Istoria Akademii Nauk SSSR;* Lipski, "Foundation."

12. Vucinich, *Science in Russian Culture*, p. 36.

13. See Fersman, *The Pacific*, ch. 1.

14. Vucinich, *Science in Russian Culture*, pp. 39–44; Lipski, "Foundation," p. 349; Weld, *History*, 1:256.

15. See "Anniversaries of Other Academies and Societies," *Notes and Records of the Royal Society of London* (1947), 4:65; Dvoichenko-Markov, "The American Philosophical Society," p. 549.

16. Vucinich, *Science in Russian Culture*, pp. 45–48; Harnack, *Geschichte* (1901), pp. 137–38; Lipski, "Foundation," pp. 349–50; Dvoichenko-Markov, "The American Philosophical Society," p. 550.

17. See previous note. Westerners were already in China as Jesuits. They served there as scientific and technical advisers and did important astronomical work for the Chinese; see Spence, *To Change China*, ch. 1.

18. Vucinich, Review of *Istoria Akademii*, p. 151; Vucinich, *Science in Russian Culture*, pp. 48–50, 59; Haumant, *La Culture française en Russie*, pp. 23–27.

19. "Hors de tout rang." PA-PV (1717), 36:318; see also p. 139; see also Vucinich, *Science in Russian Culture*, p. 67; Lipski, "Foundation," p. 350.

20. Vucinich, *Science in Russian Culture*, p. 67; Lipski, "Foundation," pp. 350–51.

21. Vucinich, *Science in Russian Culture*, p. 69; Lipski, "Foundation," p. 351.

22. Joravsky, Review of *Istoria Akademii Nauk SSSR*, p. 592.

23. For information on the composition of the Saint Petersburg Academy, see Vucinich, *Science in Russian Culture*, pp. 70–77; Lipski, "Foundation," pp. 351–52; Krassovsky and Vosper, "The Structure of the Russian Academy." See also Imperial Academy of Sciences, Saint Petersburg, *Materiali*, 10 vols. This series contains material relevant to the academy's history for the period 1716–1750, and each volume presents a budget for the various sections of the academy; 2:226–45, presents an extended discussion on "the State of the Academy" for 1732.

24. Vucinich, *Science in Russian Culture*, pp. 68–71. See also the *Protokoli* of the Saint Petersburg Academy, which reprints minutes of the meetings of the Archives-Conferences.

25. See Modzalevski, *Spisok Chlenov Imperatorskoi Academii Nauk*. This volume contains a breakdown of the membership of the Saint Petersburg Academy through 1907.

26. Vucinich, *Science in Russian Culture*, p. 70; Lipski, "Foundation," p. 352.

27. Vucinich, *Science in Russian Culture*, pp. 72–73.

28. Vucinich, *Science in Russian Culture*, p. 68.

29. On the importation of Western men of science to Saint Petersburg, see Vucinich, *Science in Russian Culture*, p. 76; Vucinich, Review of *Istoria Akademii*, p. 151; Haumant, *Culture française*, pp. 24–33; Pasquier, *Euler*, pp. 1–16; Lipski, "Foundation," p. 352; Imperial Academy, *Materiali*, vols. 1 and 2. This last source contains contracts of early academicians, materials used to recruit them, etc.

30. Joravsky, Review of *Istoria Akademii*, pp. 591–92; Vucinich, *Science in Russian Culture*, p. 78; Vucinich, Review of *Istoria Akademii*, p. 151.

31. See Vucinich, *Science in Russian Culture*, p. 81; Lipski, "Foundation," p. 352; Pasquier, *Euler*, p. 15; Menshutkin, *Russia's Lomonosov*, p. 22.

32. See Youschkevitch and Winter, *Briefwechsel*, 1:4; Vucinich, *Science in Russian Culture*, pp. 88–89; Vucinich, Review of *Istoria Akademii*, p. 151; Imperial Academy, *Materiali*, 1:504–15, 536, 568ff.

33. Youschkevitch and Winter, *Briefwechsel*, 1:5; Imperial Academy, *Materiali*, 2:776; 3:446ff., 832–53; Imperial Academy, *Protokoli*, vol. 1, October 13–18, 1738, etc.

34. Haumant, *Culture française*, pp. 35ff.; Vucinich, *Science in Russian Culture*, pp. 79–80.

35. In the polyglot record of Latin, German, French, and Russian left behind by the Imperial Academy of the 18th century, one is treated to the spectacles of Delisle working in the geographical department with five Russian and German geodesists and asking the academic chancellory for a translator so that they could understand one another; later the vice-president of the academy, de Rjevski, asked that the minutes of the Archive-Conferences be kept in French (instead of German) so that he could read them. See Imperial Academy, *Materiali*, 2:776; Imperial Academy, *Protokoli*, vol. 3, April 12, 1773.

36. Vucinich, *Science in Russian Culture*, p. 73; Youschkevitch and Winter, *Briefwechsel*, 2:1–3. Catherine's rules are reprinted in Imperial Academy, *Materiali*, 2:297–324.

37. This is the view presented by G. F. Müller in his *Nachrichten zur Geschichte der Akademie der Wissenschaften* (1765) which forms volume six of the *Materiali* of the Imperial Academy; see esp. pp. 16–23 and 73–79 for this point. His view of the initial autonomy of the academy may be somewhat exaggerated in that Müller was writing at the time of another proposed reform of the academy which also stressed greater scientific autonomy for the institution.

38. Youschkevitch and Winter, *Briefwechsel*, 2:13–15; Vucinich, *Science in Russian Culture*, pp. 77 and 92–94; Vucinich, Review of *Istoria Akademii*, p. 151; Pasquier, *Euler*, pp. 25–35.

39. Müller, *Nachrichten*, in Imperial Academy, *Materiali*, 6:269–96; Youschkevitch and Winter, *Briefwechsel*, 1:2–3; Gnuchev, *Materiali gla Istorii Ekspeditsii*, gives details for some fifty scientific expeditions carried out by the Saint-Petersburg Academy in the 18th century; some non-Russian materials relating to these expeditions are reprinted in the appendix. See also Vucinich, *Science in Russian Culture*, p. 99; Fersman, *Pacific*; Golder, *Russian Expansion;* see also Imperial Academy, *Protokoli*, vol. 2, passim.

40. Youschkevitch and Winter, *Briefwechsel*, 1:7–8; 2:26–27; Vucinich, *Science in Russian Culture*, p. 90; Lipski, "Foundation," p. 353; Imperial Academy, *Materiali*, 3:228–30. See also Andreef, *Geografischeskii Department*.

41. Vucinich, *Science in Russian Culture*, pp. 78–80.

42. Menshutkin, *Russia's Lomonosov*, pp. 25ff.; see also Imperial Academy, *Protokoli*, passim.

43. See the various "Bitteschriften" presented to the imperial crown on behalf of the academicians; Imperial Academy, *Materiali*, 5:927–29; 6:528–32, 574, 576; 7:540–46. These petitions emphasize the need for greater control for scientific academicians in running their own affairs and a wish for higher, better recognized status like members of foreign academies. The first petition cited above concludes to the Czarina Elizabeth, "Her Majesty can see that it is of small honor to the Empire and little use for us to belong to this Academy, especially compared to foreign countries where academies hold great honor

. . . Signed: The Professors and Adjuncts of the Academy." Imperial Academy, *Materiali*, 5:929.

44. Youschkevitch and Winter, *Briefwechsel*, 1:4–5; 2:27.

45. Vucinich, *Science in Russian Culture*, p. 87. This charter is reprinted in Imperial Academy, *Materiali*, 8:52–55.

46. Vucinich, *Science in Russian Culture*, pp. 90–91.

47. See Dvoichenko-Markov, "The American Philosophical Society," p. 552; Youschkevitch and Winter, *Briefwechsel*, 2:41.

48. See Kopelevich, "The Petersburg Astronomy Contest"; Youschkevitch and Winter, *Briefwechsel*, 1:11.

49. Lindroth, *Swedish Men*, p. 13; Frängsmyr, "Swedish Science," pp. 29–31, 37. Bouillier, *Institut*, pp. 279–83, discusses the rather primitive consultations between Descartes and Christina over establishing an academy in Stockholm. As indicated by citations immediately following, the literature concerning 18th-century Swedish science and scientific societies is extensive. A major source, Sten Lindroth's *Kungl. Svenska Vetenskapsakademiens Historia: 1739–1818*, unfortunately is entirely in Swedish. Sweden had other learned societies, in addition to the science societies discussed here; for an overview of them see Almhult, "Academies in Sweden," pp. 305–13.

50. Fries, *Origin and Foundation*, p. 4.

51. Rydberg, *Svenska Studieresor*, pp. 411–12.

52. Rydberg, *Studieresor*, pp. 411, 414; Liljencrantz, "Christopher Polhem"; Bring, "Polhem," pp. 26–30.

53. Von Sydow, "The Society of Sciences."

54. Lindroth, *Swedish Men*, p. 14; Bring, "Polhem," pp. 52–72; von Sydow, "The Society of Sciences"; Liljencrantz, "Polhem"; Frängsmyr, "Swedish Science," pp. 33ff.; Hildebrand, *Förhistoria*, French summary, p. 805.

55. Bring, "Polhem," p. 65; see also von Sydow, "The Society of Sciences"; Liljencrantz, "Polhem."

56. Lindroth, *Swedish Men*, pp. 14–16. A copy and a translation of the charter of the *Societas* are found in RS, "Classified Papers," vol. 17, #46, #47, #48. See also Bring, "Polhem," p. 70.

57. Lindroth, *Swedish Men*, p. 14; see also Hildebrand, *Förhistoria*, p. 805.

58. Rydberg, *Studieresor*, p. 416; Fries, *Origin*, p. 5; Hildebrand, *Förhistoria*, p. 806.

59. Fries, *Origin*, pp. 5–6.

60. Nordenmark, *Anders Celsius*, French resumé, p. 224; Nordenmark, *Pehr Wilhelm Wargentin*, French resumé, p. 311; see also Rydberg, *Studieresor*, pp. 416–17; Fries, *Origin*, p. 5.

61. Nordenmark, *Anders Celsius*, pp. 245ff.

62. Nordenmark, *Anders Celsius*, p. 247; Lindroth, *Swedish Men*, pp. 68–69.

63. Hildebrand, *Förhistoria*, p. 806.

64. Fries, *Origin*, p. 5.

65. Fries, *Origin*, p. 5; Hildebrand, *Förhistoria*, p. 806; see also Dahlgren, *Personförteckningar*, p. 210.

66. Lindroth, *Swedish Men*, pp. 16–18.

67. Lindroth, *Vetenskapsakademie Historia*, pp. 1–9; Holmberg, *Notes*, p. 5; Ramsbottom, "The Royal Swedish Academy of Science"; Hildebrand, *Förhistoria*, p. 807.

68. Hildebrand, *Förhistoria*, p. 804; Rydberg, *Studieresor*, p. 416; Lindroth, *Swedish Men*, p. 19.

69. Hildebrand, *Förhistoria*, p. 807; Ramsbottom, "Swedish Academy," p. 270.

70. Lindroth, *Vetenskapsakademiens Historia*, pp. 12–37; these details are also presented in Ramsbottom, "Swedish Academy"; Holmberg, *Notes;* and Dahlgren, *Personförteckningar.*

71. Lindroth, *Vetenskapsakademiens Historia*, pp. 41–58; Kungl. Svenska Vetenskapsakademie, *Bicentenaire*, p. 22; see also the *Protokols* (minutes) of the academy in typed MSS in the archives of the Swedish Academy of Sciences.

72. Lindroth, *Vetenskapsakademiens Historia*, pp. 136, 157, 823–24; Dahlgren, *Personförteckningar*, pp. 1–4.

73. Nordenmark, *Wargentin*, p. 331; Holmberg, *Notes*, p. 16.

74. For extensive discussion of this aspect of the academy's work, see Lindroth, *Vetenskapsakademiens Historia*, pp. 217–376; see also Hildebrand, *Förhistoria*, p. 807.

75. From 1775 the academy published a separate set of economic memoirs; see Lindroth, *Vetenskapsakademiens Historia*, pp. 130–31, 226–29; Holmberg, *Notes*, p. 9.

76. Lindroth, *Vetenskapsakademiens Historia*, pp. 141ff.; see also Nordin-Pettersson, "Prisfrågor."

77. Hildebrand, *Förhistoria*, p. 807; Fries, *Origin*, p. 7; Lindroth, *Swedish Men*, p. 19. See also Lindroth, *Vetenskapsakademiens Historia*, pp. 378ff., for a survey of the scientific work of the academy. See also the related Lönnberg, *Naturhistoriska Riksmusetts.*

78. The *Handlingar* proved to be a barrier between these two faces of the academy. Published in Swedish, they kept the economic side of the academy confined to Sweden. This situation of course was a problem for those outside of Sweden who wished to know more of the academy's scientific work. Late in the 18th century the complaint was heard from Montpellier in France: "C'est avec un regret infini que je vois que ces Mémoires seront de peu d'utilité pour moi, puisqu'ils sont écrits dans une langue que très peu de personnes entendent dans ce pais. Il est surprenant que l'Academie n'ait pas cherché à rendre plus utiles les travaux de tout les sçavans distingueés, en les donnant dans une langue plus généralement entendue. C'est assurément une perte pour les sciences & je fais des vœux pour que ces Mémoires puissent être mis à la portée des étrangers." Letter, Jean Gaussin to J. C. Wilckes, dated Montpellier, March 17, 1791, KVA-Wilckes. Two translations of the *Handlingar* were made in Europe; one in Germany by Kaestner; the other in Holland; see Lindroth, *Vetenskapsakademiens Historia*, pp. 208–14.

79. Lindroth, *Vetenskapsakademiens Historia*, pp. 823ff.; Nordenmark, *Wargentin*, pp. 314–15; Dahlgren, *Personförteckningar*, p. 4; Holmberg, *Notes*, p. 15; Ramsbottom, "Swedish Academy," p. 270.

80. Lindroth, *Vetenskapsakademiens Historia*, pp. 169ff.; Dahlgren, *Personförteckningar*, pp. 315–16.

81. Lindroth, *Vetenskapsakademiens Historia*, pp. 48–53, 378ff.; Nordenmark, *Wargentin*, pp. 313–16; Lindroth, *Swedish Men*, p. 107.

82. Lindroth, *Vetenskapsakademiens Historia*, pp. 181–83ff.; below, chapter 5.

83. The preeminent and inclusive source, from which the bulk of this interpretation is drawn, is Roche, *Siècle*. See also Mornet, *Les Origines intellectuels*, and Roddier, "Pour une histoire."

84. See Roche, *Siècle*, vol. 1, ch. 1, esp. pp. 31ff. and graphique 1. See also Desplat, *Pau*, p. 1. The various royal almanacs published by the government throughout the 18th century likewise provide important information on provincial academies (dates of foundation, organization, prizes, etc.).

85. In the first sections of his work Roche sketches three main stages to the history of the learned society movement in provincial France during the 18th century: to 1715,

1715–1760, and 1760–1789. Our periodization (1700–1750) does not conform exactly, but essentially it is Roche's second stage that concerns us here.

86. See note 84. See also Fäy, "Learned Societies"; Barrière, *Académie de Bordeaux*, p. 16; Roche, *Siècle*, 1:30–31.

87. Roche, *Siècle*, 1:75ff., and vol. 2, tables 3 and 4.

88. Roche, *Siècle*, 1:32–33, 50–54; see also vol. 2, graphiques 13, 19–21; vol. 1, ch. 1, pp. 19–20, 25, 54. Roche's first stage academies were generally known simply as *académies royales* or *académies de belles-lettres;* third stage academies usually carried the name *académies des sciences, belles-lettres et arts.*

89. In addition to Roche, *Siècle*, 1:24–29, 45–48, see Barrière, *Académie de Bourdeaux,* pp. 3, 7, 19–20; Desplat, *Pau*, p. 13; Dumas, *Histoire de l'Académie*, 1:10; Castelnau, *Mémoire historique*, p. 18; Tisserand, *Académie de Dijon*, pp. 14–15.

90. Guillon, "Ancienne Académie," p. 399. On this point see also Roche, *Siècle*, 1:34–39; Barrière, *Académie de Bordeaux*, pp. 20, 25; Castelnau, *Mémoire historique*, pp. 23–24; Tisserand, *Académie de Dijon*, pp. 18ff.; Dumas, *Histoire de l'Académie*, 1:10–11. Patrons of academies are listed in the *almanach royal.*

91. See Roche, "Milieux académiques," p. 107; *Siècle*, 1:96ff.; entries in the *almanach royal* are one sure sign of this institutional integration.

92. See Roche, *Siècle*, 1:114–21. Some academies, e.g., the marine academy at Brest, were subsidized to a much greater extent.

93. See Roche, *Siècle*, 1:324ff.; see also Berthe, *Dubois de Fosseux*, p. 129 and Guillon, "Ancienne Académie," p. 398 for evidence of dues in the academies of Arras and Orléans.

94. Société royale des sciences de Montpellier, *Histoire et Mémoires* (1766), 1:xii. See also Heilbron, *Electricity*, p. 127, where the station of the provincial academician is summarized by the epitaph "Ci git qui ne fut rien pas même académicien."

95. This theme is discussed in virtually every history of provincial academies; see, for example, Kindleberger, "The Société royale," ch. 4. Roche, *Siècle*, 1:122, summarizes: "Overall, the notion of cultural service gradually imposed itself, and academies, originally private reunions, became official institutions supported by the authorities."

96. Remarks Roche, *Siècle*, 1:124: "A veritable symbiosis existed between local powers and the learned societies."

97. Pingaud, "Documents," p. 241.

98. Barrière, *Académie de Bordeaux*, pp. 96–109; Kindleberger, "The Société royale," pp. 174–86; Gallois, "L'Académie des sciences"; Roche, *Siècle*, 1:124, 382.

99. Roche, *Siècle*, 1:369ff.; and vol. 2, graphiques 19–21.

100. Roche, *Siècle*, 1:324–55; Barrière, *Académie de Bordeaux*, p. 121; Tisserand, *Académie de Dijon*, p. 581; see also Delandine, *Couronnes académiques.*

101. Barrière, *Académie de Bordeaux*, pp. 115–37.

102. This is an important theme in Roche; see his "Milieux académiques."

103. See here chapter 4 and Taton, *Enseignement.*

104. See Dulieu, "Jean Astruc"; Tisserand, *Académie de Dijon*, pp. 171–74.

105. On the Montpellier society consult Kindleberger, "The Société royale"; Castelnau, *Mémoire historique;* LeGrand, "Chemistry in a Provincial Context."

106. See Coury, *L'Enseignement de la médecine*, pp. 41–82; Delaunay, *La Vie médicale*, pp. 290–91.

107. See Kindleberger, "The Société royale," pp. 18, 99–106. The letters-patent of the Montpellier Society can be found in AD-H, D199, #1.

108. "Le Roy ordone que la Société de Montpellier soit regardée come une extension et une partie de l'Académie des Sciences, qu'elles entretiennens commerce sur les matières qui les occupens, et que les Académiciens de l'une ayens séance dans l'autre, chacun dans son ordre, & que la Société de Montpellier envoye tous les ans à l'Acadaémie celle de toutes les Pieces qui aurons été Lües dans ses Assemblées pendans l'année qu'elle jugera la plus digne et la plus convenable, et que cette Piece sera imprimée dans les Memoires que l'Académie donera pour la même année." PA-PV (1706), 25:90.

109. See PA-PV, (1743), 62:543ff.; Kindleberger, "The Société royale," pp. 240–43.

110. Kindleberger, "The Société royale," pp. 19–155; Castelnau, *Mémoire historique*, pp. 48–49. See also BM-M, MS. 52, #87, #93, #106, and #107.

111. On the Bordeaux Academy see primarily Barrière, *Académie de Bordeaux*.

112. Roche, *Siècle*, 2:154, n. 63.

113. Barrière, *Académie de Bordeaux*, pp. 28, 30, 113.

114. Barrière, *Académie de Bordeaux*, pp. 115–37; Vivie, "Les Lauréats de l'Académie."

115. On the Lyon Academy see primarily Dumas, *Histoire de l'Académie*, 2 vols; also Ollier, "Les Deux Premiers Siècles," and Trénard, "L'Académie de Lyon."

116. Roche, *Siècle*, 1:369; 2:154, n. 63. Classes in the Lyons Academy met separately.

117. On the Dijon Academy see primarily Tisserand, *Académie de Dijon*.

118. Notable in this regard is the failed effort pushed by Guyton de Morveau to publish the *Actes* of the Dijon Academy on a trimestrial basis after the example of the Swedish Academy's *Handlingar;* see Tisserand, *Académie de Dijon*, p. 520.

119. Roche, *Siècle*, 1:369; 2:154, n. 63, p. 156, n. 73; another source is Charliat, "Académie royale de marine."

120. Charliat, "Académie royale de marine," p. 76. The Brest Academy became formally connected to the Paris Academy in 1771.

121. On the Marseille Academy see Lautard, *Histoire de l'Académie de Marseille*, 3 vols; Dassy, *L'Académie de Marseille;* Académie des sciences, lettres et beaux-arts de Marseille, *Deux Siècles d'Histoire Académique;* Roche, *Siècle*, 1:364.

122. On the Bolognese academy see Richard Rosen, "The Academy of Sciences"; Medici, *Memorie storiche*.

123. Rosen, "The Academy of Sciences," pp. 25, 33.

124. Rosen, "The Academy of Sciences," p. 29.

125. Longardi and Galdi, *Le Accademie in Italia*, p. 91.

126. "La maison du Comte de Marsigli, qui pouroit bien devenir dans la Suite une veritable Académie. On y travaille toujours." AD-H, MS. D 207, #8. See also Kindleberger, "The Société royale," pp. 263–64.

127. On this point in addition to Rosen, "The Academy of Sciences," pp. 73ff., see Denina, "De l'influence qu'a eue l'Académie de Berlin," p. 563.

128. "L'Institut de cette ville fait un des plus beaux ornemens de la Ville & de la mémoire de feu Pape Benoit XIV, par les soins duquels cette Academie se trouve dans le Lustre qui la fait admirer. Il paroit qu'on aime ici les Sciences. Dumoins y sont elles protegés." Letter Callenberg to Samuel Formey, dated Bologna, November 21, 1764; DS-Formey, V, 3, (2).

129. Maylender, *Storia*, 1:456–61.

130. Weld, *History*, 1:422–24 and 470–71; H. J. J. Winter, "Scientific Notes"; Woodward, "Paper."

131. RS-JBC, 15:126; 16:136–37; 19:72, 351, 525, etc.

132. See Horn, *A Short History of the University of Edinburgh*.

133. Hoppen, *The Common Scientist*, p. 211; Weld, *History*, 1:383.

134. See Shapin, "Property, Patronage and the Politics of Science," pp. 6–11; Royal Society of Edinburgh, *Transactions* (1788), 1:4–6; Ed., Forbes, "Address," in Royal Society of Edinburgh, *General Index*, p. 18.

135. RS-JBC, 14:242, 391–92.

136. RS-JBC, 18:251ff.

137. See Royal Irish Academy, *Transactions* (1786), "Preface," 1:xiv.

138. RS-JBC, 17:381–83; RS-CM, 3:300; RS, "Letters and Papers," Decade 1, #71 and #72. The Royal Dublin Society, the first major "economic" society of the 18th century, was founded in 1731; see here, chapter 1 and Royal Dublin Society, *Bicentenary Souvenir*, pp. 5–13.

139. See Brooke Hindle, *Pursuit*, pp. 59–74, for these details.

140. Quoted in Hindle, *Pursuit*, p. 64.

4. The Scientific Society Movement: 1750 to 1793

1. See inter alia, Goubert, *The Ancien Regime*, p. xix; Cobban, *A History of Modern France*, pp. 85–88; Krauss, *Studien zur deutschen und französischen Aufklärung*, pp. 25–27. See also Behrens, *The Ancien Regime*, ch. 3; Anderson, *Europe in the Eighteenth Century*, pp. 304–13.

2. Académie royale de Berlin, *Mémoires* (1745), 1:2.

3. "Discours prononcé . . . par M^r De Ratte," Société royale de Montpellier, *Assemblées publiques* (1743–1751), vol. 2. See also AD-H, D191; Kindelberger, "The Société royale," p. 201. De Ratte refers to the Plantade discourse of 1706 wherein the latter remarked: "Most people treat the sciences as vain, abstract, useless, and it is only by grace that they call them curious. The difficulty there is learning them, the small number of people who excell at them and the little direct link they seem to have with more ordinary and easy knowledge seems to have given birth to and established a prejudice extremely disadvantageous to them. . . . After that, will one be surprised that, in order to cultivate so vast and necessary a science which encompasses all that concerns the preservation of humanity, one establishes new learned companies?" "Discours prononcé par . . . M. de Plantade," Société Royale de Montpellier, *Assemblées publiques* (1706–1737), vol. 1. For greater public awareness of academies and societies per se, see here, chapter 1.

4. RS, "Letters & Papers," Decade I, #225.

5. Quoted in Roche, *Siècle*, 1:16.

6. Bierens de Haan, *De Hollandsche Maatschappij*, p. 273. Original English. See similar remarks by R. J. Forbes, "The Hollandsche Maatschappij," p. 898.

7. *Collection Académique, composées des Mémoires, Actes, ou Journaux des plus célèbres Académies & Sociétés Littéraires étrangères, des Extraits des meilleurs ouvrages Périodiques, des Traités particulières, & Les pièces fugitives les plus rares, Partie Etrangère*, 13 vols. Paris and Dijon, 1755–1779. (Title varies.) On the *Collection Académique*, see Kronick, *History*, pp. 212–16; Brunet, "Geúneau de Monbeillard."

8. Kronick, *History*, pp. 212–13, tells of two versions of the *Collection Académique*: the one discussed here ("partie étrangère") and a sixteen-volume "partie française" (an abridgement of the *Mémoires* of the Paris Academy of Sciences) which appeared between 1754 and 1787.

9. That there was an *Encyclopédie* and Enlightenment cast and ideology behind this publication is evident in the imitative, sixty-page "Discours préliminaire," which introduces the project in volume 1 (Dijon: François Desventes, 1755).

10. *Collection Académique* (1755), vol. 1, "Avertissement."

11. On the Göttingen Society, see Smend, "Die Göttinger Gesellschaft"; Joachim, "Die Anfänge"; Sonntag, "Albrecht von Haller"; Toellner, "Entstehung und Program der Göttinger Gelehrten Gesellschaft"; Kraus, *Vernunft*, pp. 240–51.

12. The plans were by Schmauss and Münchausen; see Smend, "Die Göttinger Gesellschaft," p. vi; Joachim, "Die Anfänge," pp. 1–3.

13. Other plans were by Mosheim, Weder, Moser, Bünaus; see Joachim, "Die Anfänge," pp. 4–30. The creation of the *Göttinger gelehrten Zeitung* (from 1735 under various titles), which the Göttigen Society later took over, was part of an effort to generate a learned group around the university.

14. Joachim, "Die Anfänge," pp. 36–37; Smend, "Die Göttinger Gesellschaft," p. vii; Kraus, *Vernunft*, pp. 240–41.

15. Joachim, "Die Anfänge," p. 37 .

16. Kraus, *Vernunft*, p. 241; Smend, "Die Göttinger Gesellschaft," p. viii.

17. Kraus, *Vernunft*, pp. 10, 242.

18. Kraus, *Vernunft*, pp. 243–44, makes the point that historical studies were a long suit of the Göttingen Society, as of other German academies. See also Smend, "Die Göttinger Gesellschaft," p. ix.

19. On the Erfurt Academy, see Kraus, *Vernunft*, pp. 251–62; Hubrig, *Die patriotischen Gesellschaften*, pp. 40–45. See also Thiele, "Die Gründung der Akademie"; Oergel, "Die Akademie nützlicher Wissenschaften"; Hammermayer, "Akademiebewegung," pp. 5–6, 46; Voss, "Die Akademien," p. 52.

20. Hubrig, *Die patriotischen Gesellschaften*, p. 40. That is not to say that the Erfurt Academy did not also concern itself with traditional questions of science, as the following report in the registers of the Saint Petersburg Academy indicates: "The Academy of Sciences of Erfurt is ardently occupying itself with botany, chemistry, and metallurgy and desires a correspondence with the Academy of Petersburg." Imperial Academy, *Protokoli*, vol. 3, January 27, 1777.

21. Hubrig, *Die patriotischen Gesellschaften*, p. 40; Kraus, *Vernunft*, p. 10.

22. The main source on the Munich Academy is Ludwig Hammermayer, *Gründungs- und Frühgeschicte*. See also Kraus, *Vernunft*, pp. 10 and 261ff.; Westenrieder, *Geschichte der Baierischen Akademie der Wissenschaften*, esp. vol. 1, and Bayerische Akademie der Wissenschaften, "Aus der Chronik."

23. Hammermayer, *Gründungs- und Frühgeschichte*, pp. 16–26.

24. Quoted in Bayerische Akademie, "Aus der Chronik," p. 11.

25. See Hammermayer, "Süddeutsch-russische Wissenschaftsbeziehungen," pp. 503–28. Hammermayer, *Gründungs- und Frühgeschichte*, pp. 200–228 presents the full picture of Munich's foreign contact and a portrait of German academies.

26. See Böhm, "Die Kurpfalzische Akademie der Wissenschaften," pp. 117–22; Walter, *Geschichte Mannheims*, 1:597–624; Kister, *Die Pflege der Naturwissenschaften*. See also Kraus, *Vernunft*, pp. 272–87; Voss, "Die Akademien," p. 53; Cassidy, "Meteorology in Mannheim."

27. Böhm, "Die Kurpfalzische Akademie," p. 118.

28. On Lamey's extended correspondence, see Walter, *Geschichte Mannheims*, p. 608.

29. See note 26 and Kraus, *Vernunft*, p. 10.

30. On the Hesse-Hambourg Society see Hubrig, *Die patriotischen Gesellschaften*, pp. 59–63; Voss, "Die Societe [sic] Patriotique"; and Hammermayer, "Akademiebewegung," pp. 27–28; see also here, chapter 6.

31. See E. Forbes, *The Euler-Mayer Correspondence*, pp. 3, 23, 106, 113. See also evidence of this society in Andreef, *Geograficheskii Departament*, pp. 154–57 ("Réponse au demandes de la Société Cosmographique de Nuremberg"); RS-JBC, 20:450.

32. See Vávra, "Die Olmützer Societas," pp. 278ff.; see also Voss, "Die Akademien," p. 52; Hammermayer, "Akademiebewegung," p. 5; "Italienische Aufklärung," p. 256; *Gründungs- und Frühgeschichte*, pp. 10–12.

33. Becker, "Abriß einer Geschichte," pp. 11, 15; Kronick, *History*, p. 105; Harnack, *Geschichte* (1901), p. 300.

34. Schmieder, *Geschichte*, pp. 9–10; Karl Hufbauer, private communication.

35. Karl Gottlob von Anton (1751–1818) was the prime mover behind the Görlitz society; for a while, after 1803, it was the *Kurfürstliche Sächsische Gesellschaft*; see Jecht, "Kurzer Wegweiser," pp. 3–5, 20.

36. Hammermayer, "Akademiebewegung," pp. 5, 67; Voss, "Die Akademien," pp. 53–54; Staszewski, "Die ersten wissenschaftlichen Gesellschaften," p. 319; Axel von Harnack, "Die Akademien der Wissenschaften," p. 865.

37. On the Danzig society, see Hammermayer, "Akademiebewegung," pp. 5, 46; Staszewski, "Die ersten wissenschaftlichen Gesellschaften," esp. pp. 312–13; Lindroth, *Vetenskapsakademiens Historia*, 2:171; Schumann, "Geschichte der Naturforschenden Gesellschaft."

38. Hammermayer, "Akademiebewegung," p. 77, refers to a short-lived society in Bratislava (Pressburg) and suggests other nebulous associations in Central Europe. See also Staszewski, "Die ersten wissenschaftlichen Gesellschaften," p. 319. For indications of failed plans by the Zaluski brothers for a society in Warsaw dating from the 1740s, see Hammermayer, "Akademiebewegung," p. 46; "Italienische Aufklärung," p. 256. Feyl, *Beiträge zur Geschichte*, p. 227, refers to a plan of 1729 for a science academy in Constantinople!

39. On the Bohemian Society, see Kraus, *Vernunft*, pp. 295–316; Teich, "The Royal Bohemian Societies of Sciences," and "Bohemia," pp. 151–53; Zacek, "The Virtuosi of Bohemia"; Hammermayer, "Akademiebewegung," pp. 24, 70; Voss, "Die Akademien," p. 53. The main source is Joseph Kalousek, *Geschichte der kön. Böhmischen Gesellschaft der Wissenschaften*.

40. On Born, see Teich, "Ignaz von Born."

41. On the Zurich Society, see Hansen, "Scientific Fellowship"; Rübel, "Geschichte der naturforschenden Gesellschaft"; Luck, *Science in Switzerland*, p. 95. Luck mentions an interesting publishing organization, the *Societas Physica-mathematico . . . Helvetica*, which issued nine volumes of *Acta Helvetica* between 1751 and 1787; see also Kronick, *History*, p. 10. On the Lausanne society, see "Histoire," "Reglemens," and "Liste des Membres," in Société des Sciences Physiques, Lausanne, *Mémoires* (1786), 2:1–13. Although without a scientific society, Geneva was not without its local community of scientific practitioners; see Montandon, "Sciences et Société." For other Swiss societies of various sorts, see Hansen, "Scientific Fellowship," table 9.

42. On this point see Fueter, *Geschichte der exakten Wissenschaften*; Jaeggli, "Recueil des pièces," p. 409; Delorme, "Académie Royale des Sciences."

43. On the Holland Society and other Dutch societies, see Bierens de Haan, *De Hollandsche Maatschappij*; R. J. Forbes, *Martinus van Marum*, esp. 1:4–7 and 3:1–66; R. J. Forbes, "The Hollandsche Maatschappij"; Heilbron, *Electricity*, p. 128.

44. For a detailed look at the system of Dutch universities in the period, consult Frijhoff, *La Société Néerlandaise*.

45. Frijhoff, *Société Néerlandaise*, pp. 24, 226, 287; Bierens de Haan, *De Hollandsche Maatschappij*, p. 275; R. J. Forbes, *Van Marum*, 1:5.

46. See J. G. de Bruijn, "Teylers Tweede Genootschap," in R. J. Forbes, *Van Marum*, 3:22–32.

47. R. J. Forbes, *Van Marum*, 1:5; 3:3.

48. On the Rotterdam society, see Kuenen, *Gedenkboek;* Singels, *Geschiedenis*, p. 21; R. J. Forbes, *Vaan Marum*, 1:5; 3:4.

49. On the Utrecht society, see Singels, *Geschiedenis;* R. J. Forbes, *Van Marum*, 1:5; 3:3.

50. R. J. Forbes, *Van Marum*, 1:5–6.

51. Consult Bataviaasch Genootschap, *Gedenkboek;* R. J. Forbes, *Van Marum*, 1:5.

52. See Quetelet, "Premier Siècle"; Lavalleye, *L'Académie royale;* L'Académie Royale des Sciences, *L'Académie Royale de Belgique*, pp. 9–39.

53. The foundation of the Danish Academy is a good instance of continuity to the academic movement from the first to the second halves of the 18th century. On the Danish Academy, see Lomholt, *Det Kongelige Danske videnskabernes Selskab;* Lomholt, "Outline"; Kronick, *History*, pp. 84–85.

54. Lomholt, *Selskab*, 3:454ff.; vol. 4, passim; Lomholt, "Outline," pp. 26–27. Hans Aarsleff, private communication.

55. On the Göteborg society, see Beckman, "Göteborgs Kungl. Vetenskaps," and Nachmanson, "Forteckning." See further Almhult, "Academies in Sweden," for more on the Göteborg society and other nonscience academies and societies in Sweden.

56. Lindroth, *Vetenskapsakademiens Historia*, 1:815–22; Bratt, *En Krönika*. See also Hammermayer, "Akademiebewegung," p. 60.

57. Consult Gertz, *Kungl. Fysiografiska Sällskapet*.

58. See Midbøe, *Det Kongelige norske videnskabers selskaps historie*, passim; Lindroth, *Vetenskapsakademie Historia*, 1:92, 205–06.

59. See Maugain, *Etude*, pp. 68–77.

60. Maugain, *Etude*, pp. 80–85.

61. Maugain, *Etude*, pp. 77–80; Pace, *Benjamin Franklin*, pp. 21, 51.

62. Pace, *Benjamin Frankliln*, pp. 51–56; Maugain, *Etude*, pp. 153–74; Académie royale des sciences de Turin, "Mémoire historique," p. ii.

63. See Pace, *Benjamin Franklin*, pp. 87–88; Longardi and Galdi, *Le Accademie in Italia*, p. 18; Accademia delle scienze, *Primo secolo*, p. 3.

64. Accademia delle scienze, *Primo secolo*, p. 3; Académie royale des sciences de Turin, "Mémoire historique," p. ii.

65. Accademia delle scienze, *Primo secolo*, p. 3; RS-JBC, 23:625; Imperial Academy, *Protokoli*, vol. 2, August 11, 1759.

66. Accademia delle scienze, *Primo secolo*, p. 3; Académie royale des sciences de Turin, "Mémoire historique," p. iii; Pace, *Benjamin Franklin*, pp. 87–88; Denina, "De l'influence qu'a eue l'Académie de Berlin," p. 565.

67. Reprinted in Accademia delle scienze, *Primo secolo*, pp. 7–14.

68. Lagrange et al., "Projet," in Accademia delle scienze, *Primo secolo*, pp. 7–9.

69. See Lagrange et al., "Projet," in Accademia delle scienze, *Primo secolo*, pp. 9–12.

70. RS, "Letters & Papers," Decade IV, #55. Contemporary, Royal Society translation.

71. RS, "Letters & Papers," Decade IV, #188. Contemporary, Royal Society translation.

72. "J'ai été assez bien reçu ici du Roi et des ministres; on m'a donné de belles es-perances; mais je n'y fais pas grand fonds. Vos lettres ont fait beaucoup d'impression à la Cour et à la Ville." Letter: Largrange to d'Alembert, dated Turin, May 30, 1764; BI, Papiers de Condorcet, MS. 876, #52.

73. "Notre Société se prépare à faire imprimer un nouveau Volume; voudriez-vous lui faire l'honneur de décorer cet Ouvrage de votre nom? Cella ferait assurément ici un grand effet, et je ne doute pas qu'il hâta beaucoup son etablissement." Letter: Lagrange to d'Alembert, dated Turin, January 26, 1765; BI, Papiers de Condorcet, MS. 876, #55.

74. "Je n'entend plus parler de la Société de Turin, je crois que toutes les belles esperances qu'on avoit s'en sont allées en fumée." Letter: Lagrange to Condorcet, dated Berlin, September 4, 1775; BI, Papiers de Condorcet, MS. 876, #15.

75. Lubimenko, Correspondance scientifique, appendix, letter #13.

76. "On m'a mandé de Turin qu'on n'y parle plus du date de l'etablissement de la Société; peut-etre a cause que le Roi a des affaires plus importantes dans la tête; ou peut etre aussi parceque depuis la ratraite du Comte Salucce on en aura abbondonné le projet." Letter: Lagrange to Condorcet, dated Berlin, September 13, 1777: BI, Papiers de Con-dorcet, MS. 876, #23.

77. Reprinted in Accademia delle scienze, Primo secolo, pp. 14–19; see also Académie royale des sciences de Turin, "Mémoire historique," p. vi. De Saluces wrote to Condorcet informing him of his selection as a foreign academician of the Turin Academy and of Lagrange's exceptional status as honorary president of the Turinese academy; see PA, Dossier Condorcet, letter dated Turin, July 30, 1783.

78. Accademia delle scienze, Primo secolo, p. 4.

79. Accademia delle scienze, Primo secolo, p. 4; Imperial Academy, Protokoli, vol. 2, May 5, 1767; PA-PV, (1773), 92:144; Kronick, History, p. 215.

80. See Accademia delle scienze, Primo secolo, pp. 14–19, and Académie royale des sciences de Turin, "Mémoire historique," pp. xii–xxii. Resident academicians were re-quired to produce one paper a year; the academy met monthly, held annual public meet-ings, and was also endowed with a cabinet, apparatus, and a library.

81. Longardi and Galdi, Le Accademie in Italia, p. 78; Accademia delle scienze, Primo secolo, p. 4.

82. Accademia delle scienze, Primo secolo, p. 5 and analytical table of the Turinese Mémoires, pp. 245–74.

83. See Académie royale des sciences de Turin, "Mémoire historique," p. xxii and Longardi and Galdi, Le Accademie in Italia, p. 78.

84. See Accademia delle scienze, Primo secolo, p. 4; Berlin academy, Mémoires 1786–1787 (1792), p. 60; AdW, "Registres 1786–1800," p. 20a.

85. Internal work of the academy centered around agricultural improvements, engi-neering projects, and economic programs for the region; the academy also served as the scientific and technical advisor to the Sardinian crown on these matters; see Accademia delle scienze, Primo secolo, pp. 5–6; Académie royale des sciences de Turin, "Mémoire historique," pp. xiiff. and xxiiff.; see also Longardi and Galdi, Le Accademie in Italia, pp. 78–90.

86. On the Georgofili see Maylender, Storia, 1:456–61; Hubrig, Die patriotischen Gesellschaften, p. 34; Cochrane, Tradition and Enlightenment, pp. 36ff.; Giuliani, "I Georgo-fili." The remarks quoted earlier (chapter 1) by Grandjean de Fouchy indicate something of the international stature of the Georgofili.

87. See Maylender, Storia, 1:96–103; Imperiale Reale Accademia de Scienze, Lettere ed Arti degli Agiate, Memorie pubblicate per commenorare, pp. 23–27.

88. See Maylender, *Storia*, 3:363–66; 5:321,469–75.

89. Maylender, *Storia*, 3:20–26.

90. Maylender, *Storia*, 5:128–30.

91. Maylender, *Storia*, 4:440–45; Moschetti, "La R. Accademia di Scienze."

92. Maylender, *Storia*, 1:141–46; Accademia di Agricoltura, Scienze, Lettere di Verona, *Anton Maria Lorgna.*

93. See appendixes and the appropriate entries in Maylender, *Storia.*

94. See Maylender, *Storia*, 2:197–200. Note that Modena was the site of another late 18th-century science academy of the "Renaissance" type, the *Accademia Rangoniana.*

95. Roche, *Siècle*, 1:55–74.

96. Barrière, *Académie de Bordeaux*, p. 69; see also pp. 129–32. Increased utilitarianism on the part of French provincial academies is likewise noted by Tisserand, *Académie de Dijon*, pp. 267 and 591ff.; Cousin, "Besançon," pp. 335, 344; Desplat, *Pau*, 95–106.

97. The Lyon Academy was the sponsor of a very early Montgolfier balloon in 1783. The "Académie de Dijon" made its ascent in 1784; balloons were also launched by the academies in Marseille, Bordeaux, and Besançon. See Tisserand, *Académie de Dijon*, pp. 349–63 and 381; Dumas, *Histoire de l'Académie*, 1:170; Cousin "Besançon," p. 332; Gillispie, *Montgolfier*, pp. 74–75, 97; Bouillier, *Institut*, pp. 127–28.

98. Barrière, *Académie de Bordeaux*, p. 134.

99. Roche, *Siècle*, 2:131; see also Roche, *Siècle*, 1:324–55. Roche, *Siècle*, 1:329, 342–43, counts over two thousand competitions in France with several tens of thousands of livres handed out in prize money; questions concerning the sciences and the arts accounted for 60 percent of the competitions overall, belles lettres 30 percent, and history 10 percent. See also Delandine, *Couronnes académiques.*

100. Roche, *Siècle*, 1:115–16, provides the following indicative figures for the Dijon Academy: 1745, 1,000 livres; 1770, 2,000 livres; 1788, 9,000 livres.

101. Barrière, *Académie de Bordeaux*, p. 96.

102. Kindelberger, "The Société royale," pp. 251–52; Castelnau, *Mémoire historique*, p. 80.

103. Tisserand, *Académie de Dijon*, pp. 518–32. These were modeled after the *Histoire et Mémoires* of the Paris Academy, although in 1781 Guyton de Morveau made the very interesting proposal to publish a quarterly review modeled on the *Handlingar* of the Swedish Academy of Sciences. One issue of the *Actes de l'Académie de Dijon* did appear in 1782, but because of problems of subscription and state control over the press, the series was suppressed shortly thereafter; see Tisserland, *Académie de Dijon*, p. 520.

104. Kindelberger, "The Société royale," pp. 202–204. AD-H, D121 (Registres), March 15, 1764. The royal letters-patent confirming this arrangement explain part of the reason: "Nous avons jugé qu'il seroit plus utile au bien public et à l'avancement des études d'en confier l'exercice à un corps Littéraire permanent, que de la faire remplir par une seule personne."

105. A royally supported school of surgery opened in Besançon in 1773 with its professors in close association with the academy there; see Cousin, "Besançon," p. 340; There were also royal chairs of mathematics and hydrography in Brest and Metz; professors of the University of Nancy had special positions within the academy of Nancy; see Roche, *Siècle*, 1:94. The Paris Academy of Sciences itself controlled the appointment of a professorship of mathematics at Rheims; see Hahn, *Anatomy*, p. 73.

106. Barrière, *Académie de Bordeaux*, pp. 97–98; Roche, *Siècle*, 1:124–31.

107. Tisserand, *Académie de Dijon*, pp. 535–94; Roche, *Siècle*, 1:129.

108. Tisserand, *Académie de Dijon*, p. 625.

109. See Kindelberger, "The Société royale," pp. 202–16; Castelnau, *Mémoire historique*, pp. 88–91. See also AD-H, D121 (Registres), March 21, 1768; D191 ("Projet pour l'Etablissement d'une Chaire de phisique [sic] Expérimentale à Montpellier"); D123 (Registres), July 31, 1780. Chaptal taught a course in chemistry; Bertholon a course in *physique expérimentale*.

110. "Ce jour la compagnie extraordinairement assemblée, Mr de Causan a exposé que les secours nécessaires pour instruire dans la physique experimentale, qui fait aujourd'hui une partie essentielle de l'éducation publique, manquent dans cette province et en particulier dans cette ville, quoique cette science soit un des principaux objets des recherches de la Société Royale, et qu'elle soit de la plus utilité aux Etudians en Medecine et en Chirurgie." AD-H, D121 (Registres), March 21, 1768.

111. Justin, *Les Sociétés royales d'agriculture*, pp. 52–130; Tisserand, *Académie de Dijon*, pp. 330–32; Roche, *Siècle*, 1:63, reports that of twenty cities with agricultural societies only seven also had academies. "However," he continues, "very often it was academicians who played a decisive role in the new institution."

112. On Masonic lodges, see Roche *Siècle*, 1:257–80; compare Fày, "Learned Societies."

113. Roche notes that 6,400 men belonged to French academies in the 18th century while some 20,000 individuals were associated with Masonic lodges. Crucially, only 5 percent of the members of ordinary academies between 1770 and 1790 were also Masons. Also, recruitment to Masonic lodges was decidedly more open to members of the Third Estate in France than was the case in academies; see Roche, *Siècle*, 1:262ff.; vol. 2, table 46.

114. See Berthelot, "Notice sur les origines," and his "Sur les publications de la Société philomatique"; Dejob, "De l'Etablissement connu sous le nom de Lycée"; Smeaton, "The Early Years of the Lycée and the Lycée des Arts"; appropriate sections in Taton's *Enseignement*; Roche, *Siècle*, 1:66–68.

115. Roche, *Siècle*, 1:68.

116. Ribeiro, *Historia dos estabelecimentos scientificos*, 2:37–61; Hammermayer, "Akademiebewegung," pp. 24, 71; Ferrâo, *Academia das Sciências*, pp. 32–42.

117. See Gillispie, *Science and Polity*, pp. 315–16; McClellan, "Un Manuscrit inédit," pp. 243–44.

118. See McClellan, "Un Manuscrit inédit," pp. 243–44; Aguilar-Pinal, *La Real Academia Sevillana*, p. 283.

119. For more on Spanish academies and scientific activity in the period, see Arias-Divito, *Las Expediciones Cientificas*, passim and pp. 16, 19; Wilson, "Scientists in New Spain," pp. 24–44, esp. 26; Rickett, "The Royal Botanical Expedition."

120. On the Barcelona Academy, see Murua y Valerdi, "Historia de la Real Academia de Ciencias y Artes."

121. See here, this chapter and chapter 2.

122. Richard Meister, *Geschichte der Akademie*, pp. 12–17. For more on Leibniz' plans, see Schneiders, "Gottesreich und gelehrte Gesellschaft"; on Hell, see Sarton, "Vindication of Father Hell," pp. 98–99.

123. Leibniz' plan failed because he was a Protestant and because war intervened as a factor; Gottsched's because he was a Protestant, too, and because his plan would have been too costly; von Petrasch's failed because, while he proposed an entirely Catholic membership, there were not enough Austrian Catholics qualified in the sciences to staff the proposed academy; finally, the Hess-Hell plan fell through because money continued to be a problem and, paradoxically, with the Jesuits officially expelled from Austria, the

ex-Jesuits Hess and Hell could not join their own proposed institution; see Meister, *Geschichte der Akademie*, pp. 12–17.

124. Meister, *Geschichte der Akademie*, p. 16.

125. Hindle, *Pursuit*, pp. 105–19. To be consulted is Oleson and Brown, *The Pursuit of Knowledge*; although the center of gravity of their book lies in the 19th century, it is informative on the early history of societies in America; see especially articles by Greene, "Science, Learning, and Utility"; Whitehill, "Early Learned Societies"; Rossiter, "Organization of Agricultural Improvement."

126. Hindle, *Pursuit*, pp. 120–21; Stearns, *Science*, pp. 662, 671; Bates, *Scientific Societies*, p. 14.

127. Hindle, *Pursuit*, p. 105.

128. Hindle, *Pursuit*, pp. 127–34; Stearns, *Science*, pp. 670–74; Bates, *Scientific Societies*, pp. 6–9. See also Ponceau, *An Historical Account* (1840); for a repudiated account of the origins of the American Philosophical Society, see Conklin, "Brief History," pp. 7–9.

129. There was no council per se, the business of the society being handled in ordinary meetings and by standing committees; see American Philosophical Society, "Early Proceedings."

130. In this area, too, the American Philosophical Society followed the tradition established by the Royal Society of London. In seeking to alleviate its financial strain, the society instituted a fine for nonattendance, hired a bill collector, and eventually brought suit against nonpaying members; see American Philosophical Society, "Early Proceeedings," p. 71 (February 14, 1772), p. 101 (April 2, 1779), p. 121 (January 16, 1784), pp. 155–56 (November 2, 1787), and p. 175 (September 18, 1789); see also Hindle, *Pursuit*, p. 140.

131. See, for example, American Philosophical Society, "Early Proceedings," p. 18 (September 20, 1768), p. 30 (February 9, 1769), p. 48 (February 2, 1770), p. 67 (November 4, 1771). See also Hindle, *Pursuit*, pp. 134, 140, 267ff.

132. Hindle, *Pursuit*, p. 140; American Philosophical Society, "Early Proceedings," pp. 34–35 (April 21, 1769), p. 45 (December 1, 1769).

133. American Philosophical Society, "Early Proceedings," p. 135 (December 2, 1785) and following; Hindle, *Pursuit*, pp. 270–71.

134. More than fifteen or twenty persons made for a large meeting, and there were complaints about attendance; see American Philosophical Society, "Early Proceedings," p. 121 (January 16, 1784) and following.

135. Hindle, *Pursuit*, pp. 169, 219–32.

136. Hindle, *Pursuit*, pp. 194–221, 327, 354–66.

137. Says Hindle, *Pursuit*, p. 7, "For some time it would be accurate merely to say that in America 'arts and sciences' were just dawning." Bell, *Early American Science*, p. 36, puts 1820 as the date science achieved an independent status in the United States.

138. Lingelbach, "Franklin," p. 12; Hindle, *Pursuit*, p. 78. Franklin had won the Copley medal of the Royal Society of London in 1753 for his work on electricity. Extraordinarily, he was elected to both the Royal Society and the Paris Academy without the usual formalities; see RS-JBC, 22:630; PA-PV, (1772), 91:353. See also Institut de France, *Index biographique*.

139. Hindle, *Pursuit*, p. 146.

140. Woolf, *The Transits of Venus*, pp. 169–74; Hindle, *Pursuit*, pp. 146–65.

141. American Philosophical Society, "Early Proceedings," pp. 62–63 (February 22, 1771); Hindle, *Pursuit*, p. 143.

142. Lingelbach, "Franklin," pp. 13ff.; Hindle, *Pursuit*, p. 138. The following note was

attached to the volumes of Transactions sent out to other institutions: "The American Philosophical Society held at Phil^a humbly desirous to cooperate wth [the society's name] in their laudable endeavors for the Advancement of useful Knowledge requests ye learned and respectable Body to accept this Vol. as the first Fruits of their Labors in this new World." See American Philosophical Society, "Early Proceedings," p. 73 (May 15, 1772); also PA-PV, 91:353 (1772).

143. See American Philosophical Society, "Early Proceedings," pp. 85–86 (January 5, 1774) for letters from the academies in Turin and Bologna. Hindle, *Pursuit*, p. 143, describes the reactions from Sweden, England, France (in the *Journal des Savants*), Venice, Florence, Berlin, and Saint Petersburg. See also Hindle, *Pursuit*, p. 165.

144. The American Philosophical Society was in contact with the Turinese Academy of Sciences, the Bolognese Institute, the Bohemian Society, the Paris Academy, the Royal Society of London, the Lyons Academy, the Göttingen Society, and the Saint Petersburg Academy. See American Philosophical Society, "Early Proceedings," passim. See also Dvoichenko-Markov, "Benjamin Franklin," and "The American Philosophical Society." Some Europeans elected to the American Philosophical Society during the period include Condorcet, Daubenton, Macquer, Lavoisier, Priestley, Linnaeus, Pallas, and Banks. Bergius, the Swede, wrote of his election to Thomas Bond, "I am greatly obliged to you for making me known to your respectable Society, and to assure you, that the Society's electing me a foreign Member flatters me as much, if not more than my admittance into any of the European Societies." See KVA-Bergius, 15:604; original in English.

145. Hindle, *Pursuit*, p. 145.

146. On the Boston Academy, see Bates, *Scientific Societies*, pp. 9–11, and his *American Academy;* Stearns, *Science*, pp. 682–86; Hindle, *Pursuit*, pp. 263–70.

147. Hindle, *Pursuit*, p. 263.

148. Hindle, *Pursuit*, p. 266. See also PA-PV, (1786), 105:369; RS-JBC, 32:293.

149. Letter: Joseph Willard to John Ewing, November 20, 1780, APS. See also Hindle, *Pursuit*, p. 267.

150. Hindle, *Pursuit*, p. 275; Bates, *Scientific Societies*, pp. 14–15; McCormick, *Experiment in Independence*, p. 62; Gray, *History of Agriculture*, 2:783. As "improving societies" these societies may also have been patterned after the Royal Society of Arts in London.

151. On the Richmond academy, see Hindle, *Pursuit*, pp. 275–77; Bates, *Scientific Societies*, pp. 11–14.

152. See PA-PV, (1788) 107:64–67 (a favorable report on the project by Lalande, Lavoisier, Tenon, and Thouin appearing in the registers of the Paris Academy).

153. Hindle, *Pursuit*, p. 276.

154. On the *Cercle* see primarily Maurel, "Une Société de pensée"; see also Hindle, *Pursuit*, p. 277. Roche, *Siècle*, sets the comparison.

155. Maurel, pp. 252, 261; *Status de Cercle des Philadelphes* (Cap François: Imprimerie Royale, 1785); see also PA-PV (1789), 108:92. The noted colonial jurist Moreau de St-Méry was the key intermediary.

156. See Maurel, p. 252, which does not quite get the full story; see further PA-PV (1789), 108:54, 87–92; see also the very interesting "Discours prononcée à l'Académie Royale des Sciences. . .par Mr. Barré de Ste Venant, Président du Cercle des Philadelphes," in PA, Pochette de séance, February 25, 1789.

157. On the Philosophical Society of Edinburgh, see articles by Emerson.

158. See Shapin, "Property, Patronage and the Politics of Science," pp. 16–24, and his "The Royal Society." For more on the Royal Society of Edinburgh, see Kendall, "Royal Society"; Campbell and Smellie, *The Royal Society;* Royal Society of Edinburgh, *Transactions* (1788), 1:5–13; Ed. Forbes, "Address," in Royal Society of Edinburgh, *General Index.* See also Dugald Steward, *William Robertson.*

159. Kendall, "Royal Society," p. 2. The rather off-beat study by Emerson, "The Edinburgh Society for the Importation of Foreign Seeds," sheds an interesting light on the phenomenon of learned societies at this time and a peculiar Scottish vantage point vis-à-vis the rest of the world.

160. The fact that the society was made up of classes gave the impression that it was modeled along the lines "of some foreign academies." See Forbes, "Address," p. 20.

161. Sir Walter Scott was an early member of the Scottish society and later was president for the twelve years between 1820 and 1832; the literary section was closed in 1827; see Forbes, "Address," p. 21.

162. See Royal Society of Edinburgh, *General Index.* Says Horn, *A Short History,* p. 71, "On the whole, Edinburgh's influence upon European scholarship and culture was exerted by the written rather than by the spoken word."

163. Royal Irish Academy, *Transactions 1* (1787), p. xiv.

164. Royal Irish Academy, *Transactions* (1787), 1:xiv–xv; Ed. Forbes, "Address," p. 16; Bonfield and Farrington, *The Royal Irish Academy.*

165. See Royal Irish Academy, *Index to Transactions* (1813), passim. The "Preface" to the first volume of Irish *Transactions* is curious in revealing the nationalistic and political overtones to the institution. The editors complain of a brain drain to England and violently attack the "vacant thoughts" and treasonable attitudes of some Irishmen; one guesses that the men involved in the academy were Protestants who wished to bring to Ireland the "civilizing" and economically uplifting benefits of a learned society; see *Transactions* (1787), 1:x–xi.

166. On these Lit. & Phil. societies, see Thackray, "Natural Knowledge in Cultural Context"; Fleure, "The Manchester Literary and Philosophical Society"; Sheenan, "The Manchester Literary and Philosophical Society"; C. L. Barnes, "The Manchester Literary & Philosophical Society"; Nicholson, "The Literary & Philosophical Society"; Robinson, "The Derby Philosophical Society"; Watson, *Newcastle-upon-Tyne.*

167. On this dissenting background and the rise of various dissenting "academies," see Parker, *Dissenting Academies;* McLachlan, *The Warrington Academy.*

168. See Shapin and Thackray, "Prosopography as a Research Tool," passim and p. 7; Shapin, "Property, Patronage and the Politics of Science," p. 4, where he remarks that the Edinburgh Society "had little to do with the industrializing context of 'Lit & Phil' science in the late eighteenth century."

169. For a sense of the scope of Lit. & Phil. societies in the 19th century, see Hume, *Learned Societies.* Only one Lit. & Phil. society received official incorporation, the Cambridge Literary and Philosophical Society, in 1832; see Hume, *Learned Societies,* p. 6; A. R. Hall, *The Cambridge Philosophical Society,* pp. 32–35.

170. See Schofield, *The Lunar Society;* Bolton, "The Lunar Society."

171. See Stimson, *Scientists and Amateurs,* pp. 116–17, 138–40.

172. Stimson, *Scientists and Amateurs,* pp. 217–18; Weld, *History,* 1:491; Allibone, *The Royal Society and Its Dining Clubs.*

173. See, for example, RS-CM, 3:172, 174; 4:103–04.

174. "La Société est un peu nombreuse, mais les associez de Vôtre merite auront toujours des prerogatives sur ceux, qui ne sont que *ad honores.*" Letter: G. M. Bosse to Samuel Formey, dated Wittenburg, July 26, 1750; DS-Formey, IV, 4, #5.

175. "Ce corps fameux est bien déchu de son ancienne gloire; aujourd'hui les savans réels dédaignent de s'y faire agréger." Letter: Jean Dechamps to Samuel Formey, dated London, July 1761; DS-Formey, IX, 23, #19.

176. "Les fonctions au reste ne sont point trop assujsttissantes, et ne prendront rien sur ma prâtique." Letter dated London, December 31, 1765. PA, Dossier Maty.

177. On Banks see Cameron, *Sir Joseph Banks;* Miller, "Sir Joseph Banks."

178. See Stimson, *Scientists and Amateurs,* ch. 11; Lyons, *The Royal Society,* chs. 6 and 7.

179. See, for example, RS-CM, 3:121–22. Nothing came of this or similar proposals for reform within the Royal Society.

180. See Weld, *History,* 2:151–70; Stimson, *Scientists and Amateurs,* pp. 186–90; Lyons, *The Royal Society,* pp. 198–200.

181. See RS-CM, 7:6,463.

182. Stimson, *Scientists and Amateurs,* pp. 176–77; RS-CM, 6:428.

183. Letter: Pringle to William Chambers, May 10, 1776; RS-CM, 6:303.

184. Weld, *History,* 2:134–37; RS-JBC, 30:120.

185. RS-JBC, 32:119–21.

5. The Communications Network of the Scientific Societies

1. See Stimson, *Scientists and Amateurs,* pp. 120–23, 125–26; Lyons, *The Royal Society,* pp. 120–22, 149. One suspects that the Newtonianism of the English (versus the Cartesianism of the Continent) as well as the conflict between Newton and Leibniz over the invention of the calculus (which turned out so badly for Leibniz in the judgment of a Royal Society committee) contributed to the seemingly alienated situation of the Royal Society; for more on these themes see A. R. Hall, *Philosophers at War,* and Richard Westfall, *Never at Rest,* ch. 14.

2. Imperial Academy, *Protokoli,* 1:5–7.

3. Valentin Boss, *Newton and Russia,* p. 96

4. Imperial Academy, *Protokoli,* 1:6–7; translated by Boss, *Newton and Russia,* p. 95; Boss (illustration 32) reproduces the original of this letter. See also report appearing in the minutes of the Royal Society, RS-JBC, 13:52–53, and Weld, *History* 1:437.

5. RS-JBC, 13:78.

6. RS-JBC, 13:82; see also JBC, 13:89.

7. RS-JBC, 13:377.

8. See RS-JBC, 13:361, 511; 14:235–36; 15, passim; 19:511; RS-LBC, 20:248–50, 252–55, 298–99; 24:16; other Letter Books, passim. See also Imperial Academy, *Protokoli,* vol. 1, passim.

9. Imperial Academy, *Protokoli,* 1:13–15, and September 17, 1737, October 12, 1739.

10. RS-LBC, 23:392–95, 422; Imperial Academy, *Protokoli,* vol. 1, April 21, 1735, February 11, and April 25, 1737.

11. PA-PV (1717), 36:318; (1721), 40:241–42; (1723), 42:135; (1728), 47:345.

12. Imperial Academy, *Protokoli,* 1:6.

13. Imperial Academy, *Protokoli,* 1:5–6.

14. Imperial Academy, *Protokoli*, vol. 1, January 31, 1735; April 14, 1735; January 14, 1737; February 14, 1737.

15. Imperial Academy, *Protokoli*, vol. 1, February 10, 1741.

16. The secretary of the Swedish Academy before Wargentin, Pehr Elvius, did attempt to establish contact with the Saint-Petersburg Academy through a traveling third party, a certain Chevalier Sagramoso, who wrote from Hamburg in July of 1748 that he had met with the director of the academy at Saint-Petersburg, Razumovskii, and that "Nous nous sommes convenus, que vous vous addresseriés à Lui même pour arranger toutes les disposistions necessaires à l'utilité reciproque des observations Astronomiques." Letter, Sagramoso to Elvius, KVA-Bergius, 8:426.

17. Imperial Academy, *Protokoli*, 1:7.

18. See Imperial Academy, *Protokoli*, vol. 1, July 27, 1731; June 2, 1735; February 10, 1737; October 31, 1737.

19. See Imperial Academy, *Protokoli*, vol. 1, October 7, 11, 28, 1734; February 11, 12, 13, 15, 16, 1735; April 14, 1735; June 17, 1737; January 20, 1738; June 8, 1738, July 1, 1743.

20. Imperial Academy, *Protokoli*, vol. 1, March 16, 1739.

21. Imperial Academy, *Protokoli*, vol. 1, June 2, 1735; July 7-8, 1735. The rather offbeat connection between these two academies began through the offices of a Dr. Sanchez, an Imperial physician, who may have been Portuguese.

22. Imperial Academy, *Protokoli*, vol. 1, July 14, 1735 (the *Commentarii* and nine other volumes sent); September 11 and 15, 1740.

23. For other communications back and forth, see Imperial Academy, *Protokoli*, vol. 1, December 20, 1736; October 17, 1737, July 7, 1738, August 22, 1739, and January 9, 1741. An interesting insight into the real conditions under which these exchanges took place can be seen in the fact that (in one instance at least) it took ten months for a letter to pass from Lisbon to Saint Petersburg. As a footnote to the international dimension of the Saint Petersburg Academy in the first half of the century, one might add that the academy was receiving information on the natural history of Pennsylvania, Carolina, and Virginia via correspondents in London. See Imperial Academy, *Protokoli*, vol. 1, October 12, 1739; December 8, 1740.

24. Weld, *History*, 1:428; see also RS-CM, 3:9-11.

25. See RS-LBC, passim. Such an ability to communicate better with the learned of the Continent must have helped overcome the isolation of the Royal Society. In a revealing communication of 1706 to the Montpellier *Société royale des sciences*, Luigi Marsigli of the Bolognese *Accademia degli Inquieti* wrote of England, "I do not presently see that there is anybody in [London] who is distinguishing himself by new works. It is true that the English have this custom of writing in their own language which is not very practical in foreign countries." Kindelberger, "The Société royale," pp. 263-64. See also AD-H, D207, #8. At the time, of course, Newton was president of the Royal Society, and his *Opticks* had just appeared in 1704.

26. See Stimson, *Scientists and Amateurs*, pp. 135-38; Lyons, *The Royal Society*, p. 149-53; Jacquot, "Sir Hans Sloane"; H. Brown, "Buffon and the Royal Society of London," pp. 143-44; "Voltaire and the Royal Society of London," p. 30; Andrade and Martin, "The Royal Society and its foreign relations [sic]," pp. 74-75.

27. Lyons, *The Royal Society*, pp. 151-52 and RS-CM, 3:2-7. See also certificates of election to the Royal Society reproduced in Brown, "Buffon," and "Voltaire."

28. "Et pour ce qui concerne M. Pitot, je suis si bien persuadé de ses Talens que je souhaitterois le voir Membre de Notre Corps. Mais il y a cette condition requise préable-

ment à l'election de tout Etranger, c'est qu'il lui faut etre recommandé par deux ou trois Membres de son voisinage, à moins qu'il n'ait communiqué des Observations à la Société. Ainsi, Monsieur, dès que vous trouverés à propos de m'envoyer un Certificat de ce scavant, signé de quelques uns de nos Membres à Paris, j'y concourerai avec plaisir tant par moi-meme, que par mes amis." Letter: Sloane to Duhamel, London, March 6, 1740, PA, dossier Sloane.

29. Foreign membership in the period 1700–25 averaged 55 men; for the period 1725–50 the average is 125. See Lyons, *The Royal Society*, appendix 3; see also Stimson, *Scientists and Amateurs*, p. 137.

30. RS-JBC, 12:464; 13:35–36, 223–24, 285–87, 458, 514. The last entry reads: "The President Communicated a Letter from Ericus Benzelius Junr. dated at Gottenburg June 6th: 1730, Wherein he Speaks of the Transactions of the Society at Upsal for the Year 1729 as nearly Compleated for publication. He Sends Inclosed the Articles of the Constitution of that Society." See also RS-LBC, 18:327–28, 395–96, 475; "Classified Papers," vol. 17, #46, #47, and #48.

31. See RS-JBC, 15:192; 16:323; 17:32–33; RS-LBC, 25:100. The Council Minutes of the Royal Society for June 21, 1742 (3:300) read: "It being represented by Dr. Mortimer that it might be of Service to promote correspondencies to Send Presents of the Philosophical Transactions to all Such persons who send their works of the like kind to the Society, it was ordered. . .That the like present be sent to Mr Celsius and to Mr Triewald at Upsal so long as they continue the Present of theirs."

32. See RS-LBC, 26:409; "Letters & Papers," Decade 1, #93 (1741); Decade 2, #159 (1750). See further, this chapter.

33. See RS-JBC, 12:268.

34. See RS-JBC, 12:308–10, 540–41; 16:158, etc.; RS-LBC, 19:299–300; 22:63–72; 23:41–42.

35. See RS-JBC, 13:373; 14:158, 243; 16:186; RS-LBC, 19:382; 24:145–46, 214–16; 26:34.

36. RS-JBC, 16:304.

37. RS-LBC, 24:397. On contemporary English connections to the land of Port, see Thomson's brief, "Anglo-Portuguese Scientific Relations."

38. Quoted from RS-JBC, 13:203. See also RS-JBC, 12:540; 13:34; RS-LBC, 18:122–24, 320–22; 19:298.

39. RS-JBC, 13:643ff.

40. RS-JBC, 12:413–15. See also notice "Invitatio ad Observationes meteorologicas communi consilo instituendas," *Philosophical Transactions*, (1723), 32:422–27; Lyons, *The Royal Society*, pp. 148–49. This was reprinted in the *Acta Eruditorum* from Leipzig, and it is noteworthy that in his French index to the *Phil. Trans. (Table des mémoires*, Paris: Piget, 1739, p. 29) F. de Brémond placed Jurin's announcement at the head of his section on meteorology. See also Middleton, *The Experimentors*, pp. 108, 255, for a report of a 17th-century meteorological project centered around the *Accademia del Cimento*.

41. A good indication of the importance of meteorology to the Royal Society is found in the analysis of papers appearing in the *Philosophical Transactions* of the society in the 18th century presented by Thomas Thompson. He finds that meteorology was the fourth most popular subject after medicine and surgery (478 papers), astronomy (416), and zoology (298); there were 281 meteorological reports, most "bare diaries." See Thompson, *History of the Royal Society*, pp. 458, 552.

42. See RS-JBC, 12:507 (sent "in the name of the Director of their Academy"); 13:34, 203. The Berlin *Societas* was already engaged in a similar meteorological project of its own.

43. See RS-JBC, 13:35–36, 78, 223; 14:2–3.

44. See RS-JBC, 12:540–41; 13:38, 46–47.

45. RS-JBC, 13:290; see also 12:623.

46. See Weld, *History*, 1:434. See also RS-CM, 3:24; RS-JBC, 14:327, where one reads, "a Thermometer of the Standard of those sent abroad to other places should be sent to M^r de l'Isle at Petersburg in order to compare his observations there with others made here in England."

47. RS-JBC, 14:15.

48. Jablonski of the Berlin Academy wrote to the Royal Society in 1726 that he was "disappointed in his hopes of Establishing the like Undertaking at Petersburg and Konigsburg and other places." RS-JBC, 13:34. In a way it is unfortunate that the Royal Society and the academy at Saint Petersburg were not in closer contact over this program of meteorological research because the Saint Petersburg Academy was regularly receiving meteorological reports sent in from many points of the Russian Empire; see Imperial Academy, *Protokoli*, passim.

49. For reports out of the Paris *Mémoires*, see RS-JBC, 12:178, 283, 288, 551–52; 13:220, '234, 273; for Paris prize questions, see 13:216–17, 358; 14:280; 15:172, 312.

50. See PA-PV, (1703), 22:47,267.

51. See RS-LBC, 18:349–51, 488–90; PA-PV, (1707), 26:155, 172, 182, etc.

52. For example, Bélidor asked of the Royal Society, "un brevet ou Certificat, comme m'en ont Envoié les academies roiales de France & de prusse par la Voie de leur Secretaires." Even Woolhouse asked the Society "to put the names of the other academies of which I am a certified member" on the next published list of F.R.S.; see RS-LBC, 18:350,400.

53. See Brown, "Buffon" and "Voltaire"; McKie, "Fontenelle et la Société Royale de Londres."

54. McKie, "Fontenelle," pp. 334–38.

55. RS-JBC, 14:243.

56. For Sloane communications, see PA-PV, (1709), 28:241, 403; (1722), 41:33; (1726), 45:1, 135; (1727), 46:349; (1737), 56:208; (1742), 61:2, 9–10; For Mortimer communications, see PA-PV, (1734), 53:272; (1735), 54:23–27; (1737), 56:93; (1745), 64:25–27. For Folkes, see PA-PV, (1743), 62:2, 18; (1746), 65:315–16.

57. "On a lû une Lettre de Mr Sloane à M^r L'Abbé Bignon, par la quelle il le remercie de l'avoir fait entrer dans L'Académie des Sciences, et promet différentes Observations ou curiositez." PA-PV, (1709), 28:241.

58. RS-LBC, 18:489.

59. See RS-JBC, 15:48–49, 90–92. Godin made this offer while attending the Royal Society, and "thanks were ordered for the great civility of these communications." Regarding a slightly abortive 17th-century precursor to this exchange, see A. R. and M. B. Hall, "Les Liens publics."

60. RS, "Letters & Papers," Decade 1, #128; see also #200.

61. RS-JBC, 15:179 and 248; see also RS-LBC, 22:92–94.

62. RS-JBC, 17:454–56; RS-CM, 3:298.

63. This first account is pieced together from the "Eloge de Brémond" by Dortous de Mairan in Paris Academy, *Mémoires* 1742 (1745), pp. 189–94; M. Demours, "Préface du

Traducteur," *Transactions philosophiques de la Société Royale de Londres, 1737, 1738* (Paris, 1759), pp. ix–xlvii; and Scudder, *Catalogue.*

64. Demours, "Préface," p. xxxiii. All Brémond's volumes were printed with the "approbation et privilège du Roi."

65. See letter of September 1, 1737, from Bignon to the Royal Society, where he says, "enfin nous avons engagé un jeune homme nommé M. de Bremond à mettre [vos Transactions philosophiques] en françois." RS-LBC, 24:4–5.

66. Paris Academy, *Mémoires* (1742/45), "Histoire" section, pp. 191–92. See original éloge in PA-PV, (1742), 61:477–82, here p. 479; Paul, *Science and Immortality*, pp. 114, 21.

67. See PA-PV, (1741), 60:123.

68. Demours, "Préface," p. xxxv.

69. See PA-PV, (1743), 62:209, where one reads, "Vers la fin de l'assemblée Mr. l'abbé de Gua qui s'etoit chargé avec Mr. Demours, Medecin, de continuer la Traduction des Transactions Philosophiques de la Société Royale de Londres, sous la même forme que ce que nous en avoit deja donné M. de Brémond, demande à la compagnie qu'elle veüille bien le dispenser de ce travail." See also Demours, "Préface," pp. xxxv–vi.

70. Reported in RS-JBC, 18:333.

71. See RS-LBC, 24:4–5; 25:252–53, 354. This last communication is from a correspondent in Brest, who writes Mortimer about this translation, "Le public les recoit avec beaucoup d'applaudissement." See also RS-JBC, 16:418.

72. See RS-JBC, 17:155.

73. See AD-H, D188 (Registres), February 8, 1742; Alain Brieux, *Histoire des Sciences: Livres anciens* [Book Catalogue, June 1983], #11896.

74. Demours took up the *Phil. Trans.* project again in 1749 and in 1759–60 he published several further volumes; see Demours, "Préface," p. xxxvii. These later developments are beyond our present focus. That the project continued to draw the attention of French governmental authorities, however, is evident from the following report of 1749 appearing in the records of the Royal Society: "Dr. Mortimer further acquainted the council that he had lately received a Letter from the Rev^d Mr. Stoeber acquainting him, that the Chancillor of France, having been a chief encourager of a Translation of the philosophical Transactions into French desir'd that the persons employed in that Translation might have leave to Copy such of the papers laid before the Society as were Originally in the French Language." RS-CM, 4:23.

It should be noted that in this period the Royal Society was also in one kind of contact or another with the French academies in Montpellier and Bordeaux (minimal), the Paris Surgeons Academy (regular), the *Academia Naturae Curiosorum Leopoldina* (receiving its *Acta*), the *Real Academia de Medicina* of Seville (passing), and a private "Society of Physicians" at Nuremberg, with which it maintained a regular "commercium litterarium." Jesuits in China were also in contact with the Royal Society. See RS-JBC and RS-LBC, passim; McClellan, "International Organization," pp. 177–79.

75. See Kindelberger, "The Société royale," pp. 240–45; McClellan, "International Organization" p. 190.

76. PA-PV, (1719), 38:247 (Bordeaux); (1737), 56:64 (Béziers); (1746), 65:1–2, 53, 56, 67, 140–45, 204, 245 (Toulouse).

77. Roche, *Siècle*, 1:19. Says Roche, "In this context, the birth of provincial societies always took place in a more or less clearly tense relation between the capital and the regions of France." See also Barrière, *Académie de Bordeaux*, p. 28.

78. See PA-PV, (1701), 20:27, 220; (1702), 21:207; (1713), 32:3.

79. For the period 1740 to 1748 see PA-PV, (1740), 59:120; (1741), 60:40, 110; (1742), 61:309; (1743), 62:259, 523; (1748), 67:339, 490.

80. See Imperial Academy, *Protokoli*, vol. 1, May 27, 1734; May 6, 1735; March 6, 1736; July 20, 1736; August 9, 1736; October 14, 1736; August 16, 1737, June 22, 1738; December 5, 1740.

81. For more on Delisle, see Chapin, "Delisle"; Boss, *Newton and Russia*, 133ff.; Nevskaja, "Delisle,"

82. Imperial Academy, *Materiali*. 1:126. The *need* for such a clause reflects the perils a man like Delisle faced in going to Russia and his caution.

83. See PA-PV, (1727), 46:350; (1729), 48:170; (1741), 61:311. RS-JBC, 13:234, 253, 377; 14:327; AD-H, D120, March 24, 1746; Imperial Academy, *Protokoli*, vol. 1, February 8, 1734; May 6, 1735; August 9, 1736; June 3, 1737; September 16, 1737; July 8, 1737; July 1, 1737; vol. 2, October 26, 1744.

84. Because of the sources, this survey of inter-institutional interactions is not complete, but it does cover the major academies. For indications of less significant activity, see McClellan, "International Organization," p. 202.

85. Youschkevitch and Winter, *Briefwechsel*, 1:2.

86. Imperial Academy, *Protokoli*, vol. 1, February 22, 1740; on Saint Petersburg pensions to foreign honoraries, see Lubimenko, *Correspondance scientifique*, pp. 38–39; Pavlova, "Lalande."

87. RS-JBC, 18:424.

88. The Royal Society did begin receiving prize circulars from the Berlin Academy from 1745; see RS-JBC, 12:413–15.

89. See Imperial Academy, *Protokoli*, vol. 1, June 3, 1737; July 3, 1737; May 29, 1740. The Paris prize questions were ordered published in the Saint Petersburg *Zeitung*.

90. See Imperial Academy, *Protokoli*, vol. 1, September 7, 1739.

91. "M. le Prince de Cantemir envoye à la Compagnie le 7e et 8e Tome des Mémoires de l'Académie de Petersbourg, de la part de cette Académie." "En conséquence de ce qui été remarqué dans l'assemblée précédente des mémoires envoyez par l'Académie de Petersbourg, la Compagnie a jugé à propos d'envoyer à cette Académie les quatre derniers volumes de nos Mémoires." PA-PV, (1742), 61:141, 150.

92. "J'ai lu une Lettre de Mr. *Mortimer* Secrétaire de la Société Royale de Londres, par laquelle il me mande que cette Compagnie propose d'envoyera à l'Académie les Transactions Philosophiques, et d'accepter en échange ses Mémoires: Cette proposition a été acceptée, et l'Académie m'a chargé de lui faire réponse en conséquence." PA-PV, (1749), 68:263.

93. PA, Dossier Mortimer, letters dated August 2, 1748 and November 12, 1748. Mortimer does, however, speak of his hopes of closer liason betwen the Paris and London institutions with the conclusion of the War of Austrian Succession.

94. RS, "Letters & Papers," Decade 2, #116 (contemporary translation); see also RS-JBC, 20:296–97.

95. RS-CM, 4:35; see also RS-JBC, 20:334. The Royal Society did have to pay duty on the case.

96. RS-CM, 4:35; see also RS-JBC, 20:343–45; PA-PV, (1750), 69:358, where one reads, "Mr. de Buffon a lu une Lettre de Monsr. Folkes Président de la Société royale de Londres, par laquelle, au nom de cette Compagnie, il remercie l'Académie de l'Exemplaire complet de ses Ouvrages qu'elle lui a envoyé."

97. See RS-CM, 4:93, 98, 106–07, 130; RS-JBC, 20:561; 21:70; PA-PV, (1753), 72:634.

Exchanges included the *Connaissance des Temps* as well as other productions of the Academy.

98. RS-CM, 4:35–36. It is to be noted that the *Transactions* from 1746 only were sent, the Swedish Academy "having the preceding Volumes." RS-JBC, 20:346. See also RS, "Letters & Papers," Decade 2, #159 (back of letter of July 5, 1750), which speaks of the initiative for an "exchange of the philosoph. Trans. as they come out for the Acta of the Royal Academy of sciences of Sweden" as having come from the Royal Society of London.

99. See Stimson, *Scientists and Amateurs*, p. 169; Lyons, *The Royal Society*, pp. 151, 179; RS-CM, 4:49–88; McClellan, "The Scientific Press," p. 429. This change in direct control held little import for the *Transactions* themselves and merely made official the de facto relations between the society and its journal from the 1660s.

100. RS-CM, 4:121. (Punctuation differs in original.) Note the present to the Hanoverian Göttingen Society of Sciences, which had been founded just the year before.

101. RS-CM, 4:93; see also pp. 132–33.

102. RS-CM, 4:161–62.

103. Nordenmark, *Wargentin*, p. 312.

104. RS-JBC, 20:453–56.

105. See PA-PV, (1748), 67:339 and (1750), 69:367 and 439. The latter two reports read: "Mr. de L'Isle a dit que l'Académie Suedoise de Stockholm desiroit être en correspondance avec la Compagnie; et que M. l'Envoyé de Suede l'avoit chargé de s'informer si le Présent qu'elle se proposoit de faire de ses Mémoires imprimés à l'Académie, lui seroit agréable: On a chargé Mr. de l'Isle de répondre que l'Académie les recevroit avec plaisir, et envoyeroit par la suite un Exemplaire de ses Volumes à celle de Suede." "Mr. de l'Isle a lu une Lettre de Monr. *Wargentin* de l'Académie de Suede, par laquelle il remercie au nom de cette Compagnie, de la Correspondance et des Volumes que l'Académie leur a accordé; sur quoi il a été déliberé qu'on envoyeroit tous les ans à MM de l'Académie de Suede la connoissance des Temps, le volume courant de l'Académie, et le volume courant des Sçavants Etrangers."

106. KVA-Bergius, 8:353–54 (an official communication to the Saint Petersburg Academy); 7:713–14 (letter from Zanotti to Wargentin, Bologna, May 1, 1752). See also Imperial Academy, *Protokoli*, vol. 2, March 15, 1751; May 24, 1751.

107. Quoted in Nordemark, *Wargentin*, pp. 315–16. Curiously, Wargentin's overtures were rebuffed by the Danish Academy of Sciences; see Lomholt, *Det Kongelige Danske Videnskabernes Selskab*, 2:173. The Danish Academy itself, however, was in contact with other societies, e.g., the Saint Petersburg Academy from 1752.

108. See RS-JBC, 18:424, 474; PA-PV, (1748), 67:361; Imperial Academy, *Protokoli*, vol. 2, February 13, 1749; July 5, 1751, etc.

109. AdW, Registers, 1746–1766, September 17, 1750, December 14, 1752, etc. See also DS-Formey, 31, 23 (correspondence with Strube de Piermont).

110. Nowhere in the records of the French, English or Russian societies is there any indication of reciprocal exchanges with the Berlin Academy; the same holds true for the Swedish Academy (Sten Lindroth, private communication and *Vetenskapsakademiens Historia*, 2:172). The Berlin scientific community was not cut off from the outside, however. Youschkevitch and Winter (*Briefwechsel*) have shown the close connection to the Saint Petersburg Academy, and Formey of the Berlin Academy was in contact with Sweden and Wargentin (although not that frequently); see DS-Formey, 6, 2 (Catteau) and 12, 12 (Ferner); KVA-Wargentin, Formey to Wargentin, February 14, 1756. The Berlin Academy itself, however, does seem cut off.

111. AD-H, D121 (Registers of the Montpellier Society), April 26, 1770; AD-CD, Registers of the Dijon Academy, 5, June 15, 1770.

112. "Mr. D'aigrefeuille auquel l'Académie a envoyé aussi un exemplaire de ses mémoires exprimoit la sienne dans les termes les plus forts et parlant du volume dont la Société Royale de Montpellier faisoit present à l'Académie il disoit 'la réciprocité du present qu'elle vous offre doit vous faire connoitre combien elle desire d'entretenir une correspondance suive qu'elle cultivera avec soin.['] Cette correspondance est egalement le vœu de l'Academie, qui sent tous les avantages qu'elle peut en retirer et il a été arrête que le Secretaire dans la Lettre qu'il ecriroit à Mrs de la Société Royale de Montpellier leur temoignueroit que nous n'epargnerons rien pour entretenir cette utile correspondance." AD-CD, Registers of the Dijon Academy, 5, June 15, 1770, pp. 195v–196.

113. Letter from the Turinese Society to the American Philosophical Society dated Turin, June 30, 1773; APS, "Early Minutes," p. 157. Much the same letter was sent again in 1787 after the Turinese Society was upgraded into an academy and began publishing its *Mémoires;* see APS Turin letter, shelf list 1787.

114. We find a convenient echo of this view in a committee report à propos of a proposed union between the Paris Academy and the Marine Academy at Brest: "Que quelques avantages qu'ait déja produit une correspondance entre quelques membres des deux compagnies, il y en auroit encore de plus grands a espérer, si cette correspondance pouvoit avoir lieu entre les deux Compagnies mêmes." See PA-PV, Pochette de séance, February 13, 1771; see also PA-PV, (1771), 90:52–54.

115. RS, "Letters & Papers," Decade 3, #157; PA-PV, (1767), 86:62, 134; Bierens de Haan, *De Hollandsche Maatschappij,* p. 275. The overture to the Royal Society of London merits transcription from the contemporary translation: "High Nobly-born Lord! We have been very agreably informed by our very worthy Member Mr. Allamand Professor at Leyden that your High-Praise-worthy Royal Society of Sciences in London, is disposed, as it does with others, to enter into a Correspondence with our Dutch Society establish'd at Harlem, so that we are to communicate with your High-Praise-worthy Society our Transactions whenever any part of them shall come out, while we likewise are to receive those publish'd by your Society. As Noble as the Views which your famous Society hath in such a Proposal, So certain is it, that the keeping up a good Correspondence between learned Societies, cannot but be of very great benefit to the Reign of Truth. It is therefore with the utmost Satisfaction that we enter into this Correspondence . . . [We] assure you, that we shall contribute all that lies in our power to the maintaining so excellent and useful a correspondence; and in general, that we shall, on all occasions, endeavor to shew, that we are at least Lovers of the useful Arts and Sciences."

116. "J'ai lû une Lettre de M. Desrocher, Sécretaire de l'Académie de Bruxelles, envoye à l'académie, le recueil de ses ouvrages: il été délibéré que l'académie lui enverroit les volumes courants." PA-PV, (1777), 96bis;397.

117. RS-CM, 6:73.

118. McClellan, "International Organization," appendix 3.

119. See Imperial Academy, *Protokoli,* vol. 3, March 14, 1774 and January 18, 1779, for lists of the Saint Petersburg Academy; see PA, 1784 Dossier, "État de distribution du Volume de 1781 des mémoires de l'Académie des Sciences," for the same for the Paris Academy. The 1779 distribution list of the Saint Petersburg Academy, for example, included foreign and domestic members of the academy, various Russian and European royalty, and "the academies of Paris, Brussels, Berlin, Stockholm, Copenhagen; the so-

cieties of science of London and Göttingen, the Imperial University of Moscow, and the municipal library of Riga."

120. Compare Kronick, *History*, p. 216; "Scientific Journal Publication," pp. 42–44.

121. On this prize see Maindron, *Fondations*, pp. 32–33.

122. Special fliers were sent to the Berlin and Saint Petersburg academies and the Royal Society of London. See Imperial Academy, *Protokoli*, vol. 3, October 16, 1775; AdW, "Registres 1766–1786," p. 181 (October 26, 1775); RS-JBC, 28:349 (January 18, 1776). The Swedish Academy of Sciences was also contacted; KVA-Wargentin, de Fouchy to Wargentin, August 8, 1775; quite conceivably other institutions were reached also.

123. RS-JBC, 28:349.

124. Imperial Academy, *Protokoli*, vol. 3, September 3, 1784.

125. PA-PV (1784), 103:175; Berlin Academy, *Mémoires* (1784/1786), 39:48; KVA-Wilckes, Parker letter, June 24, 1784. Other institutions may well have been contacted.

126. KVA-Wilckes, Parker letter, June 24, 1784.

127. AdW, "Registres 1766–1786," p. 9, (October 29, 1767). Still, no exchange took place until 1769, as the following report in the minutes of the Berlin Academy indicates: "M. Sulzer a rapporté que la Société Royale de Londres ayant envoyé à l'Académie la suite des Transactions Philosophiques, il parrisait convenable de lui donner nos Mémoires complets. l'Académie y a consenti; & l'envoi se fera en conséquence." AdW, "Registres, 1766–1786, p. 58a (July 6, 1769).

128. See Imperial Academy, *Protokoli*, vol. 3, May 30, 1782; PA-PV, (1786), 105:150; AdW, "Registres 1786–1800," p. 41a (February 7, 1788); Lindroth, *Vetenskapsakademiens Historia*, 2:172. The Berlin Academy did present twenty-four volumes of its *Mémoires* to the visiting Swedish king in 1771; see AdW, "Registers 1776–1786," pp. 92–92v (April 26, 1771).

129. See Harnack, *Geschichte* (1901), pp. 269–301, for these details.

130. See AdW, "Registres 1746–1766," p. 231a (January 12, 1764). There is an ironic twist to Frederick's naming members to the academy because he would first ask for a report from the academy about a prospective candidate and then order his admittance into the organization; "Ce à quoi," in the typical words of the Berlin minutes, "l'Académie s'est respectueusement conformée." See AdW, "Registres, 1766–1786," p. 183; see also AdW, I:III:3, "Acta Betr. die Aufnahme der Mitglieder 1744–1785."

131. See AdW, I:III:3, "Acta betr. die Aufnahme der Mitglieder 1744–1785"; and AdW, I:VII:19, "Versendung der Mémoires an auswärtige Akademieen, 1786–1788."

132. See AdW, I:VII:19, "Versendung der Mémoires an auswärtige Akademieen, 1786–1788," pp. 4–5 (Memorandum dated June 4, 1786).

"La Société n'ayant pas repondu, ou White est un frippon qui n'a pas remis l'Exemplaire destiné à la Société, ou la Société ne se souci pas de nos Mémoires."

[Note:] "La Société R. de Londres, et surtout Mr. Magellan, son secrétaire ou du moins un de ses principaux membres correspond avec Mr. Bernoulli et nous a envoyé tant ses propres ouvrages que ceux d'autres de ses confrères. Elle nous envoie ses *Transactions* ainsi c'est un échange avantageux pour nous."

"Envisageant la chose en general, je remarquerai . . . [sic]

"B. que l'empressement à envoyer nos produits aux Académies étrangères, est une espece d'aveu d'inferiorité que ne convient pas."

[Note:] "B. Oui, à moins que ce ne soit un échange."

"C. que les petites Académies d'Italie, auxquelles on a envoyé les Mémoires de préférence, n'ont guère de droit à cette distinction."

[Note:] "C. J'excepterais les trois qui ont repondu, et qui sont plus de célébrité, celles de Padoue, et de Turin surtout, et celle de Siène en vertu de l'ordre du Roi 1777."

"D. Que les Académies de Petersbourg, Stockholm & Paris jouissent d'une toute autre reputation."

[Note:] "L'échange a lieu entre celle de Pétersbourg et la notre; et celle de Paris vient elle-meme de la proposer à Mr. Bernoulli; et nous donnera encore la description des Arts et métiers.

133. On two occasions at least the Berlin Academy launched entrepreneurial ventures in England. See AdW, I:VII:78, "Acta betr. die nach England zum Verkauf versandten Landkarten, 1773–1780" and I:VII:17, "Acta. betr. den Verkauf der Mémoires in Londen [sic] und Berechnung darüber mit dem Kaufman White und Sohn, wie auch Überreichung der Mémoires an die Londoner Societät 1787–1788." Remarks by Merian and Jean III Bernoulli in this last set of memoranda again reflect the caution of the Berlin Academy in entering into relations and exchanges with other scientific institutions: [Merian:] "La Société Royale de Londres nous a constamment envoyé les vols. de ses *Transactions*, de même que d'autres ouvrages qui se sont publiés en son nom, ou par ses associés. Elle demande quelques voll de nos mémoires qui lui manquent; et que j'estime que nous ne pouvons pas lui refuser." [Bernoulli note:] "Je ne manquerai pas de faire votre Commission pour [les volumes] qui manque dans notre collection des *Transactions Philosophiques*, mais pour avoit meilleure grace à la demander, il faudroit ne plus différer de satisfaire la Société Royale, sur la note qu'elle nous a envoyée."

134. See AdW, "Registres 1786–1800," pp. 44 (March 13, 1788), 223a (September 10, 1795), and passim.

135. PA- PV, (1759), 78bis:770. See also Lindroth, *Vetenskapsakademiens Historia*, 1:187ff.; and Baër correspondence, KVA-Bergius, 8:456–96; KVA-Wargentin. His correspondence with Wargentin during this period amounted to approximately 140 letters.

136. Representatives of other scientific societies, like Bäer, were also granted the special privilege of attending and being seated in the meetings of the Paris Academy from time to time, e.g., men from the Royal Society of London, the Montpellier Society, the Paris Surgeons Academy, the Saint Petersburg Academy, and the American Philosophical Society. See, for example, PA-PV, (1754), 73:241; (1756), 75:466; (1757), 76:331, 510. Such courtesies tended to reinforce the ties that existed between the Paris Academy and these other institutions. Benjamin Franklin of the American Philosophical Society played an exceptional role in the Paris Academy, where he not only attended meetings but also was named to committees, wrote reports, and generally participated in the business of the academy as an active member; see for example, PA-PV, (1778), 97:166–68; (1783), 102:160.

137. Youschkevitch and Winter, *Briefwechsel*, vol. 1, letter #86. While in Berlin Euler still played an active role in the Saint Petersburg Academy. He assisted in editing the Russian *Commentarii*, suggested possible prize questions, recruited persons for imperial Russian service, and instructed students sent from Saint Petersburg; see Youschkevitch and Winter, *Briefwechsel*, 1:5–6, 11, 20–22, 24, and letters #51, #60, #67, #69; 2:21, 36, and letter #99.

138. Lubimenko, *Correspondance scientifique*, pp. 38–39.

139. Pavlova, "Lalande," p. 743.

5. COMMUNICATIONS NETWORK OF SOCIETIES

140. Lubimenko, *Correspondance scientifique*, p. 35.

141. American Philosophical Society, "Early Proceedings," p. 255 (1797).

142. Based on Institut de France, *Index Biographique* and Royal Society of London, *Record* (1940). Jackie McClellan did the study. Members with dual locales were assigned 1/2 membership in each locale.

143. Figure 6 integrates the common membership of the Royal Society of London and the Paris Academy of Sciences over the whole period, 1699–1793. In the final analysis, this approach was felt to be more revealing of the scope of their common members than any other. In order to have some sense of changes over time, however, three *sondages* of this data were taken for the years 1720, 1750, and 1780. The results are as follows.

Locale:	1720:	1750:	1780:
Paris:	6	25	17
London:	6	5	7
French Provinces:	0	2	1
British Isles:	1	1	0
Italy:	2	6	3
Germany/Austria:	2	3	2
Switzerland:	1	5	4
Netherlands:	2	1	2
Belgium:	0	0	2
Sweden:	0	1	3
Spain:	1	2	2
Saint Petersburg:	0	0	2
Lithuania:	0	0	1
Philadelphia:	0	0	1
Peru:	0	1	0
Totals:	21	52	47

To be noted especially: the rise in the total number of common members between 1720 and 1750; the large increase in the number of Paris members in 1750 and a decline thereafter; the virtually constant number of London members.

144. Connections through common members of ten leading scientific societies are explored further in McClellan, "International Organization," pp. 331–39 and appendix 4.

145. Tisserand, *Académie de Dijon*, pp. 635–36; Barrière, *Académie de Bordeaux*, p. 112.

146. See Roche, *Siècle*, 1:55,69. Twenty-eight different academies were involved in the first of these efforts.

147. Roche, *Siècle*, 1:69–70; Tisserand, *Académie de Dijon*, p. 636.

148. Quoted in Roche, *Siècle*, 1:70–71.

149. Barrière, *Académie de Bordeaux*, p. 111.

150. E.g., Sapte of the Toulouse Academy, La Tourette of the Lyon Academy, Le Cat of the Rouen Academy. See PA, individual dossiers for these men.

151. PA, Pochette de Séance 1751 and February 13, 1771; PA-PV, (1772), 90:52–55. See also BI, Papiers de Condorcet, MS. 871, pp. 159–60 ("Modèle d'une Acte d'association entre l'Académie des sciences et celle de Marine").

152. Baker, *Condorcet*, pp. 48–55, explores this story in full; see also his, "Les Débuts de Condorcet"; Bouillier, *Institut*, p. 90, indicates that Turgot himself engaged provincial academies over his proposals for the grain trade.

153. See BI, Papiers de Condorcet, MS. 870, pp. 161–65 (proposal to Maurepas); pp. 166–73 (proposal to Malesherbes); pp. 174–75 (response of Seguier, Nîmes).

154. See Roche, *Siècle*, 1:71; Barrière, *Académie de Bordeaux*, p. 111; Tisserand, *Académie de Dijon*, p. 138; Baker, *Condorcet*, p. 51; Hammermayer, "Akademiebewegung," pp. 28–29.

155. "M. De Buffon annonçoit que le moment n'est pas encore arrivé ou ce projet conçu par M. Condorcet pourra avoir son exécution." AD-CD, Dijon Academy Registres VI, pp. 125–26v; see also pp. 202v–03.

156. "Votre projet ne pouvoit qu'être très avantageux et très honorable pour les académies de province . . . Vous pouvez donc être assure Monsieur, que dès que cette association sera établie, notre academie s'engagera volontiers à lui remplir les conditions." Letter: Morveau to Condorcet, Dijon, July 23, 1775; BI, Papiers de Condorcet, Cote MS. 876, #33.

157. BI, Papiers de Condorcet, MS. 870, pp. 174–75; MS. 876, #34.

158. "Je ne crois pas que cette derniére reflection puisse être lüe en pleine Académie; il faudra la modifier . . . il faut garder plus de ménagement." BI, Papiers de Condorcet, MS. 870, p. 175; Guyton de Morveau made the same point; see note 156.

159. See Baker, *Condorcet*, pp. 51–54; Barrière, *Académie de Bordeaux*, p. 112; Roche, *Siècle*, 1:71–72.

160. See Baker, *Condorcet*, pp. 54–55; Hankins, *Jean d'Alembert*, p. 144.

161. Tisserand, *Académie de Dijon*, p. 237; Roche, *Siècle*, 1:69, 72.

162. See Berthe, *Dubois de Fosseux*, pp. 151–52, 159, 161–62, 168–69, 175, 199–201, 292.

163. Berthe, *Dubois de Fosseux*, passim and appendix 1 ("Les Companies savantes en relation avec l'Académie d'Arras), pp. 385–402.

164. Berthe, *Dubois de Fosseux*, pp. 165–66; Roche, *Siècle*, 1:68–69.

165. Berthe, *Dubois de Fosseux*, pp. 180–81 and appendix 2 ("Les Lettres collectives"), pp. 407–22, and 192n, 198.

166. Berthe, *Dubois de Fosseux*, ch. 6 title; see also Roche, *Siècle*, 1:69.

167. APS, Archives: 1780, December 15, (Lewis Nicola Proposal); 1790, February 5, (Committee Report on By-Laws).

168. Letter: Lorgna to Banks, dated Verona, May 1, 1783 in Dawson, *The Banks Letters*, p. 555; see also Hammermayer, "Akademiebewegung," p. 31.

169. Hubrig, *Die patriotischen Gesellschaften*, pp. 30–32 and n. 165; see also AdW, "Registres, 1786–1800," p. 183 (December 5, 1793).

170. Hubrig, *Die patriotischen Gesellschaften*, pp. 66–67.

171. Economic societies in Germany and in Switzerland likewise initiated plans for union at this same time; see Hammermayer, "Akademiebewegung," p. 29. The prize question of 1787 of the *Zeeuwsch genootschap der Wetenschappen* on how Dutch societies might cooperate is another indication of this tendency to unite institutional efforts; see R. J. Forbes, *Van Marum*, 3:3–4.

172. Hubrig, *Die patriotischen Gesellschaften*, p. 67.

173. The culmination of this theme in our story was not reached until 1900, when, upon the proposal of the Royal Society of London, the International Association of Academies was established in Paris. See Stimson, *Scientists and Amateurs*, pp. 232–33; Axel von Harnack, "Die Akademien der Wissenschaften," pp. 871–74; Andrade and Martin, "Foreign Relations," pp. 78–79; Schroeder-Gudehus, "Division of Labor and the Common Good," pp. 6–8; see also Bouillier, *Institut*, passim and ch. 16.

174. Kronick, *History*, pp. 143, 153, 163; see also his, "Scientific Journal Publication,"

p. 32; McClellan, "The Scientific Press," p. 431. That delays in the publication of society proceedings were a problem for the active man of science in the latter part of the 18th century is evident in the remarks of the Saint Petersburg academician Paul Henri Fuss, who said in 1782 of a 1780 paper of his, "It is very discouraging for an author to see his productions remaining buried thus in the storerooms." See Imperial Academy, *Protokoli*, vol. 3, August 22, 1782.

175. A lesser-known Dutch series, *Uitgezogte Verhandelingen uit de nieuwste werken van de Societeiten der Weetenschappen in Europaen van andere geleer de Mannen*, 10 vols., Amsterdam, 1756–65, is another indication of the importance of scientific societies and interest in disseminating their researches; see Bolton, *Catalogue*, #4602, p. 557; Scudder, *Catalogue*, #781, p. 50.

176. Letter: Lalande to Wargentin, March 8, 1774; KVA-Wargentin; original printed, McClellan, "The Scientific Press," p. 431.

177. See 450-page Ruffey manuscript, BM-D, MS. 995, also AD-CD , Dijon Academy Registres 5:52–52v; François Delandine, *Couronnes académiques* (1787). See also *Journal Encyclopédique*, October 1780, "Project d'une table chronologico-synoptique des prix de toutes les académies," pp. 328–31. A letter from Domaschneff, director of the Saint Petersburg Academy, to Formey in Berlin likewise reveals this desideratum: "Depuis les fondations des Académies, chacunes d'elles cherche par Esprit de son institution à interresser toute la Republique des Sçiences et de Lettres, à la resolution de certaines Questions qu'elles leurs proposoient pour des prix. Plusieurs de ces Questions ont été resoluës. Mais il n'existe pas a moins que je sache un Livre qui Contiendroit la Collection de toutes les prix . . . [portions of letter destroyed] . . . Une Collection come cella en auroit formé l'Histoire des progrez des Sciences et de la marche qu'on leur assignoit." Letter dated Saint Petersburg, January 18/27, 1776, DS-Formey, 9, 38, #1.

178. Hahn, *Anatomy*, p. 62. See also PA-PV, (1747), 66:44.

179. See *Savants Etrangers* (1751), "Préface," 1:iii. The point is made that the *quality* of the work of foreigners was not up to par at the beginning of the century.

180. *Observations sur la physique, sur l'histoire naturelle et sur les arts* (1773), "Avis," 1:iii–iv; see Kronick, *History*, pp. 107–10, for full translation, and McClellan, "The Scientific Press," pp. 433–34.

181. For full details consult McClellan, "The Scientific Press," passim; for the literature on this well-studied journal, see p. 427.

182. Kronick, *History*, p. 158. See also Guerlac, *Lavoisier*, pp. 58–64.

183. Guerlac, *Lavoisier*, p. 61. The *Philosophical Transactions* of the Royal Society and the *Handlingar* of the Swedish Academy were published quarterly or trimestrially, and delays in publication were not such a factor with these institutional journals. Language barriers were a problem for both, however.

184. *Observations*, January 1773, "Avis," pp. v–vi.

185. *Observations*, 1779, p. 88; see also January 1773, p. vii.

186. Letter: Rozier to Wargentin, Paris, August 8, 1775; KVA-Wargentin; original printed, McClellan, "The Scientific Press," p. 445.

187. For this aspect of Rozier's activity, see McClellan, "The Scientific Press," pp. 444ff.

188. For the generally antipathetical attitude of academies toward independent journalists, see McClellan, "The Scientific Press," pp. 441–42.

189. The major source on the Hesse-Hamburg society and its effort to coordinate the activities of other societies is now Voss, "Die Societe [sic] Patriotique." See also Hubrig,

Die patriotischen Gesellschaften, pp. 59–64; Hammermayer, "Akademiebewegung," pp. 27–28; McClellan, "International Organization," pp. 431–36.

190. Voss, "Die Societe [sic] Patriotique," pp 199, 204, 210, 218; Hubrig, *Die patriotischen Gesellschaften*, pp. 62n., 63n.

191. Voss, "Die Societe [sic] Patriotique," pp. 201–03; Hammermayer, "Akademiebewegung," p. 28.

192. Voss, "Die Societe [sic] Patriotique," pp. 197, 199–200. The Swedish were especially well represented among the German society's members, and it had especially close contact with the *Patriotiska sällskapet* in Stockholm. See also Hubrig, *Die patriotischen Gesellschaften*, pp. 59–60.

193. Voss, "Die Societe [sic] Patriotique," pp. 198–99, 201.

194. Voss, "Die Societe [sic] Patriotique," pp. 207–09.

195. "J'ai l'honneur de vous adressé de la part du chef comité de la Société patriotique de Hesse Hombourg et de S. A. R. Monseigneur le Landgrave, Chef et Protecteur de la dite Société un projet de correspondance qu'il desireroit établir avec l'Académie Royal des Sciences et Belles-Lettres de Berlin . . . Son but est de se dévouer au service de toutes les autres sociétés, de les lier entre elles par une correspondance aisée et rapide, d'attirer toutes les connoissances à un dépot commun pour les verser ensuite plus promptement et également dans le sein de toutes les Nations . . . Il est certain que l'eloignement et le défaut de communication entre les savants et les associations qu'ils ont formées dans les differents Etats de l'europe, nuit aux progrès des Sciences et des Arts qu'ils cultivent [.E]n les rèunissant par une correspondance suivie et en faisant circuler leurs découvertes, ce seroit les mettre á, même d'épurer leurs travaux au creuset de la discussion, de l'experience et de gout, et faire participer egalement toutes les nations aux avantages qui doivent en résulter; ce seroit par conséquent étendre la masse des connoissances et celle du bonheur. Voilà qui a été la base de notre institut et le motif qui nous a engager à propager ses rameaux dans toutes les parties de l'Europe. À present que cette extension a lieu, nous desirerions que les academies et les autres Sociétés voulussent bien nous aider à perfectionner un ouvrage entrepris pour leur propre satisfaction." Covering letter from L'Armbrüster to Formey, dated Hamburg, March 20, 1780; DS-Formey, 1, 39. See also official proposal, "Projet de Correspondance de la Société Patriotique de Hess-Hombourg," in AdW, MS. I:V:5b, "Literarischer Briefwechsel oder Schreiben verschiedener Gelehrter an die Akademie aus den Jahren 1751–1788," pp. 271–73.

196. Voss, "Die Societe [sic] Patriotique," pp. 201, 208, and passim; Hammermayer, "Akademiebewegung," p. 27; Hubrig, *Die patriotischen Gesellschaft*, p. 63.

197. Voss, "Die Societe [sic] Patriotique," pp. 212–13, 217; he points out, however, some overlap with the *Société royale de médecine* of Paris.

198. Imperial Academy, *Protokoli*, vol. 3, August 17, 1780.

199. Voss, "Die Societe [sic] Patriotique," p. 209.

200. Cousin, "L'Académie des sciences, belles-lettres et arts de Besançon et ses relations extérieurs," p. 59.

201. See de la Blancherie, *Correspondance générale*, pp. 14–15.

202. See de la Blancherie, *Correspondance générale*, pp. 8–13. He explains, pp. 8–9, the purposes of this Assembly as follows: "This assembly has three purposes: The first is to serve as a rendez-vous and meeting point for communication for all savants, men of letters, artists, amateurs, and distinguished national and foreign travelers who find themselves in this capital. The second, to bring together beneath their eyes books, paintings, mechanical pieces, specimens of natural history, sculpture models and all sorts of ancient

and modern works, the existence, or value, or author of which one wishes to make known promptly. Finally the third, to secure for me the means to extend correspondences and contacts in all parts of the world with regard to all the subjects of the sciences and the arts."

203. See de la Blancherie, *Correspondance générale*, pp. 19–25. He says, p. 20: "In publishing notices of the works of foreigners in the form of a gazette, I can acquire a new manner of exciting their interest in correspondence, a way of gaining their support, and a means that will dispense them from writing me and me replying to them about many things."

204. De la Blancherie, *Correspondance générale*, p. 25.

205. De la Blancherie, *Correspondance générale*, pp. 15, 18.

206. "Nous avons assisté aux assemblées hebdomadaires indiquées sous le nom de rendés vous de la république des Lettres; nous y avons vû des savans, des artistes et des amateurs de présque touts les parties de l'Europe, nous avous vû dans ses registres les preuves d'une correspondance qu'il n'a pû former qu'avec beaucoup de tems et de peines, et nous avons été témoin d'une activité et d'un zèle qui sont très rares, et qui ne peuvent être que très utiles au progrès des sciences et des arts . . . Plus il sera encouragé, plus il deviendra utile, soit aux françois, soit aux etrangers à qui il desire d'epargner les embarras d'une correspondance à laquelle beaucoup de gens de Lettres sont très peu propres, qui fatigue beaucoup les autres et qui leur fait perdre beaucoup de tems, faute d'avoir à leur portee les moyens, les relations et les secours que M. Delablancherie a sû se procurer. On ne sauroit trop favoriser les correspondances qui sont un des grands moyens d'accélérer les progrés des connoissances humaines, en conséquence nous croyons que le projet de M. Delablancherie mérite d'être encouragé et que l'Académie ne pourra voir avec plaisir le succès de cet établissement." PA-PV, (1778), 97:166–68 (May 23, 1778). The report was authored by Lalande and signed by Franklin, LeRoi, and Condorcet.

207. See de la Blancherie, *Correspondance générale*, pp. 8, 16, 25, 27–30; PA-PV, (1779), 98:32, 100, 153, 230.

208. Imperial Academy, *Protokoli*, vol, 3, May 27, 1777; September 28, 1778; February 22, 1779; March 11, 1779, etc.

209. See Imperial Academy, *Protokoli*, vol. 3, June 10, 1779; June 21, 1779.

210. AD-CD, Dijon Academy Registers, 10:34–35v (March 11, 1779); see also p. 91v (July 29, 1779).

6. A Record of Common Endeavors

1. Weld, *History*, 1:350.

2. On these expeditions, see Chapin, "Expeditions of the French Academy of Sciences, 1735"; Brunet, *Maupertuis*,1:9–33; 2:89–163; Maury, *Académies*, pp. 73–75; Bertrand, *Académie*, pp. 119–26; Hahn, *Anatomy*, pp. 90–91.

3. On discussions within the academy attendant to these expeditions, see McClellan, "International Organization," pp. 183–84 and PA-PV, vols. 51–54, passim. See also Newton, *Principia*, book 3, prop. 18–19. These were not the first expeditions of the Paris Academy of Sciences in the 18th century; botanical expeditions, notably, were sent out in the 1710s, lead by Antoine de Jussieu and Tornefort; see Bertrand, *Académie*, pp. 108ff.

4. PA-PV, (1735), 54:23–28.

5. Imperial Academy, *Protokoli*, vol. 1, February, 11, February, 14, March 21, 1737; see also Boss, *Newton and Russia*, pp. 132–33; Nevskaja, "Delisle," p. 312.

6. Imperial Academy, *Protokoli*, vol. 1, January 29, 1737, May 6, 1737, November, 10, 1738; RS-JBC, 15:168, 173–74, 193–94, 277, 332, 352–53; RS-LBC, 21:467; 22:1, 61–62, 313, 360–61; 23:3–4, 124–25, 249–52ff., 334–36; 24:22–24. The results of the expeditions, disputed initially, soon confirmed Newton.

7. See Woolf, *The Transits of Venus*, pp. 9–16. On the basis of data available to him, Newton put the astronomical unit at 70 million miles (*Opticks*, Query 21); that the astronomical unit was still a pressing question, see the Delisle proposal from Saint Petersburg to observe the 1740 passage of Mercury, "invisible in Europe." He proposed three observing stations, one in Russia, one in Peking, and one in Peru, "where the French astronomers perhaps are still." He concludes: "One will have one of the best occasions to date to determine the distance from the sun to the earth, which is not yet entirely certain." See Imperial Academy, *Materiali*, 4:313–14.

8. This chapter is heavily endebted to Woolf's excellent volume, which so well captures the spirit of science and the scientific enterprise of the classical 18th century. For Woolf's account of the LaCaille expedition and the 1753 Mercury transit, see *The Transits of Venus*, pp. 35–53.

9. See Woolf, *The Transits of Venus*, pp. 36, 45–46, 48–49. About private correspondence, Woolf says (p. 45), "Private correspondence . . . played a large role in the discussion of scientific problems like those connected with the transits . . . The quality of the letters in detail and continuity is of a very high order, and the frequency with which they were exchanged must come as a shock to modern man so pressed for time."

10. See Woolf, *The Transits of Venus*, pp. 35, 37, 46–47. Notes Woolf (p. 46), "It is easy to be overly impressed by the penned interchange of ideas and of opinions in the process of formation. In the long run, the solid foundation upon which the working machinery of astronomy evolved in these years is to be found in the memoirs and notices, great and small, which were published under the auspices of scientific societies, by the patronage of royal friends or by the subsidy of men of wealth." Of the centrality of academies and societies to his subject, Woolf has communicated privately, he is "absolutely convinced."

11. Woolf, *The Transits of Venus*, pp. 36–37.

12. The verb "contract" ("contracter") is Delisle's; see KVA-Wargentin, Delisle to Wargentin, December 18, 1750. See also Lindroth, *Vetenskapsakademiens Historia*, 1:393–98; *Swedish Men*, 107–08; Nordenmark, *Wargentin*, pp 320–21.

13. PA-PV, (1750), 69:408, 411, 442.

14. "On a déliberé sur le voyage proposé par Mr. de l'Isle, L'academie dit qu'elle croyoit suffisant de prier MM. des Académies de Stockholm et d'Upsal de faire les Observations correspondantes à celle de Mr. l'abbe de la Caille; et il a été résolu que je leur en écrivons en conséquence." PA-PV, (1750), 69:463. Delisle himself wrote to Wargentin, sounding disappointed: "Notre Académie, qui a en l'avantage de s'associer avec la vôtre de Suede, et qui a souhaité que je n'entreprisse le voyage de Suede vous a fait ecrire par son Secretaire pour vous prier de vouloir bien faire les observations correspondantes à celles de Mr. de la Caille." Letter, dated Paris, February 19, 1751; KVA-Wargentin.

15. "L'Académie s'est flattée que Messieurs les Astronomes de Stockholm et d'Upsal voudroient bien prendre par à un travail aussi utile en faisant de leur coté les observations correspondantes à celles de Mr. l'abbé de la Caille." Letter, January 20, 1751; KVA-Wargentin. De Fouchy included samples of the form to be used for the observations.

16. ". . . pour [les] exhorter à vacquer avec toute l'assiduité possible aux observations correspondantes à celles de Mr. l'abbé de la Caille." PA-PV, (1751), 70:201.

17. ". . . par laquelle il promet au nom de cette Compagnie de faire faire les Observations correspondantes à celles de Mr. l'abbé de la Caille." PA-PV, (1751), 70:315.

18. ". . .ravie des dispositions que vous avez faites pour nous procures dans les plus d'endroits qu'il est possible en Suede des observations correspondantes à celles de Mr de la Caille." Letter dated March 1, 1751; KVA-Wargentin.

19. Lindroth, *Swedish Men*, p. 108.

20. See Imperial Academy, *Protokoli*, vol. 2, September 13, 1751, June 12, 1752, March 29, 1753, January 7, 1754, June 17, 1754, January 16, 1755, August 11, 1756, November 28, 1756.

21. "Mon devoir et envers Votre Illustre Academie et envers Vous, Monsieur, est à present de Vous donner avis que j'ai continué jusques ici avec beaucoup de soin et d'exactitude les observations correspondantes à St. Petersbourg . . . l'Academie Imp. de St. Petersbourg etant donc infiniment redevable à Votre Ill. Academie des Sçiences du Conseil qu'Elle a pris de prolonger le sejour de Mr. l'Abbe de la Caille au Cap, il sera de mon devoir de Vous communiquer Monsieur les observations que je vais faire içi, en vous suppliant de les presenter à Votre Ill. Corps. Etant donc actuellement en etat de continuer içi les observations correspondantes avec la dernière exactitude autant qu'il plaira à Votre Ill. Academie de laisser M. l'Abbé de la Caille au Cap de Bonne Esperance, je vous supplie, Monsieur, de me faire la grace de me marquer environ jusqu'à quel temps Vous croyez que Mr. l'Abbé de la Caille y continuera ses observations." Letter dated Oesel, August 14/25, 1752; PA, Dossier Grischov. The letter goes on to say that Grischov was having forwarded the second volume of the *Novi Commentarii* and other books for the Paris Academy.

22. PA-PV, (1752), 71:557; Youschkevitch and Winter, *Briefwechsel*, vol. 2, #223 (letter from Schumacher to Euler, March 13/24, 1753).

23. See Woolf, *Transits of Venus*, p. 37 and note. (Apparently Maupertuis was anxious to have Lalande in Berlin to help perform extra astronomical work for the Academy there.) See also Berlin *Mémoires* 1759 (1766), p. 481; AdW, "Registers 1746–1766," p. 88a (January 19, 1752); Youschkevitch and Winter, *Briefwechsel*, vol. 2, #188; PA-PV, (1752), 71:86.

24. "M. le Comte de Maillebois Président dit que le Roy avoit accordé 2500# au Sieur de la Lande Eléve de Mr. de l'Isle pour aller à Berlin faire des observations correspondantes à celles de Mr. l'abbé de la Caille: Il a aussi invité l'Académie à prendre des mesures pour en faire faire à Malthe et à Tripoli." PA-PV, (1751), 70:447. There is no indication that stations in Malta and Tripoli were ever established.

25. Woolf, *The Transits of Venus*, pp. 40, 45; Rosen, "The Academy of Sciences," p. 120; PA-PV, (1751), 70:187; (1753), 72:451; RS-JBC, 20:394; 21:179; AD-H, D120 (Registers of Montpellier Society), March 24, 1746, February 4, 1751; D129 (Astronomy), pp. 72–98.

26. PA-PV, (1752), 71:559; (1754), 73:424, 457.

27. Quoted in Woolf, *Transits of Venus*, p. 47. The American colonies already knew of the event and the preparations for it, having indirectly obtained a copy of Delisle's *Avertissement* bound for Quebec. Only one report was forthcoming from the Americas for the 1753 Mercury transit, from Antigua; Woolf, pp. 43–45 and Hindle, *Pursuit*, pp. 82–84.

28. KVA-Bergius, 7:728 (Grischov to Wargentin), pp. 713–14 (Zanotti to Wargentin); KVA-Wargentin, Mortimer to Wargentin, May 11, 1750.

29. Youschkevitch and Winter, *Briefwechsel*, vol. 2, #188.

30. PA-PV (1758), 77bis:936–37; see also Maindron, *Fondations*, pp. 24–25. I am grateful to Craig Waff for sharing his researches on the Lauragais prize initiative.

31. "Il a été question de la proposition faite par M. le Cte de Lauragais d'un prix de 2400# destiné a celuy qui auroit donné la solution la plus élégante et sans aucune approximation du Problème des 3 Corps. et Comme L'Intention de M. de Lauragais est de ne point exclure les académiciens de Concourir et que par Conséquent ce prix ne peut estre adjugé dans La forme ordinaire on a déliberé sur la maniere dont les pièces devoient estre Jugées pour éviter tout soupcon de partialité sur quoy il a été proposé d'engager La Societé R. de Londres et Lacad[e] de berlin a nommer chacune trois Juges qui avec pareil nombre de L'Acad[e] formeroient un Comité de neuf personnes et que si quelqu'un de ces Juges vouloit travailler pour le prix, il seroit tenir de la Déclarer pour qu'on peut nommer un autre en sa place. Ce projet a été approve et a été Décidé qu'avant cette Deliberation fut enregistrée . . . [on attendrait] L'acceptation des 2 Compagnies étrangères." PA, Pochette de Séance, 1758, "Suite du 6 Décembre, 1758, Mrs. les Mathématiciens ét ant demurés aprés la séance." See also PA-PV (1758), 77bis:936–37. Lauragais also proposed a companion prize in chemistry dealing with the extraction of mercury; the results of this proposal are unknown; see PA-PA (1758), 77bis:934.

32. Kopelevich, "The Petersburg Astronomy Contest"; Waff, "Universal Gravitation." The precise problem was formulated as follows: "Given three bodies, launched in a vacuum each with a certain speed and direction, mutually attracting in direct proportion to their masses and inversely to the square of the distance, determine rigorously and without approximation the curves described by each of these three bodies." This problem is known to be insoluble, a state of affairs recognized by Clairaut when he wrote in the *Journal des Sçavans* (August, 1759, p. 546): "Aussi lorsque M. le Comte de Lauragais, dont l'Académie connoît le zèle & les lumières, a choisi cette espèce de solution pour le sujet d'un des prix qu'il propose aux Sçavans, j'ai été des premiers à reconnoître les obligations que les Mathématiques lui auront, soit par cette solution même si elle a lieu, soit par les découvertes qui seront dûes aux efforts qu'on fera pour la trouver."

33. Kopelevich, "The Petersburg Astronomy Contest," p. 656; Youschkevitch and Winter, *Briefwechsel*, vol. 2, #136

34. See PA-PV (1759), 78bis: 527–28; see also PA, Pochette de Séance, 6 December 1758. Wrote de Fouchy to Florentin: "Comme ces deux Comp[ies] sont précisément celles des nations avec les quelles La france est en Guerre actuellement L'Acad[e] a cru ne devoir rien faire sur cet article sans prendre les ordres du Roy et elle m'a chargé d'avoir Lhonneur de Vous en écrire et de Vous envoyer la Copie de la lettre que je dois écrire à ce sujet aux secretaires de Ces Académies."

35. "J'ai, Monsieur, rendu compte au Roy du Projet de la Lettre à écrire à Mrs. les Secretaires des Academies de Londres et de Berlin. Le Roy a fort approuvé la Zèle qui porte M. le Comte de Lauragais à fonder un prix qui ne peut que contribuer à donner de l'émulation pour le progrès des sciences, mais Sa M[té] pense qu'il convient de remettre à la Paix à écrire les Lettres proposées à Mrs. les secretaires de ces deux Académies Etrangeres." PA, Pochette de Séance, 28 July 1759; see also PA-PV (1759), 78bis:614. This outcome contrasts somewhat with the more optimistic internationalism of Gavin de Beer's *The Sciences Were Never At War.*

36. See *The Transits of Venus*, pp. 53–149.

37. Woolf, *The Transits of Venus*, pp. 135–41.

38. Woolf, *The Transits of Venus*, p. 63; see also his appendix 5, pp. 209–11.

39. Woolf, *The Transits of Venus*, pp. 54–57, 62–63; PA-PV, (1760), 79:257, 265–66, 421–27.

40. Woolf, *The Transits of Venus*, pp. 95–130.

41. ". . . de M . . . [sic] secrétaire de l'académie de Petersbourg, par laquelle il démande de la part de cette Compagnie si quelu'un des astronômes de l'academie voudroit aller en Siberie observer le passage de Vénus sur le soleil qui doit arriver en 1761. Mrs. Pingré et de Chappe se sont offerts." PA-PV, (1760), 79:239.

42. Thus Baër: "L'Académie de Petersbourg ne se fiant pas à la capacité de ses membres et ayant prié celle d'icy de lui envoyer quelqu'un aux depens de l'Imperatrice pour observer en Siberie le passage de Venus, l'on vient de nommer l'Abbe de Chapte . . ." Letter dated Paris, July 15, 1760; KVA-Bergius, 8:487. Thus Fouchy: ". . . [l'Académie] auroit cru manquée à son objet si elle avoit retalé de Concourir du Zèle que vous temoignez pour l'avancement des sciences." PA, Pochette de Séance, 1760.

43. Youschkevitch and Winter, *Briefwechsel*, vol. 2, #129.

44. See Imperial Academy, *Protokoli*, vol. 2, February 9, 1761; Woolf, *The Transits of Venus*, pp. 119, 122.

45. Woolf, *The Transits of Venus*, pp. 118–19, 134–36.

46. Woolf, *The Transits of Venus*, pp. 134–37, 141; Lindroth, *Vetenskapsakademiens Historia*, 1:399–404; Nordenmark, *Wargentin*, p. 322; KVA-Bergius, 18:487–88 (Baër to Wargentin); KVA-Wargentin, Delisle to Wargentin, January 27, 1761.

47. Woolf, *The Transits of Venus*, p. 141.

48. RS-JBC, 23:894–99 (June 5, 1760); Woolf, *The Transits of Venus*, pp. 72–73.

49. RS, "Letters & Papers," Decade IV, #21. Contemporary translation. Nothing came of the proposed French station on Cyprus; see Woolf, *The Transits of Venus*, p. 64.

50. RS-JBC, 23:918; RS-CM, 4:228–30 (June 26, 1760).

51. Woolf, *The Transits of Venus*, pp. 130–33.

52. RS-JBC, 24:20, 33.

53. Woolf, *The Transits of Venus*, pp. 138–40; other Englishmen observed at Calcutta and Madras.

54. Woolf, *The Transits of Venus*, pp. 94–95; Hindle, *Pursuit*, p. 99.

55. Woolf, *The Transits of Venus*, pp. 135, 137–39; Rosen, "The Academy of Sciences," p. 122; AD-H, D120 (Registers of the Montpellier Society), April 9, 1761, May 7, 1761; D129 (Astronomy), pp. 130–34.

56. See Winter, *Die Registres der Berliner Akademie* for the period.

57. For the case of the Royal Society of London and communications it received regarding the plans of the Swedish and Russian academies, see RS-JBC, 24:43–44, 95–96, 98; RS, "Letters & Papers," Decade 4, #41, #59.

58. "Le Pere Pingré va partir egalement au premier jour pour se rendre pour le meme sujet sur les cotes occidentales d'Afrique. Il devoit se rendre à St. Helene, mais on a appris que la Société de Londres y envoyoit & cela suffit." Baër to Wargentin, Paris, July 15, 1760; KVA-Bergius, 8:487–88.

59. Quoted in Woolf, *The Transits of Venus*, p. 60.

60. RS-CM, 4:245–46 (July 10, 1760); reprinted in Weld, *History*, 2:11–13; see also Woolf, *The Transits of Venus*, pp. 81–85. Woolf's interpretation of this letter is slightly different from mine in that he sees only "apprehension about national prestige . . . patriotism, honor," as being expressed in it; see *The Transits of Venus*, pp. 81–82.

61. Because of the original emphasis, this single quotation is taken from the original copy of the Council Minutes, 4:239–40.

62. See PA-PV, vols. 80–82 (1761–63), passim; one might note the existence of a station in Peking.

63. Weld, *History*, 2:19; RS-JBC, vol. 24.

64. AD-H, D129 (Astronomy), pp. 135–40.

65. Delisle to Wargentin, dated Paris, May 3, 1762; KVA-Wargentin.

66. Woolf, *The Transits of Venus*, pp. 90–91, 103, 112–13. The Royal Society and the Paris Academy were both informed of these incidents and entertained some correspondence over them.

67. ". . . il est satisfaisant pour ceux qui s'interessent au progrès des sciences dapprendre que dans un Temps de guerre, ou les depenses enormes quelle entraine sembleroient eloigner toutes celles que les connoissances qui pourroient ne paroitre que de pure curoisité exigent, tous les astronomes de l'Europe sont en mouvement pour rendre chacun dans les postes les plus avantageux, pour tenter lentreprise, la plus courageuse, la plus utile, et la plus subtile . . . quelle gloire pour les astronomes dentreprendre pour la [découverte] dune verité, des voyages, que les autres nentreprennent que pour ammasser des Tresors." PA, "Mémoire sur la conjonction de Vénus," p. 1, inserted in PV, vol. 79 (1760), between pp. 471 and 472.

68. "Tout le mode est persuadé en angleterre que la parall. du⊙ est de 8″1/2, mais M. Pingré la soutient toujours ici de 10″1/2. Je crois que la question ne sera décidé qu'en 1769." Letter, Lalande to Wargentin, dated Paris, October 10, 1763; KVA-Wargentin. Lalande's semi-official trip from the Paris Academy to England and the Royal Society was undertaken in part "to renew the Union and Correspondence which ought to subsist between the two bodies." See RS-JBC, 24:653; PA-PV, (1763), 82:52.

69. Woolf, *The Transits of Venus*, pp. 150–51.

70. See Taylor, *The Haven-Finding Art;* Bertrand, *Académie*, pp. 188–98; Weld, *History*, 1:506–08; Waters, "The Problem of Longitude."

71. Maindron, *Fondations*, notes the 1716 Regents prize in France worth 100,000 livres; the British longitude award dates from 1714 and was worth 20,000 pounds; see Taylor, *Haven*, pp. 253–54; E. G. Forbes, *The Euler-Mayer Correspondence*, p. 14. A Dutch prize worth 30,000 florins dates from 1626; see Drake, *Galileo at Work*, p. 307.

72. Taylor, *Haven*, pp. 260–63.

73. Weld, *History*, 1:508; RS-CM, 4:309–13, esp. 312 (June 25, 1761); using eclipses of the moons of Jupiter was proposed.

74. See PA-PV, (1763), 82:174–76.

75. See previous note and PA-PV, (1763), 82:252–53.

76. PA-PV, (1766), 86:211; Imperial Academy, *Protokoli*, vol. 2, August 22, 1763.

77. Taylor, *Haven*, pp. 261–62. For his lunar tables Tobias Mayer of Göttingen shared £3,000 of the prize; see E. G. Forbes, *Euler-Mayer*, p. 19; Weld, *History*, 1:508.

78. This story seems not to have been told previously. On Chabert's background and an earlier geodetic expedition in Canada, see Brooks, "M. de Chabert"; Montagne, "L'Expédition hydrographique." Chabert seems associated with the effort to establish a station on Cyprus for the Venus transit of 1761; see Woolf, *The Transits of Venus*, pp. 64, 67.

79. PA-PV (1765), 84:225, 256, 381; Imperial Academy, *Protokoli*, vol. 2, June 20, 1765, July 1, 1765; KVA-Wargentin, De Fouchy to Wargentin, May 23, 1766.

80. For fullest report see Imperial Academy, *Protokoli*, vol. 2, June 20, 1765. Chabert's work at this time was concentrated in the Mediterranean, and he ultimately published a general nautical atlas of the Mediterranean in 1791; see Brooks, "Chabert," p. 342.

81. PA-PV (1765), 84:255, 381.

82. Imperial Academy, *Protokoli*, vol. 2, July 1, 1765.

83. PA-PV, (1765), 84:286.

84. For a full account of the 1769 transit, see Woolf, *The Transits of Venus*, pp. 150–97. The 1761 observations had discovered certain features of the transit (involving apparent and real contact—the black-drop effect) that made men better prepared to observe the phenomenon more acurately in 1769; see Woolf, *The Transits of Venus*, pp. 193–95.

85. Woolf, *The Transits of Venus*, pp. 148, 182–87.

86. Weld, *History*, 2:33–35; RS-CM, 5:292–95, 303.

87. See RS-CM, 5:130–31 and following.

88. Woolf, *The Transits of Venus*, pp. 163–67; Weld, *History*, 2:35–42; RS-CM, 5:184, 187–90, 192–209.

89. Stearn, "A Royal Society Appointment with Venus," p. 65.

90. Stearn, "A Royal Society Appointment with Venus"; Wooley, "Captain Cook and the Transit of Venus"; Beaglehole, *The Exploration of the Pacific*.

91. Woolf, *The Transits of Venus*, p. 188.

92. Woolf, *The Transits of Venus*, pp. 155, 157–60.

93. Woolf, *The Transits of Venus*, p. 153; Nordenmark, *Wargentin*, pp. 322–23; Lindroth, *Vetenskapsakademiens Historia*, 1:404–08.

94. Woolf, *The Transits of Venus*, pp. 183–84, 189.

95. Discussion began with a proposal by the academicians of Saint Petersburg in 1764; see Imperial Academy, *Protokoli*, vol. 2, October 21, 1764.

96. Woolf, *The Transits of Venus*, pp. 182–83, 188–89; Imperial Academy, *Protokoli*, vol. 2, October 19, 1767, and subsequently.

97. "Messieurs, vous savez que les principales Academies de l'Europe font des grands préparatifs pour l'observation du prochain passage de Venus; il conviendroit, ce me semble, que notre Academie y prit quelque part." AdW, "Registers 1766–1786," p. 19 (November 26, 1767).

98. See AdW, "Registers 1766–1786," pp. 33–34 (June 23, 1768, July 17, 1768).

99. "[Le Secrétaire] a rapporté que S. M. n'avoit pas répondu á la Lettre de l'Académie concernant le voyage pour observer le passage de Venus." AdW, "Registers 1766–1786," p. 35a (August 25, 1768).

100. Hindle, *Pursuit*, pp. 146–67.

101. Woolf, *The Transits of Venus*, p. 175.

102. American Philosophical Society, "Early Proceedings," December 18, December 30, 1768, February 9, 1769; see also "Address to Assembly," APS Archives, 1769.

103. Hindle, *Pursuit*, p. 153; American Philosophical Society, "Early Proceedings," p. 30 (January 7, 1769).

104. Woolf, *The Transits of Venus*, pp. 176, 183; see also DS-Formey, 16, 39 (letter from Danish Academy).

105. Rosen, "The Academy of Science," p. 122; Castelnau, *Mémoire historique*, p. 66; Woolf, *The Transit of Venus*, p. 188.

106. Woolf, *The Transits of Venus*, pp. 184–85.

107. RS, "Letters & Papers," Decade 5, #1. Other examples of this type could be cited.

108. RS-CM, 5:195 and 208.

109. Imperial Academy, *Protokoli*, vol. 2, October 5, 1767; October 8, October 19, 1767; RS-JBC, 26:417–19 (January 17, 1768); Woolf, *The Transits of Venus*, p. 180.

110. See Woolf, *The Transits of Venus*, p. 190; PA-PV, (1768), 87:253.

111. See Woolf, *The Transits of Venus*, p. 180; Lubimenko, *Correspondance scientifique*, Euler-Lalande correspondence.

112. Woolf, *The Transits of Venus*, p. 169; RS-CM, 5:367 (December 15, 1768).

113. RS, "Letters & Papers," Decade 5, #133; PA-PV, (1766), 85:297; (1768), 87:55.

114. Woolf, *The Transits of Venus*, 175.

115. See Imperial Academy, *Protokoli*, vols, 2–3.

116. Imperial Academy, *Protokoli*, vol. 3, January 24, 1771, September 9, 1771.

117. Woolf, *The Transits of Venus*, pp. 189–92, 196, 197; Woolf notes (p. 197), "for all ordinary purposes the dimensions of the solar system were established by the achievement of eighteenth-century astronomy in availing itself of the opportune transits of Venus."

118. Woolf, *The Transits of Venus*, pp. 193–94, points to several problems that hindered more accurate results: longitudinal precision, the black-drop effect, weather conditions, instrument flaws, observational psychology.

119. Woolf, *The Transits of Venus*, p. 197.

120. RS-JBC, 22:162.

121. Khrgian, *Meteorology*, pp. 68–69; Imperial Academy, *Protokoli*, vol. 2, November 3, 1768.

122. Berlin Academy, *Mémoires* (1771/1773), 27:36; (1778/1788), 33:71; Imperial Academy, *Protokoli*, vol. 2, October 5, 1769.

123. This evident from a letter Cotte wrote to Rozier and his journal. Cotte complains that he is not receiving all the observations he would like and that "the various public papers inform me that meteorological observations are taken very exactly at Lyon, Dijon, Montpellier, Toulouse, etc., at Stockholm, Berlin, Petersburg, and in several cities in Germany." *Observations* (1776), 7:93.

124. Lalande to Wargentin, letter dated Paris, December 13, 1771; KVA-Wargentin. On Cotte himself, see Taylor, "Cotte."

125. On the Royal Society of Medicine and its research program see Hannaway, "Medicine, Public Welfare and the State," esp. pp. 146–61 and her "Société royale de Médecine" See also Desaive et al. *Médecins, Climat et Epidémies*; Gillispie, *Science and Polity*, pp. 222–30ff; McClellan, "International Organization," pp. 379–82.

126. On this aspect of the society's program, see Hannaway, "Medicine, Public Welfare and the State," pp. 146–49; Desaive et al., *Médicins, Climat et Epidémies*, pp. 11–12, 24–25; Rosen, *From Medical Police to Social Medicine*, esp. pp. 201–19. The Baconian program is outlined in the prefatory material to the first volume of the *Histoire et Mémoires* of the *Société royale* for 1776(1779); Vicq d'Azyr was the moving force behind these and other programmatic statements of the *Société royale*. See Hannaway, "Medicine, Public Welfare and the State," pp. 148–52. See also Coleman, "Health & Hygiene in the *Encyclopédie*," for the deeper roots of the enterprise of the *Société royale*. One recalls in this connection the climatological perspective expressed by Montesquieu in his *Esprit des Lois*.

127. Desaive et al., *Médecins*, pp. 29, 139, 206; Société royale de médecine, *Mémoires* (1776), 1:105, for list of correspondents.

128. Hannaway, "Medicine, Public Welfare and the State," p. 160.

129. Société royale de médecine, *Mémoires* (1776/1779), 1:34; Roche, *Siécle*, 1:283–85; AD-CD, Dijon Academy Registres, 9:61v, 67, 68v.

130. Hannaway, "Medicine, Public Welfare, and the State, pp. 155–56; Société royale de médecine, *Mémoires* (1776/1779), 1:33–35.

131. Roche, *Siécle*, 1:284, for remark on academic movement.

132. *Observations* (1774), 3: 239.

133. On the complex history of 18th-century meteorological instruments, see Middleton, *Invention of Meteorological Instruments*, chs. 1–2; Daumas, *Les Instruments scientifiques*, ch. 4.

134. Daumas, *Instruments scientifiques*, p. 280.

135. On the Meteorological Society of Mannheim, see above, chapter 4; Khrgian, *Meteorology*, pp. 74–76; Frisinger, *The History of Meteorology*, pp. 10–11; Walter, *Geschichte Mannheims*, pp. 619–20; McClellan, "International Organization," pp. 382ff. The present discussion has benefitted considerably from a draft of Cassidy, "Meteorology in Mannheim," and from correspondence that author has so graciously entertained. The new society was the brainchild of the Abbé Johann Jakob Hemmer, member of the Mannheim Academy interested in electricity and meteorology and in charge of the academy's physical sciences cabinet; he was appointed permanent secretary of the new Meteorological Society.

136. See Societas Meteorologicae Palatinae, *Ephemerides*, (1781/1783), 1:1–124.

137. Societas Meteorologicase Palatinae, *Ephemerides*, (1781/1783), 1:23–42.

138. "L'Académie reçu les ordres particuliers de soigner le choix des Stations & celui d'Observateurs savans & etendus dans cette partie. Elle expédia en consequence des Lettres d'invitation, accompagnées du Plan imprimé pour coopérer, à ce grand & vaste institut, á diverses Académies, Universités & Corps littéraires, & leur offrit, suivant la volonté de S. A. S. tous les Instrumens gratis avec cette condition cependant que tous ces corps s'engageroient de les conserver soigneusement, de choisir de bons observateurs, de les continuer les obervations sans interruption, & d'envoyer annuellement leurs travaux à l'Académie," a copy to be found in AdW, L:XII:9, "Acta betr. die von Herrn Fontaine aus Mannheim eingesandten *Ephemerides Palatinae*, 1786–1799, pp. 3–4. This publicity effort was the result of the Mannheim Society contracting the publishing of its *Ephemerides* to the bookseller Fontaine. This document complains of the unsystematic and interrupted meteorological observations of the past and says that "all the observations that have been made cannot be mutually compared because of the differences in instruments."

139. Walter, *Geschichte Mannheims*, p. 620.

140. See Imperial Academy, *Protokoli*, vol. 3, November 6, 1780; a declination instrument was also supplied as part of the package of instruments.

141. See Imperial Academy, *Protokoli*, vol. 3, January 18, 1781 and earlier, January 15, 1781 and November 6, 1780.

142. See subsequent communications, Imperial Academy, *Protokoli*, vol. 3; see also Lubimenko, *Correspondance scientifique*, letters #1744, #1845. The Mannheim "Avertissement" (cited n. 138) remarked: "The Society also has great hopes that the Imperial Academy of Petersburg will establish stations in all the points of the vast Empire of Russia." That was in 1785.

143. Imperial Academy, *Protokoli*, vol. 3, June 17, 1781.

144. Societas Meteorologicae Palatinae, *Ephemerides*, passim; Hammermayer, "Süddeutsch-russische Wissenschaftsbeziehungen," p. 523; Mannheim "Avertissement."

145. Khrgian, *Meteorology*, p. 75; Societas Meteorologicae Palatinae, *Ephemerides* (1781/1783), 1:23–40.

146. Fifty individuals from thirty-seven locales participated in the project, sending in reports from sixty stations. The project was also the occasion for collecting and distribut-

ing non-meteorological data. See Cassidy, "Meteorology in Mannheim," appendix and map; Societas Meteorologicae Palatinae, *Ephemerides*, (1781/1783), vol. 1, passim; see also Mannheim "Avertissement," p. 3a.

147. Societas Meteorologicae Palatinae, *Ephemerides* (1781/1783), 1:23–24; Berlin Academy, *Mémoires* (1781/1783), 36:46–50; adW, "Registres 1766–1786," p. 301a (February 15, 1781).

148. AdW, "Registres 1766–1786," p. 302 (February 22, 1781); p. 305a (April 27, 1781).

149. Berlin Academy, *Mémoires* (1781/1783), 36:51.

150. Berlin Academy, *Mémoires* (1794/1799), vol. 46.

151. Westenrieder, *Geschichte*, 1:258.

152. Societas Meteorologicae Palatinae, *Ephemerides* (1781/1783), 1:23ff.

153. Kington, "Meteorological Observing in Scandinavia," p. 223; Societas Meteorologicae Palatinae, *Ephemerides* (1781/1783), 1:23, 29–30; Lindroth, *Vetenskapsakademiens Historia*, 1:455.

154. See letter from Wargentin reprinted in Societas Meteorologicae Palatinae, *Ephemerides* (1781/1783), 1:29–30 and *Ephemerides*, passim; see also Mannheim "Avertissement," p. 3a.

155. Societas Meteorologicae Palatinae, *Ephemerides* (1781/1783), 1:23ff.; *Ephemerides*, passim; Lomholt, *Danske Videnskabernes Selskap*, 3:152–53; Kington "Meteorological Observing in Scandinavia," p. 223.

156. Kington, "Meteorological Observing in Scandinavia," p. 226; Societas Meteorologicae Palatinae, *Ephemerides* (1783/1785), vol. 3; (1784/1786), vol. 4, and passim.

157. Kington, "Meteorological Observing in Scandinavia," p. 224; see also Societas Meteorologicae Palatinae, *Ephemerides* (1784/1786), vol. 4 and passim; DS-Formey, 34, 25, # 6: letter Jacob Wilse to Formey, dated Spydeberg, March 30, 1783.

158. Societas Meteorologicae Palatinae, *Ephemerides* (1781/1783), 1:42; AD-CD, Dijon Academy Registres, 11:268v (November 15, 1781).

159. See AD-CD, Dijon Academy Registres, 13:47–48 and D131 (letter: Hemmer to Maret, dated Mannheim, April 12, 1784).

160. Societas Meteorologicae Palatinae, *Ephemerides* (1781/1783), 1:23; AD-H, MS. D192 ("Adresse de l'Académie Palatine à la Société Royale des Sciences, 1781").

161. Societas Meteorologicae Palatinae, *Ephemerides* (1781/1783), 1:23ff. and *Ephemerides*, passim; see also Mannheim "Avertissement," p. 3.

162. Societas Meteorologicae Palatinae, *Ephemerides* (1781/1783), 1:23, 30–31, 41; PA-PV (1781), 100:17, and Pochette de Séance, February 3, 1781.

163. Societas Meteorologicae Palatinae, *Ephemerides* (1781/1783), 1:23 ff., and *Ephemerides*, passim; Mannheim "Avertissement"; Quetelet, "Premier Siècle," p. 16; Académie Royale des Sciences, *Académie*, pp. 38–39.

164. Societas Meteorologicae Palatinae, *Ephemerides*, passim; Walter, *Geschichte Mannheims*, p. 621; Mannheim "Avertissement."

165. Societas Meteorologicae Palatinae, *Ephemerides*, passim; Mannheim "Avertissement," p. 3a.

166. Societas Meteorologicae Palatinae, *Ephemerides* (1781/1783), 1:23; RS-JBC, 30:205 (February 22, 1781); Dawson, ed., *The Banks Letters*, Mannheim entry, #575, Latin letter dated February 19, 1781.

167. Hindle, *Pursuit*, pp. 347–48; Societas Meteorologicae Palatinae, *Ephemerides* (1782/1784), vol. 2.

168. Mannheim, "Avertissement," p. 3a.

169. Walter, *Geschichte Mannheims*, p. 621; Westenrieder, *Geschichte*, p. 258; for the distribution of the *Ephemerides*, see, for example, Imperial Academy, *Protokoli*, vol. 3, July 4, 1785, August 22, 1785; RS-JBC, 32:395 (February 8, 1787); AdW, "Registres 1766–1786," p. 324 (March 7, 1782).

170. That is not to say that academies and societies were the sole institutional contacts of the Mannheim Society, which also enlisted universities, schools, colleges, and individuals. See Societas Meteorologicae Palatinae, *Ephemerides* (1781/1783), 1:23, 44, 48, and *Ephemerides*, passim.

171. Letter: Fontaine to Merian, Mannheim, August 12, 1786 in AdW, I:XII:9, "Acta betr . . . *Ephemerides Palatinae*," p. 2; see also Böhm, *Kurpfälzische Akademie*, p. 122.

172. See Dawson, ed., *The Banks Letters*, Blagden Correspondence, #75 (p. 59), and RS-JBC, 32:66; Lyons, *The Royal Society*, p. 215. This episode is explored in some detail in the unpublished Battut, "Relations," pp. 58–64.

173. See Bertrand, *Académie*, pp. 114–19, 126–29; Hahn, *Anatomy*, p. 160. The general map of France was one of the projects of the 17th-century Paris Academy, suggested by Colbert; it had been actively worked on by Picard, Cassini I in 1701 and 1713–18 and by Cassini (III) de Thury in 1733 and again from 1750. Triangulation was at the heart of measurements taken to determine the shape of the earth in Lapland and Peru in the 1730s.

174. Banks to Bladgen, October 13, 1783, in Dawson, ed., *The Banks Letters*, Blagden Correspondence, #75 (p. 59).

175. "M. de Thury a dit, qu'il avoit reçu des Lettres d'Angleterre par les quelles on proposoit de former des triangles semblables à ceux de la méridienne de France et de les joindre à ces derniers, par le pas de Calais si l'Académie vouloit bien y conduire une chaine de triangles qu'on lieront par des signaux à ceux d'Angleterre, ce que l'Académie a accepté et a prié M. Cassini d'écrire pour pouvoir concerter l'éxecution de ce projet qui donneroit, avec une entière certitude la position respective des deux capitales et des deux observatoires." PA-PV, (1784), 103:129.

176. See RS-CM, 7:73–74; JR-JBC, 32:66; Lyons, *The Royal Society*, p. 215.

177. Hahn, *Anatomy*, p. 160; PA-PV (1788), 107:67–68.

178. RS-CM, 7:169 (June 29, 1787), p. 178 (November 8, 1787); RS-JBC, 34:400; PA-PV (1788), 107:67–68.

179. Lyons, *The Royal Society*, p. 215; Weld, *History*, 2:186–89; RS-JBC, 32:113–22; 35:438–48.

180. Lyons, *The Royal Society*, pp. 185–86; Weld, *History*, 2:30–31; RS-CM, 4:343–45 ff.; 5:13, 29–30, 64–66.

181. Hahn, *Anatomy*, p. 63; PA-PV, passim; RS-JBC, passim.

182. Weld, *History*, 2:359.

183. See Maindron, *Fondations*, pp. 44–46; Berlin Academy, *Mémoires* 40(1785/1787), vol. 40, "Histoire," pp. 36–43.

184. Berlin Academy, "*Mémoires* (1785), 40:42–43.

185. See PA-PV (1785), 104:72–73. This is a commission report of the Paris Academy on the proposal written by Condorcet; others signing the report were Borda, Duséjour, and Laplace.

186. "Cette question nous a parue importante de la manière dont elle est exposée dans le programme, à elle appartient la Logique et à la science des combinaisons, autant qu'à la jurisprudence et à la politique: et la science des combinaisons peut être regardée comme une branche de mathématiques. Nous croyons que l'académie ne doit pas considérer

l'objet de ce prix comme étranger à ses occupations." PA-PV (1785), 104:72; see also PA, Dossier Condorcet.

187. ". . . toutes les grandes académies de l'Europe ne doivent pour leur propre intérêt comme pour celui des sciences, que prétendre à l'égalite entre elles." PA-PV (1785), 104:73.

188. "Cette manière de se procurer le résultat des jugements combinés de trois compagnies, nous a parue ingénieuse . . . Cette méthode dont c'est ici le premier exemple peut être utile; par ce qu'il y a plusieurs questions et celle ci nous paroit etre du nombre pour lesquelles il seroit bon de se procurer les Lumières reunis de notions les plus éclairées." PA-PV (1785), 104:73.

189. Berlin Academy, *Mémoires* (1785), 40:37–43.

190. Berlin Academy, *Mémoires* (1785), 40:42–43; there is mention (p. 38) that in any event "the Academy cannot enter into these engagements with being authorized by the King."

191. See Imperial Academy, *Protokoli*, vol. 3, March 21, 1785.

192. Hahn, *Anatomy*, pp. 162–64; PA-PV (1789), 108:207; Bertrand, *Académie*, pp. 417–18; Maury, *Académies*, pp. 321–23.

193. Hahn, *Anatomy*, pp. 162–63, and consult Zupko, *French Weights and Measures*.

194. Hahn, *Anatomy*, pp. 162–64, 255–57.

195. Hahn, *Anatomy*, p. 163.

196. See PA-PV (1790), 109:85.

197. Hahn, *Anatomy*, p. 164.

198. PA-PV (1791), 109:314.

199. Crosland, "The Congress on Definitive Metric Standards."

200. "Il a été décidé que l'Académie ecrivait une lettre circulaire à toutes les Compagnies savantes de l'Europe pour leur faire part des moyens qu'elle se propose de mettre en usage pour fixer une unité de mesure et de les inviter à y concourir de leur côté. l'Académie invite egalement ceux de MM. les Academiciens qui voyageraient pour leurs affaires particulieres de se concerter avec les compagnies savantes des pays où ils se trouveraient sur l'adoption de la mèsure universelle et les opérations necessaires pour y parvenir." PA-PV (1791), 109:332.

7. Scientific Societies and the Making of the Scientist

1. See Introduction. The recent volume edited by Geison, *Professions and the French State, 1700–1900*, casts the discussion of this problem in a new light. Geison and his contributors make the state central to the development of medical, technical, and scientific professions in France, as opposed to the traditional "independence" from the state of professions in the United States and England. Fox's chapter, "Science, the University, and the State," covers science in the 19th century, but the professionalization of science in the 18th century is not treated directly in Geison's volume. Using the case of pre-Revolutionary barber-surgeons, Gelfand in his chapter, "A 'Monarchical Profession,'" argues for premodern professional forms existing in the 18th century. The case concerning science is different, but still, based on Gillispie's previous work and on what has been presented here, the parallel notion of a premodern, state-based profession for ancien régime science receives strong support. The role of the state relative to the scientific

societies not only in France but throughout Europe has been treated at length already, and this chapter provides some working details for a science professionalized in the context of 18th-century scientific societies. These details support the larger historiographical points made by Geison et al. See also MacLachlan, "Scientific Professionals in the Seventeenth Century." for a worthwhile contrast to the extent of professionalism in the 18th century.

2. See Ross, "Scientist: The Story of a Word."

3. *Encyclopédie* (1751), vol 1, article "Académicien." A somewhat free translation.

4. AD-H, MS. D123 (Registers of the Montpellier Society), June 28, 1777; see also Hahn, *Anatomy*, p. 72, for other, typical privileges reserved for academicians.

5. Quoted in Académie Royale des Sciences, *Index biographique*, p. 9. Their wives were to be called "Madame."

6. Imperial Academy, *Protokoli*, vol. 3, September 12, 1785.

7. One might mention in this connection a study of the backgrounds of 680 British[*] "scientists" of the 18th century in Hans, *New Trends in Education*, esp. pp. 31–36.

8. Without extensive documentation the point can ony be asserted. In two related articles, "Scientific Careers in Eighteenth-century France" and "Scientific Research as an Occupation in Eighteenth-century Paris," Roger Hahn argues effectively that contemporary science was not a profession by current standards or in ways that 18th-century usage would especially recognize. That leaves open the possibility that it was effectively a profession, i.e., that in the world of 18th-century science training was required, positions secured, careers pursued, standards exercised, research undertaken, and communities defined along lines we would judge de facto as professional. The social context may have been dramatically different, but the practice seems remarkably the same. Compare Hahn, *Anatomy*, passim and pp. 72–73; Duveen and Hahn, "Laplace's Succession to Bézout's Post"; Gillispie, *Science and Polity*, pp. 84–85ff.

9. On the 1785 reform and careers within the academy, see Hahn, *Anatomy*, pp. 98–101; McClellan, "The Académie Royale des Sciences," pp. 546–50. On the increasing age of Paris academicians and contemporary complaints, see Hahn, *Anatomy*, p. 97n.; McClellan, "International Organization," pp. 479–80.

10. See Roche, *Siècle*, 1:96–114. Says Roche (p. 107), "In its totality, the provincial milieu excluded professionalism of the intellect or functionalization of the Parisian type."

11. See Lindroth, *Swedish Men;* Frängsmyr, "Swedish Science in the Eighteenth Century."

12. Kopelevich, *Vozniknovenie*, pp. 176–229; Griffiths, Review, p. 155.

13. Imperial Academy, *Protokoli*, vol. 3, December 1, 1774.

14. Imperial Academy, *Protokoli*, vol. 3, December 1, 1774; see also March 17, 1774, March 24, 1774, for other instances of this regulation. This notion that work had to be submitted to an academy as a condition of candidacy finds echo in late regulations of the Montpellier Society of Sciences. A commission report on adjuncts of 1787 questioned "s'il ne seroit pas également important, vû l'accroissement des Sciences depuis 1706 de donner plus d'etenduë aux preuves de capacité des Aspirans aux places d'Adjoint et de dresser en conséquence un project, qui seroit soumis à l'examen de la société." AD-H, MS. D123 (Registers of the Society), August 30, 1787; see also earlier statements of this regulation, MS. D119 (Registers of the Society), August 17, 1730.

15. See Imperial Academy, *Protokoli*, vol. 3, February 13, 1783 and earlier December 18, 1775, January 15, 1776.

16. For a detailed view of another Saint Petersburg career, see Home, "Science as a Career in 18th-Century Russia: The Case of F. U. T. Aepinus."

17. On the work and careers of Euler and Lagrange, see Youschkevitch, "Euler"; Itard, "Lagrange."

18. Letter: Euler to Teplov, Berlin, July 15/26, 1763, in Youschkevitch and Winter Briefwechsel, 2:434–35, /348; see also Calinger, "Frederick the Great," p. 246.

19. Reprinted in Accademia della scienze di Torino, Primo secolo, p. 93.

20. The elitist and exclusive nature of learned societies precipitated some causes célèbres and violent reaction at the time of the French Revolution. See Gillispie, Science and Polity, ch. 4; "The Encyclopédie and the Jacobin Philosophy of Science"; Hahn, Anatomy, pp. 107–08 and ch. 5; Roche, Siècle, 1:102, where he says, "In effect, you were an academician because the group recognized you as such."

21. For various examples, see Hahn, Anatomy, p. 97n.; Roche, Siècle, 1:96ff., 301 ff.; Rosen, "The Academy of Sciences," pp. 80–83; Imperial Academy, Protokoli, vol. 3, December 9, 1779, January 10, 1780; etc.

22. See, for example, Dijon Academy, "Règlement de Police Académique," for May 14, 1764, in AD-CD, MS. "Table Générale des Registres de l'Académie."

23. "Vous en avez fait imprimer quelques uns avec le titre de corresondant, ce qui compromet toujours un corp Academique lors que un ouvrage est foible ou reprehensible a certains egards telle est par example votre lettre imprimée. Pour vous conformer plus exactement a nos status ayez l'attention Monsieur quand vous voudrez donner au public quelque ouvrage avec la qualité de notre correspondant de communiquer le manuscrit de cet ouvrage a L'Academie pour obtenir préalablement son aprobation ou du moins son agrement." BM-D, MS. 1627, p. 128. As Hahn, Anatomy, esp. pp. 135–58, makes clear, the abilities of the Paris Academy to make and enforce judgments on work submitted to it brought on a whole tradition of criticism of it as an institution. A report by Bailly appearing in the records of the Paris Academy provides one indication of why some people felt as they did: "J'ai lu par ordre de l'academie un mémoire qui contient quarante propositions absolument unintelligibles, et qui ne méritent en aucune façon l'attention de l'Academie." PA-PV (1770), 89:204.

24. For examples see Imperial Academy, Protokoli, vol. 3, March 12, 1781, May 28, 1781; PA-PV (1714), 33:57–58 (where the individuals in question were "un dangereux Exemple"); AD-H, MS. D117 (Registers of the Society), June 9, 1707.

25. "Mr. Peyre ayant pris la qualité d'Associé Chymiste de la Société Royale des Sciences dans une affiche imprimée et publié par un Operateur vendeur d'Orvietran et autres drogues qui est actuellement dans cette ville, il a été délibéré que dans l'Assemblée suivante, qui seroit à cette effect convoquée extraordinairement, le dit Mr. Peyre seroit admoneté par Mr. le President de la Séance qui lui représenteroit le tort qu'il s'est fait dans cette occasion en prenant le titre d'Associé qu'il n'a pas étant simplement Adjoint, et en le prenant dans une pareille circonstance, En même tems, on l'avertira que s'il ne tient á l'avenir une autre conduite, on sera forcé de le rayer du catalogue des Academiciens . . . En consequence de la délibération prise dans la dernière séance du 20 de ce mois, Mr. Brun sous-Directeur . . . a admoneté Mr. Peyre et lui a representé la grieveté de la faute qu'il a commise en prenant la qualité d'Academicien dans une affiche distribuée dans toute la ville par des Charlatans et Bateleurs de profession. Mr. Peyre a répondu que cette qualité lui avoit été donnée à son insçû, et qu'il en avoit témoigné hautement son mécontentement. Mr. Brun lui a dit que ce désaveu diminuoit sa faute à la verité, mais ne

l'excusoit pas entièrement, et qu'en donnant une approbation authentique a des compositions préparées par des Charlatans, il avoit visiblement oublié ce qu'exigeoit de lui la dignité du Corps, dont il a l'honneur d'etre membre. il a promis de se comporter plus sagement et plus décemment à l'avenir. De plus la compagnie usante du droit d'inspection et de dicipline qu'elle a sur chacun de ses membres, leur défend de s'immiscer à l'avenir en aucune manière soit directement, soit indirectement dans tout ce qui concerne la composition ou la distribution des drogues preparées par les Charlatans et Bateleurs de profession, se réservant de statuer contre ceux qui contreviendroient à cette déliberaton." AD-H, MS. D121 (Registers of the Society), August 20, 1761, August 27, 1761. See also, Kindelberger, "The Société royale," pp. 70–71. For a similar incident affecting the Montpellier academician and doctor Barthez, see AD-H, MS. D191, August 23, 1764.

26. "Monsieur, Il faut que je vous avertisse qu'a cause de la mauvaise conduite que le Sr. De L'Isle tient vers moy, j'ai defendû à tous les academiciens et Professeurs icy presens de n'avoir plus aucun Commerce des Lettres avec Luy, ni luy communiquer la moindre chose qui regarde les Sciences. Si j'auray le plaisir d'apprendre de vous Monsieur que vous avez rompû tout le commerce avec ce Mechant, qui portera sans doute la peine de son impertinence, je vous temoigneray ma reconnoissance dans toutes les occasions qui se presenteront." Letter: Razumovskii to Formey, dated Saint Petersburg, July 30, 1748, in DS-Formey, 27, 13. Separately J. B. d'Anville, first geographer to the French king and adjunct of the Paris Academy, did write back to Razumovskii, saying: "I have no connections here with whoever is suspect by the Academy and my own interest does not wish that I engage in any new contact or amelioration." Letter: August 5, 1751, reprinted in Andreef, *Geograficheskii departament*, p. 169; see also DS-Formey, 25, 35, #123, for another letter relating to this incident.

27. RS-CM, 5:105–07, 140–45; Weld, *History*, 1:509–14; 2:567. Ironically, Canton won the Copley medal of the Society in 1765.

28. Harnack, *Geschichte* (1901), p. 254; see pp. 250–57 for the full story.

29. Reprinted in Berlin Academy *Mémoires* (1750), 6:523; see this volume, passim, and Berlin Academy *Mémoires* (1757), 13:453ff., for other documents pertaining to the König case.

30. Johnson, *Professions and Power*, pp. 18, 37, 45.

31. RS-CM, vol. 5:146 (December 19, 1765); 4:302 (February 26, 1761); Lyons, *The Royal Society*, pp. 165–66; Weld, *History*, 2:43.

32. RS-CM, 5:148 (December 26, 1765).

33. Lyons, *The Royal Society*, pp. 343–44; see also RS-CM, 7:162 (March 8, 1787).

34. Maury, *Académies*, pp. 46–49. Note Delisle's letter to Wargentin in 1748: "Je vous ai pris pour mon correspondant principal en Suede. C'est un usage dans nôtre Academie de Paris que lorsque l'on choisit quelque correspondant—et connu de reputation, le Secretaire de l'Academie lui expedie des lettres de correspondance; vous trouverez icy inclus celles qui j'ay demandées pour vous & qui m'ont été accordées par le Suffrage de toute l'Academie." Delisle to Wargentin, April 12, 1748; KVA-Wargentin.

35. The first list of correspondents was published in the *Connaissance des Temps* for 1748, pp. 213–19. On the growth of the correspondent class of the Paris Academy, see McClellan, "The Académie Royale des Sciences," p. 547.

36. PA-PV (1747), 66:348; (1752), 71:577; (1753), 72:36, 167, and esp. 169–70.

37. PA-PV (1762), 81:151–52; (1767), 86:223; McClellan, "The Académie Royale des Sciences," pp. 551–52.

38. "Il paroissoit très important de s'occuper de cet objet, d'autant plus que la correspondance ayant été accordée depuis quelque temps à un grand nombre de personnes, on avoit senti la necessité d'en regler la forme et les loix." AD-H, MS. D121 (Registers of the Society), June 28, 1769. Kindelberger, "The Société royale," pp. 254ff.

39. "On ne recevra pour Correspondans que ceux qui auront donné à la Société Royale une idée avantageuse de leurs connoissances." AD-H, MS. D121 (Registers of the Society), July 13, 1769. Ten rules were passed for the government of correspondents.

40. "De plus il a été arrêté que Mrs. les Correspondans seroient avertis que leur titre ne doit être exprimé que par la phrase française *Correspondant de la Société Royale des Sciences de Montpellier*, ou par cette phrase Latine, qui en est la traduction, *Regiae Scientiarum Societatis Monspeliensis Correspondens*; les mots *Socius* et de *Socius Extraneus* que quelques-uns ont employé, ayant toûjours été usage pour distinguer les différens ordres qui composent le Corps proprement dit de la Société, il a fallu l'exemple de l'Academie Royale des Sciences employer un mot plus propre, quoique peu Latin." AD-H, MS. D121 (Registers of the Society), July 13, 1769. This same kind of concern for the appropriate use of academic titles is evident in a letter Lexell wrote from Paris to the Saint Petersburg Academy: "Mr. Rome has sent something to the Society of Medicine here, and he names himself as member or associate of the Petersburg Academy in his memoir. I wish to know if he really is such. It is strange that even the most reasonable of this nation have the folly to want to seem more than they are." Lubimenko, *Correspondance scientifique*, appendix, letter #24. On the subject of titles, see also Roche, *Siècle*, 1:301–2; Tisserand, *Académie de Dijon*, pp. 84–87.

41. "Il paroissoit necessaire de fixer le nombre des Correspondans, parceque ce nombre restant illimité, et augmenté indéfiniment, les places en étoient moins considerées; la Société royale etant interessée à donner à l'etat de Correspondant toute la consistance dont il est susceptible, en imitant sur cet objet important l'académie des Sciences de la Capitale." AD-H, MS. D123 (Registers of the Society), February 19, 1777; a circular letter was sent to all correspondents of the Montpellier Society informing them of the new regulations and requirements for them to maintain annual contact with the Society. See AD-H, MS. D202, #41 and D205, #37, for some responses to this letter.

42. "S'il est intéressant pour les progrès des sciences d'etendre des correspondances utiles, il est dangereux de prodiguer les marques d'estime d'une Compagnie savante et de les affoiblir en les multipliant." AD-H, MS. D123 (Registers of the Society), May 21, 1787; see also January 10, 1780.

43. "L'effet de sa deliberation qui a gueri plusieurs de ses Membres d'une Letargie indecente en prouve la sagesse et la necessité. D'ailleurs elle n'a fait que suivre l'exemple de l'Aademie des sciences de Paris qui a fait le meme reglement . . . Lorsque notre Academie exige que les associes mettent chaque Année quelque ouvrages en fond commun ce n'est pas un statut particulierement fait pour elle cest la loy fondamentalle de toutes les societes litteraires. Une academie est une société destinée a cultiver les sciences et les lettres, dont tous les membres doivent s'empresser de remplir cette destination et travailler à luy être utile. en effet une Academie composée de simlpe titulaires qui borneroient leurs vues au seul honneur d'etre inscrits sur son tableau seroit un corps paralytique et inutile dans la republique des lettres." BM-D, Ruffey Correspondence, MS. 1627, pp. 125–27.

44. Imperial Academy, *Protokoli*, vol. 3, March 15, 1781, November 29, 1781, August 22, 1782, February 5, 1784.

45. Imperial Academy, *Protokoli*, vol. 3, February 5, 1784. Another sign of the professional standards of academies and socities is their refusal to accept papers on the squaring of the circle, trisection of an angle, perpetual motion, and the like. The Paris Academy formally ruled out papers of this kind in 1775; the Berlin Academy did likewise in 1775 and again in 1777; the Royal Society of London and the Montpellier Society of Sciences enacted similar rules at least as early as 1749. See PA-PV (1775), 94:126; Berlin Academy *Mémoires* (1775), p. 64; (1777), p. 41; RS-JBC, 20:110 (May 4, 1749); AD-H, MS. D120, May 8, 1749. For further indications of academies and societies tightening up on standards, see Heilbron, *Electricity*, pp. 73–77.

46. Berlin Academy, *Mémoires* (1775/1777), p. 17.

47. See Roche, *Siècle*, 1:301ff., for an extended discussion of the *cumul* (or multi-memberships) in French provincial academies. Roche writes (p. 301), "It was a means of mounting the rungs of fame, to be indexed in the catalogue of members of the Republic of Letters. The multiplication of titles was an effective investment in publicity; it permitted, being known, to be received by the local elites in the provinces or in foreign countries."

48. "Eh! qui pourroit blamer mon ambition à cet egard! Ces honneurs Littéraires sont la seule recompense digne d'un homme de Lettres . . . Cela pourroit sans dout contribuer à augmenter dans nos Provinces ma reputation naissant." DS-Formey, 29, 8, #2, #6.

49. "Je ne saurais vous exprimer combien je serai sensible si vos illustres académiciens veuillent bien m'accorder le title d'*Associé* de votre célébre Académie de Berlin. Je suis déjà Associé de dix Académies." Letter dated November 11, 1786; DS-Formey, 29, 9, #6. Samoilowitz enclosed testimonial letters from these ten institutions. Later in 1790 he repeated his request to Formey and the Berlin Academy, saying: "I am already an Associate of twelve academies . . ." See letter dated Moscow, December 19/30, 1790.

50. "J'ai l'honneur d'etre d'une vingtaine de sociétés academiques." AD-H, MS. D206 (Correspondences), Béquillet to de Ratte, letter dated Dijon, April 1, 1777; Roche, *Siècle*, 1:302–04. See also Lubimenko, *Correspondance scientifique*, appendix, letter #30, for another example of this type.

51. "Si cela l'enrichit pas, au moins cela encourage et fait plaisir." DS-Formey, 22, 12, #3a, letter dated May 2, 1785. Bergius of the Swedish Academy made a similar statement upon his election to the American Philosophical Society: "the Society's electing me a foreign member flatters me as much, if not more, than my admittance into any of the European societies." Letter to Thomas Bond, dated Stockholm, October 8, 1771; KVA-Bergius, 15:604–05; original in English.

52. Lalande made strenuous efforts to get himself elected to the Swedish academy, as can be seen in his correspondence with Wargentin; see KVA-Wargentin, Lalande letters for the period 1763-64. In a letter dated Paris, October 10, 1763, Lalande broached the subject by saying: "Il y a longtemps mon cher confrere que je desire d'augmenter la confraternité qui unit, en obtenant une place d'associé etranger dans votre academie royale des Sciences de Stokolm." Similarly, Formey angled for a place as honorary foreign member of the Saint Petersburg Academy; see DS-Formey 31, 23, #10, correspondence with Strube de Piermont. Not everyone struggled so to achieve titles; Hufbauer, *German Chemical Community*, p. 182, relates the story of one Andrade who turned down memberships as "learned charlatanry."

53. "C'est la Société royale des sciences qui m'ouvert la porte de celle d'ici, et les deux ensembles, celles de Berlin et de Stockholm." AD-H, MS. D203 (Correspondences),

#29, letter dated Paris, February 10, 1761; see also earlier letter of Deparicieux, #28, dated February 18, 1746.

54. "Docteur en medecine de l'Université de Montpellier, Membre de la Société royale de Medecine de Paris, de l'Academie imperiale des curieux de la nature et de celle de Berlin. Ces titres academiques et un Scavant ouvrage . . . ont déterminé le choix de réuni les suffrages de la Société." AD-H, MS. D123 (Registers of the Montpellier Society), July 10, 1783.

55. Imperial Academy, *Protokoli*, vol. 3, January 31, 1774.

56. "Dr. Lüis Godin, qui n'a pas besoin d'autre recommendation pour donner une haute idée de son sçavoir, et de sa grande pratique dans les Sciences mathématiques, que d'être connu pour un des dignes membres de votre illustre et Royale académie." PA-PV (1746), 65:54.

57. "Nous avons été chargés m. D'alembert et moi de rendre Compte à l'académie si M: le Comte de Malvesi est dans le cas de pouvoir etre admis au nombre de ses Correspondants. Voici ce que nous pensons à cet égard. Mr. le Comte de Malvesi est membre de l'institut de Bologna, l'une des plus celebres Sociétés savantes de l'europe. Cette Consideration seule nous paroitroit suffisante pour devoir determiner L'académie à luy accorder le titre qu'il demander." PA, Dossier Malvezzi, report of May 5, 1773.

58. "Si donc cette illustre compagnie me l'accorde, ayez, je vous prie, la bonté de me le faire connaître le plutôt qu'il vous sera possible, afin que je puisse en user à la tête des ouvrages, que je vais mettre au jour." AD-H, MS. D207 (Correspondence) #18, letter dated May 27, 1783.

59. In Rozier, *Nouvelles Tables des Articles* (1775), 1:cxxi-cxxviii.

60. For the related example of Lorenz Crell, see Hufbauer, *German Chemical Community*, pp. 75–81.

61. In this regard it is noteworthy that academies and societies gave out parchment certificates of membership, which the Saint-Petersburg Academy (Imperial Academy, *Protokoli*, vol. 3, October 25, 1775) specifically referred to as "diplômes académiques." See samples of the Paris Academy, PA Dossier La Tourette; the Montpellier Society in AD-H, MS. D191; the Dijon Academy in AD-CD, MS. D129. See also RS-CM, 4:216–17 re the diplomas of the Royal Society of London. Exact specifications inscribed on membership certificates was of extraordinary concern to some; one correspondent of the Paris Academy insisted his *lettres de correspondance* be redrafted because his titles were improperly noted; see PA, Dossier Klinkemberg. Belidor of the Paris Academy made a similar request of the Royal Society through Jurin. "Je vous prie, Monsieur, tres instamment, si cela n'est point Contraire à vos formalités ordinaires, de me donner un brevet or Certificat, comme m'enont Envoié les academies roiales de France & deprusse par la Voie de leur Secretaires, dans lequel il soit Exposé que la Société roiale d'un consentement unanime, m'a receûe pour associé; et d'y aposer le Sceau de la Société, si elle en a un, si je vous demande cela, ce n'est pas que la lettre que vous m'avez fait la grace de M'ecrire, ne suffise; mais c'est que je regarderai ce que jai l'honneur de vous demander comme une marque d'honneur, qui sera egalement bien recue de la Cour et de mes amis dans les occasions, ou j'aurai interêt de le produire, parce qu'en france C'et l'usage d'en avoir pour les moindres dignités." Letter dated December 14, 1726; RS-LBC, 18:399–402, quote on pp. 400–01.

62. Imperial Academy, *Protokoli*, vol. 3, June 30, 1785. These mercantilistic remarks

are similar to ones made earlier in 1765 with regard to Euler's return to Saint Petersburg: "He could even help procure good subjects for the Academy because he has a lot of friends who would let themselves be engaged more easily if they knew he was in Saint-Petersburg." Letter: P. Dolgorouki to the Chancellor, Voronchof, 1765, in *Zapiski Imperatorskoi Akademii Nauk*, 6:86–7.

63. "Il me parait avantageux et meme nécessaire pour l'Académie de conserver ce Savant du premier order . . . Il ne s'agit que de menanger un peu l'état hypocondre et l'amour propre de ce grand géometre, sans lequel notre académie n'auroit aucun mathematicien celèbre et ne le retrouveroit aussi que difficilement en Allemagne et dans le reste de l'Europe." AdW, I:III:60, "Acta betr. die Dimission de M. de la Grange, 1787–1792," pp. 1–2.

64. ". . . a grand besoin d'un genie." Letter: Chambrier d'Oleires to Formey, Turin, February 19, 1785, DS-Formey, 7, 1, #34a.

Epilogue: The End of an Era

1. Roche, *Siècle*, 1:387, marks the decline from 1787. See also Kindelberger, "The Société royale," pp. 280ff., which would seem a representative instance of pre-1789 decline; Hahn, *Anatomy*, pp. 167–72; Berthe, *Dubois de Fosseux*, p. 202; McClellan, "The Académie Royale des Sciences," p. 551.

2. Hahn, *Anatomy*, pp. 159–66; Barrière, *Académie de Bordeaux*, pp. 23–24; Tisserand, *Académie de Dijon*, pp. 250–66; Kindelberger, "The Société royale," pp. 283ff.; Hannaway, "Medicine, Public Welfare, and the State," ch. 8.

3. On the closing of the academy and the progression to the institut, see Hahn, *Anatomy*, 7–8; Fayet, *La Révolution française*, passim and pp. 130–31. Outram, "The Ordeal of Vocation," presents a provocative interpretation of the effects of the Terror on individual scientists at the time.

4. Officially the Dijon Academy reopened in 1815; the Bordeaux Academy in 1828; the Lyon Academy in 1814. See Tisserand, *Académie de Dijon*, p. 250; Barriére, *Académie de Bordeaux*, p. 23; Dumas, *Histoire de l'Académie*, 1:372ff. Comments Dumas (1:388), "Since that time the Academy has been rightly considered as essentially a local institution." See also Savarit, "Les Académies de province au travail," pp. 464–69. He notes, "The academicians of our provinces in their works of such diversity seem to be the faithful mirrors of the thoughts and activities of their regions" (p. 464).

5. Fayet, *La Révolution française*, pp. 230ff.; Hahn, *Anatomy*, pp. 161–62; Crosland, *The Society of Arcueil*, pp. 25–40; Gillispie, "Science in the French Revolution," p. 95; Gillispie, *Science and Polity*, pp. 50–73.

6. Quetelet, "Premier Siècle," p. 15.

7. Harnack, *Geschichte* (1901), pp. 408–10.

8. Accademia delle scienze, *Primo secolo*, p. 6; Rosen, "The Academy of Sciences," pp. 136, 151–59.

9. Teich, "The Royal Bohemian Society of Sciences," p. 174, for example, speaks of the "deadening effect" of the French Revolution on the science society of Prague.

10. Bierens de Haan, *De Hollandsche Maatschappij*, p. 276.

11. Accademia della scienze, *Primo secolo*, p. 6.

12. Hahn, *Anatomy*, pp. 309–12. Hahn (p. 310) sees such institutional developments as

an effort to "soften the impact of France's military aims by forcing world attention onto its cultural merit." See also Crosland, *Society of Arcueil*, pp. 14–17.

13. For these details see Lindroth, *Vetenskapsakademiens Historia*, 2:6–7, 571–75; *Swedish Men*, p. 22; Frängsmyr, "Swedish Science," pp. 40–42; Ramsbottom, "The Royal Swedish Academy of Science," p. 371; Jorpes, *Jac. Berzelius*, pp. 84–87.

14. Becker, "Arbiß einer Geschichte," p. 15.

15. See Harnack, *Geschichte* (1901), pp. 297–300, 361–450, for further discussion of this period of reform in the Berlin Academy; see also Turner, "Prussian Universities and the Research Imperative," ch. 4.

16. Vucinich, *Science in Russian Culture*, p. 184; see pp. 184-293 for these details.

17. Stimson, *Scientists and Amateurs*, pp. 177–78, 190–93; Miller, "Sir Joseph Banks"; Lyons, *The Royal Society*, ch. 6; Weld, *History*, 2:298–303.

18. On the reform of the Royal Society, see Stimson, *Scientists and Amateurs*, pp. 205–17; Lyons, *The Royal Society*, ch. 6. On further changes in the organization of science in England see Ashby, *Technology and the Academies*; Miller, "Between Hostile Camps."

19. See Kellner, "Alexander von Humboldt"; Weld, *History*, 2:439–46; Théodoridès, "Humboldt and England," pp. 47–48, 53; Sharlin, *The Convergent Century*, pp. 11–12; Lyons, *The Royal Society*, p. 275; De Terra, *Humboldt*, pp. 205, 287.

20. The Paris Observatory and Gauss at Göttingen likewise coordinated observations. The (then) newly founded British Association for the Advancement of Science played a large role in this project along with the Royal Society; see Kellner, "Alexander von Humboldt," p. 255; Pancaldi, "Scientific Internationalism and the British Association," in MacLeod and Collins, eds., *The Parliament of Science*, pp. 145–69, and esp. 152–56.

21. The magnetic survey of the 19th century had its antecedents in the 18th century. The Mannheim Meteorological Society sought geophysical information in sending out declination intruments, and the records of the Royal Society of London report that on the occasion of presenting their map of magnetic variations, William Mountaine and James Dodson "renew their invitation to the ingenious in all nations to assist them with observations." RS-JBC, 22:618 (1757); the Mountaine-Dodson survey for the years 1700–1756 reported involved 50,000 magnetic observations. For more on the background to the magnetic project of the 1840s, see Weld, *History*, 2:428–38. On early 19th-century internationalism, see Merz, *History*, p. 252. So much has the 18th-century tradition of scientific society cooperation faded that Kellner can attribute the "origins of the organization of international scientific collaboration on a large scale" to Humboldt and the 19th-century magnetic survey; see Kellner, "Alexander von Humboldt," p. 252.

22. See appendixes; note the Temple Coffee House Botany Club of London as an early "specialized" grouping.

23. On the Linnean Society, see Gage, *A History of the Linnean Society*; Brockway, *Science and Colonial Expansion*, pp. 65–69.

24. Unlike the Linnean Society, the Geological Society emerged in conflict with the Royal Society of London; see Rudwick, "The Foundation of the Geological Society of London"; Woodword, *The History of the Geological Society*, pp. 25–30. On other British specialized societies, see Brockway, *Science and Colonial Expansion*, pp. 71–72.

25. For the further growth of specialized societies, see McClellan, "Scientific Associations."

26. On specialized scientific journals and the post-18th-century scientific press, see Thornton and Tully, *Scientific Books*, pp. 284–93; Price, *Science Since Babylon*, pp. 164–69.

27. Concerning these and related journals, see Thornton and Tully, *Scientific Books*, pp. 284–85; Kronick, *History*, pp. 105–06, 150, 163, 203–04, 264; McClellan, "The Scientific Press," pp. 436, 438; McKie, "Periodicals," pp. 130–31; Hahn, *Anatomy*, pp. 174–75; Hufbauer, *The German Chemical Community*, passim and pp. 62–82. Lalande's failed effort of the 1760s to transform the *Connaissance des Temps* into a specialized "annals of astronomy" is an interesting indication of the timing of specialization in the 18th-century press; see Marguet, "La Connaissance des Temps," pp. 136–37.

28. See Hahn, *Anatomy*, pp. 304–08; Crosland, ed., *Science in France in the Revolutionary Era*, passim; Crosland, *The Society of Arcueil*, esp. 3–4. For a fine feel for an early 19th-century career in French science, see Crosland, *Gay-Lussac*, especially the role of the *Institut* in his science, pp. 159–66; see also Shinn, *Ecole Polytechnique*.

29. See Hahn, *Anatomy*, chs. 9–10 and conclusions; Gillispie, *Science and Polity*, p. 187. Concludes Hahn of the academy (p. 318), "Today it is a glorious relic of the past, more akin to a Hall of Fame than an Olympic stadium. Time and the very nature of the growth of science, which the Academy has so successfully stimulated, were its undoing. Age, wisdom and ceremony now prevail where once youth, creativity and debate reigned supreme."

30. Ben-David, *The Scientist's Role*, ch. 7; Ben-David and Zloczower, "Universities and Academic Systems in Modern Societies"; Turner, "Prussian Universities"; see further, Schnabel, *Deutsche Geschichte in neunzehnten Jahrhundert*, and Merz, *History of European Thought*, pp. 159–72.

31. On the Royal Institution see Berman, *Social Change and Scientific Organizations*; Martin, "The Origins of the Royal Institution." See also Williams, ed., *Album of Science*, passim and ch. 1.

32. See Beer, "Coal Tar Dye Manufacture"; Cardwell, *The Organization of Science in England*; Fox and Weisz, eds., *The Organization of Science and Technology in France*; Gilpin, *France in the Age of the Scientific State*; and Bernal, *Science in History*, 2:535–54. All offer valuable further insights into the topology of science in the 19th century. If time or space permitted, a separate word would be said about the movement over mechanics institutes in 19th-century Britain and elsewhere; see Sharlin, *The Convergent Century*, p. 197; Mason, *A History of the Sciences*, ch. 35.

33. Morrell and Thackray, *Gentlemen of Science*, pp. 44ff.; Williams, ed., *Album of Science*, p. 6; Bernal, *Science in History*, 2:550; Mason, *A History of the Sciences*, pp. 444ff.

34. Morrell and Thackray, *Gentlemen of Science*; Orange, "The Beginnings of the British Association, 1831–1851," in MacLeod and Collins, eds., *Parliament*, pp. 43–64; Sharlin, *The Convergent Century*, pp. 201–02.

Bibliography

Aberhalden, Emil, ed. *Bericht über den Verlauf der Feier der 250 Wiederkehr der Tages der Erhebung der am 1 Jan. 1652 gegrundeten Academia Naturae Curiosorum zu 'Sacri Romani Imperii Academia Caesareo-Leopoldiana Naturae Curiosorum' durch Leopold I.* Halle, 1937.

Académie des sciences, belles-lettres et arts de Lyon. *Le Deuxième Centenaire de l'Académie, 1700–1900.* Lyon: A. Rey, 1900.

Académie des sciences, lettres et beaux-arts de Marseille. *Deux Siècles d'Histoire Académique, 1726–1926.* Marseille, 1926.

Académie royale des sciences de Turin. "Mémoire historique." *Mémoires de l'Académie* 1 (1784), Preface.

Académie Royale des Sciences, des Lettres et des Beaux-Arts de Belgique. *L'Académie Royale de Belgique depuis sa Fondation (1772–1922).* Brussels, 1922.

——— *Index biographique des membres, correspondants et associés de l'Académie royale de Belgique de 1769 à 1963.* Brussels: Palais des Académies, 1964.

Accademia delle scienze di Torino. *Il primo secolo della R. Accademia delle scienze di Torino: Notizie storiche e bibliographifche, 1783–1883.* Turin: Paravia, 1883.

Accademia di Agricoltura, Scienze, Lettere di Verona. *Anton Maria Lorgna: Memorie pubblicate nel secondo centenaria della Nascita.* Verona, 1935.

Adrian, Lord. "Academies of Science in the Modern World." *Proceedings of the American Philosophical Society* (1956), 100:326–30.

Aguilar-Pinal, Francisco. *La Real Academia Sevillana de Buenas Letras en el Siglo XVIII.* Madrid: C.S.I.C., 1966.

Akademia Nauk, Leningrad. See Imperial Academy, Saint Petersburg.

Allan, D. G. C. *William Shipley: Founder of the Royal Society of Arts.* London: Hutchinson, 1968.

Allibone, T. E. *The Royal Society and Its Dining Clubs.* Oxford: Pergamon Press, 1976.

Almhult, Arthur. "Academies in Sweden." *Baltic and Scandinavian Countries* (1937), 3:305–19.

Amburger, Erik. *Die Mitglieder der deutschen Akademie der Wissenschaften zu Berlin, 1700–1950.* Berlin: Akademie-Verlag, 1950.

American Philosophical Society. "Early Proceedings from the Manuscript Minutes of its Meetings, 1744–1838." *Proceedings of the American Philosophical Society* (1885), 22:1–711.

Anderson, M. S. *Europe in the Eighteenth Century 1713/1783*. London: Longman, 1970. (Original edition 1961.)

Andrade, E. N. da C. *A Brief History of the Royal Society*. London: The Royal Society, 1960.

Andrade, E. N. da C. and Martin, D. C. "The Royal Society and its foreign relations [sic]." *Endeavour* (1960), 19:72–82.

Andreef, A. I. *Geograficheskii Departament Akademii Nauk XVIII Veka*. Moscow, 1946.

Antony, H. D. "Scientific Books of the 18th Century and the Emergence of Modern Science." *Proceedings of the Royal Institution of Great Britain* (1958–1959), 37:334–43.

Arias-Divito, Juan Carlos. *Las Expediciones Cientificas Españolas durante el siglo XVIII*. Madrid: Ediciones Cultura Hispanica, 1968.

Armitage, Angers. "The Pilgrimage of Pingré, an Astronomer-Monk of Eighteenth-Century France." *Annals of Science* (1953), 9:47–63.

Arnim, M. *Mitglieder—Verzeichnisse der Gesellschaft der Wissenschaften zu Göttingen 1751–1927*. Göttingen: Dieterich, 1928.

Ashby, Sir Eric. *Technology and the Academies: An Essay on Universities and the Scientific Revolution*. London: Macmillan, 1958.

Aucoc, Leon. *L'Institut de France: Lois, Statuts et Règlements concernant les anciennes académies et l'Institut, de 1635 à 1829*. Paris: Imprimerie Nationale, 1889.

Baker, Keith Michael. *Condorcet: From Natural Philosophy to Social Mathematics*. Chicago and London: University of Chicago Press, 1975.

—— "Les Débuts de Condorcet au Secrétariat de l'Académie royale des sciences (1773–1776)." *Revue d'Histoire des Sciences* (1967), 20:229–88.

Barnes, C. L. "The Manchester Literary & Philosophical Society." In *The Soul of Manchester*, pp. 1–11. Manchester, 1931.

Barnes, Eugene B. "The International Exchange of Knowledge in Western Europe, 1680–1689." Ph.D. dissertation, University of Chicago, 1947.

Barnes, Sherman B. "The Editing of Early Learned Journals." *Osiris* (1936), 1:155–72.

Barrière, Pierre. *L'Académie de Bordeaux: Centre de culture internationale au XVIIIe siècle, 1712–1792*. Bordeaux and Paris: Editions Bière, 1951.

Bartholmèss, Christian. *Histoire philosophique de l'Académie de Prusse depuis Leibniz jusqu'à Schelling, particulièrement sous Frédéric-le-Grand*. 2 vols. Paris: Librairie de Marc Duclous, 1850.

Bataviaasch Genootschap van Kunsten en Wetenschappen. *Gedenkboek, 1778–1878*. Batavia, East Indies: Ernst, 1879.

Bates, Ralph S. *The American Academy of Arts and Sciences, 1780–1940*. Boston: American Academy, 1940.

—— *Scientific Societies in the United States*. New York: Wiley, 1945; 2d ed., New York: Columbia University Press, 1958; 3d ed., Boston: M.I.T. Press, 1965.

Battut, Jean Michel. "Les Relations entre l'Académie royale des sciences et la Royal Society of London à la veille de de la Révolution." Typescript, Maîtrise d'Histoire, Université de Paris VII, Faculté des Lettres et Sciences Humaines, 1978.

Bayerische Akademie der Wissenschaften, Munich. "Aus der Chronik der Bayerischen Akademie der Wissenschaften." *Jahrbuch 1956.* Munich: Verlag der Akademie, 1956.

Beaglehole, J. C. *The Exploration of the Pacific,* 3d ed. Stanford, Calif.: Stanford University Press, 1966.

Beall, Otho T., Jr. "Cotton Mather's early 'Curiosa Americana' and the Boston Philosophical Society of 1682." *William & Mary Quarterly,* 3d series (1961), 18:360–72.

Becker, Kurt. "Abriß einer Geschichte der Gesellschaft Naturforschender Freunde zu Berlin." In Gesellschaft Naturforschender Freunde, *Sitzungsberichte* (1973), 13:1–58.

Beckman, Nat. "Göteborgs Kungl. Vetenskaps -och Vitterhets- Samhällets, 1778–1928: En Historik." In Göteborgs Kungl. Vetenskaps -och Vitterhets- Samhällets, *Handlingar,* Fjärde Földjen (1928), 32(1):1–52.

Beer, John J. "Coal Tar Dye Manufacture and the Origins of the Modern Industrial Research Laboratory." *Isis* (1958), 49:123–31.

Behrens, C. B. A. *The Ancien Regime.* New York: Harcourt, Brace & World, 1967.

Bell, Whitfield. *Early American Science: Needs and Opportunities for Study.* Williamsburg, Va.: Institute of Early American History and Culture, 1955.

Ben-David, Joseph. "The Profession of Science and Its Powers." *Minerva* (1972), 10:362–83.

——— "The Scientific Role: The Conditions of its Establishment in Europe." *Minerva* (1965), 4:15–54.

——— *The Scientist's Role in Society: A Comparative Study.* Englewood Cliffs, N.J.: Prentice-Hall, 1971.

——— "The State of Sociological Theory and the Sociological Community: A Review Article." *Comparative Studies in Society and History* (1973), 15:448–72.

Ben-David, Joseph and Awraham Zloczower. "Universities and Academic Systems in Modern Society." *European Journal of Sociology* (1962), 3:45–84.

Berman, Morris. *Social Change and Scientific Organizations: The Royal Institution, 1799–1844.* Ithaca, N.Y.: Cornell University Press, 1978.

Bernal, J. D. "Les Rapports scientifiques entre la Grande-Bretagne et la France au XVIIIe siècle." *Revue d'Histoire des Sciences* (1956), 9:289–300.

——— *Science in History.* 4 vols. Cambridge, Mass.: M.I.T. Press, 1971; original ed., 1954.

Berthe, Leon-Noël. *Dubois de Fosseux, Secrétaire de l'Académie d'Arras et son bureau de correspondance.* Arras: C.N.R.S., 1969.

Berthelot, M. "Notice sur les origines et sur l'histoire de la Société Philomatique." *Mémoires publiés par la Société Philomatique à l'occasion du centenaire de sa fondation* (1888), pp. i–xvii.

—— "Sur les publications de la Société Philomatique et sur ses origines." *Journal des Savants* (1888), pp. 477–93.

Bertrand, Joseph. *L'Académie des Sciences et les Académiciens de 1666 à 1793.* Paris: Hetzel, 1869; reprint Amsterdam: Israël, 1969.

Biagi, Luigi. *L'Accademia di Belle Arti di Firenze.* Florence: LeMonnier, 1941.

Bierens de Haan, J. A. *De Hollandsche Maatschappij der Wetenschappen.* Haarlem, 1952.

Bierman, K.-R. and Gerhard Dunken. *Biographischer Index der Mitglieder: Deutsche Akademie der Wissenschaften zu Berlin.* Berlin: Akademie Verlag, 1960.

Bigourdan, G. *Sociétés savantes de Paris au XVIIe siècle et les origines de l'Académie des sciences.* Paris, 1917.

Birch, Thomas. *The History of the Royal Society of London.* 4 vols. London, 1756–1757. New York: Johnson Reprint, 1968.

Birembaut, Arthur. "L'Académie royale des sciences en 1780, vue par l'astronome suédois Lexell (1740–1784)." *Revue d'Histoire des Sciences* (1957), 10:148–66.

Böhm, Ludwig von. "Die Kurpflazische Akademie der Wissenschaften." In *Mannheim und der Rhein-Neckar Raum*, pp. 117–22. Mannheim: Bibliographisches Institut, 1965.

Bolton, Henry Carrington. *A Catalogue of Scientific and Technical Periodicals, 1665–1895*, 2d ed. Washington, D.C.: Smithsonian Institution [Miscellaneous Publications, 40], 1897.

—— "The Lunar Society." In Bolton, *The Scientific Correspondence of Joseph Priestley*, pp. 195–219. New York: privately printed, 1892; New York: Kraus Reprint, 1969.

Bonfield, Caitlin and Anthony Farrington. *The Royal Irish Academy and Its Library: A Brief Description.* Dublin, 1971.

Boreau, A. "Histoire de l'Académie des sciences et belles-lettres d'Angers, 1685–1789." *Mémoires de la Société académique de Maine et Loire* (1861), 9:1–43.

Boss, Valentin. *Newton and Russia: The Early Influence, 1698–1796.* Cambridge, Mass.: Harvard University Press, 1972.

Bouchard, Marcel. *L'Académie de Dijon et le premier discours de Rousseau.* Paris: Société des Belles-Lettres, 1950.

Bouillier, Francisque. *L'Institut et les académies de province.* Paris: Hachette, 1879.

Bowen, Catherine Drinker. *Francis Bacon: The Temper of a Man.* Boston and Toronto: Little Brown, 1963.

Bratt, Einar. *En Krönika on svenska glober.* Stockholm: Almqvist & Wiksells, 1968.

Braudel, Fernand. *Capitalism and Material Life 1400–1800.* New York: Harper & Row, 1975.

Bring, Samuel E. "A Contribution to the Biography of Christopher Polhem." In *Christopher Polhem: The Father of Swedish Technology*, Wm. A. Johnson, tr., pp. 1–105. Hartford, Conn.: Trinity College, 1963.

Brockway, Lucile H. *Science and Colonial Expansion: The Role of the British Royal Botanic Gardens.* New York and London: Academic Press, 1979.

Bronk, Detlev W. "Joseph Priestley and the Early History of the American Philosophical Society." *Proceedings of the American Philosophical Society* (1943), 86:103–07.

Brooks, R. C. "M. de Chabert and the 1750 Louisbourg Observatory." *Journal of the Royal Astronomical Society of Canada* (1979), 73:333–48.

Brown, Harcourt. "L'Académie de Physique de Caen, 1666–1675." *Mémoires de l'Académie nationale des sciences, arts et belles-lettres de Caen* (1939), 9:117–208.

—— "Buffon and the Royal Society of London." In Ashley Montagu, ed., *Studies and Essays Offered to George Sarton*, pp. 136–66. New York: Schuman, 1946.

—— "Maupertuis philosophe: Enlightenment and the Berlin Academy." *Studies on Voltaire and the Eighteenth Century* (1963), 24:255–69.

—— *Scientific Organizations in Seventeenth-Century France: 1620–1680*. Baltimore: Williams & Wilkins, 1934; reprint, New York: Russell & Russell, 1967.

—— "Voltaire and the Royal Society of London." *University of Toronto Quarterly* (1943), 13:25–43.

Brown, John H. "The Publication and Distribution of the *Trudy* of the Free Economic Society [St. Petersburg] 1765–1796." *The Russian Review* (1977), 36:341–50.

Brown, Theodore M. "The College of Physicians and the Acceptance of Iatromechanism in England, 1665–1695." *Bulletin of the History of Medicine* (1970), 44:12–30.

Bruinj, J. D. de. "Teylers Tweede Genootschap." In R. J. Forbes, ed., *Martinus van Marum*, 3:22–32.

Brunet, Pierre. "Guéneau de Montbeillard." *Mémoires de l'Académie des sciences, arts et belles-lettres de Dijon* (1925), pp. 125–31.

—— *Maupertuis, Étude Biographique*. 2 vols. Paris: Librairie Scientifique Albert Blanchard, 1929.

Buche, Joseph. "Notice sur la Société d'Émulation et d'Agriculture de l'Ain (1755–1899)." *Annales de la Société d'Émulation et d'Agriculture de l'Ain* (1899), 32:88–98.

Buck, A. "Die humanistischen Akademien in Italien." In Hartman and Vierhaus, eds., *Der Akademiegedank im 17. und 18. Jahrhundert*, pp. 11–26.

Butterfield, Herbert. *The Origins of Modern Science 1300–1800*, rev. ed. (1957). New York: Free Press, 1965.

Buttet, X. "Notice sur l'Académie de Soissons." *Bulletin de la Société Archéologique, historique et scientifique de Soissons* (1913–1921), 30:79–249.

Calinger, Ronald S. "Frederick the Great and the Berlin Academy of Sciences (1740–1766)." *Annals of Science* (1968), 24:239–49.

Cameron, Hector Charles. *Sir Joseph Banks, The Autocrat of the Philosophers*. London: Batchworth Press, 1952.

Campbell, N. and R. M. S. Smellie. *The Royal Society of Edinburgh, 1783–1983*. Edinburgh: Royal Society of Edinburgh, 1983.

Cantor, G. N. "The Eighteenth Century Problem." *History of Science* (1982), 20:44–63.

Cardwell, D. S. L. *The Organization of Science in England, A Retrospect.* London: Heinemann, 1957.

Carutti, Domenico *Breve Storia della Accademia dei Lincei.* Rome: Salviucci, 1883.

Cassidy, David C. "Meteorology in Mannheim: The Palatine Meteorological Society, 1780–1795." Unpublished.

Cassirer, Ernst. *The Philosophy of the Enlightenment.* F. C. A. Koelln and J. P. Pettegrove, trs. Boston: Beacon Press, 1955.

Castelnau, Jacques. *Mémoire historique et biographique sur l'ancienne Société royale des sciences de Montpellier.* Montpellier: Boehm, 1858.

Castres [France], Ville de. *L'Académie de Castres, 1648–1670: Son Œuvre et son rayonnement.* Castres: Musée Goya, 1970.

Cesarotti, Melchoir. "Riflessioni sopra i doveri Accademici: Verità, Novità, Utilità." Accademia di Scienze, Lettere ed Arti di Padua, *Saggi scientifici e letterarj* (1786), vol. 1.

Chaillou des Barres. "L'Histoire des anciennes sociétés littéraires et scientifiques à Auxerre." *Bulletin de la Société des sciences historiques et naturelles de L'Yonne* (1851), 5:177–205.

Chapin, Seymour L. "The Academy of Sciences During the 18th Century: An Astronomical Appraisal." *French Historical Studies* (1968), 5:371–404.

—— "Les Associés Libres de l'Académie Royale des Sciences: Un Projet Inédit pour le Modification de leurs Status (1788)." *Revue d'Histoire des Sciences* (1965), 18:7–13.

—— "Delisle." In C. C. Gillispie, ed., *Dictionary of Scientific Biography.* 4:22–25.

—— "Expeditions of the French Academy of Sciences, 1753," *Navigation* (1952), 3:120–22.

Charliat, P.-J. "L'Académie royale de Marine et la Révolution nautique au XVIIIe siècle." *Thalès* (1934), 1:71–82.

Cobban, Alfred. *A History of Modern France.* Vol. 1: *Old Régime and Revolution, 1715–1799.* 3d ed. Baltimore: Penguin Press, 1963.

Cochrane, Eric W. *Tradition and Enlightenment in the Tuscan Academies.* Chicago: University of Chicago Press, 1961.

Cohen, I. Bernard. *Franklin and Newton: An Inquiry into Speculative Newtonian Experimental Science.* Philadelphia: American Philosophical Society, 1956.

Cohen, John, et al. "Natural History of Learned and Scientific Societies." *Nature* (1954), 173:328–33.

Cohendy, Michael. "Les Constitutions de l'Académie des sciences, belles-lettres et arts de Clermont-Ferrand." *Mémoires de l'Académie des sciences, belles-lettres et arts de Clermont-Ferrand* (1871), 13:15–26.

Coleman, William. "Health & Hygiene in the *Encyclopédie*: A Doctrine for the Bourgeoisie." *Journal for the History of Medicine and Allied Sciences* (1974), 29:399–421.

Colonjon, Henry de. "Société Académique et Patriotique de Valence." *Bulletin de la Société départementale d'archéologie et de statistique de la Drome* (1866), 1:90–99.

Confino, Michael. "Les Enquêtes économiques de la Société libre d'économie de Saint-Pétersbourg, 1765–1820." *Revue Historique* (1962), 227:155–80.

Conklin, Edwin Grant. "Brief History of the American Philosophical Society." American Philosophical Society, *Yearbook for 1950*, pp. 5–34.

Conradi, Edward. "Learned Societies and Academies in Early Times." *Pedagogical Seminary* (1905), 12:384–426.

Corlieu, A. *L'Ancienne Faculté de Médecine de Paris.* Paris: Delahaye, 1877.

Coury, Charles. *L'Enseignement de la médecine en France dès ses origines à nos jours.* Paris: Expansion Scientifique Française, 1968.

Cousin, Jean. *L'Académie des sciences, belles-lettres et arts de Besançon.* Besançon: Jean Ledoux, 1954.

——— "L'Académie des sciences, belles-lettres et arts de Besançon au XVIIIe siècle et son œuvre scientifique." *Revue d'Histoire des Sciences* (1959), 12:327–44.

——— "L'Académie des sciences, belles-lettres et arts de Besançon et ses relations extérieurs au XVIIIe siècle." *Actes du IIIe Congrès National de la Société Française de Littérature Comparée*, Dijon 1959, pp. 55–64. Paris: Didier, 1960.

Crombie, A. C. *Medieval and Early Modern Science.* Garden City, N.Y.: Doubleday, 1959.

Crosland, Maurice P. "The Congress on Definitive Metric Standards, 1788–1794: The First International Scientific Conference?" *Isis* (1969), 60:226–31.

——— *Gay-Lussac: Scientist and Bourgeois.* Cambridge: Cambridge University Press, 1978.

——— *The Society of Arcueil: A View of French Science at the Time of Napoleon I.* Cambridge, Mass: Harvard University Press, 1967.

Crosland, Maurice P., ed. *The Emergence of Science in Western Europe.* New York: Science History Publications, 1976.

Crosland, Maurice P., ed. and commentator. *Science in France in the Revolutionary Era, Described by Thomas Bugge with Extracts from Other Contemporary Works.* Cambridge, Mass.: M.I.T. Press, 1969.

Dainville, François. "Foyers du culture scientifique dans la France méditerranéenne du XVIe au XVIIIe siècles." *Revue d'Histoire des Sciences* (1948), 1:289–300.

Dahlgren, E. W. *Kungl. Svenska Vetenskapsakademien Personförtechningar 1739–1915.* Stockholm: Almqvist & Wiksells, 1915.

Darnton, Robert. *The Business of Enlightenment: A Publishing History of the Encyclopédie, 1775–1800.* Cambridge, Mass.: Belknap Press, 1979.

——— *Mesmerism and the End of the Enlightenment in France.* Cambridge, Mass.: Harvard University Press, 1968.

— "The Rise of the Writer." *New York Review of Books*, (1979), 26:26–29.

Dassy, Louis Toussant. *L'Académie de Marseille: Ses Origines, ses publications, ses archives, ses mémoires.* Marseille: Ballatier-Feissat, 1887.

Dawson, Warren R., ed. *The Banks Letters.* London: British Museum, 1958.

Daumas, Maurice. *Les Instruments scientifiques aux XVIIe et XVIIIe siècles.* Paris: Presses Universitaires de France, 1953.

de Beer, Sir Gavin. "H. B. de Saussure's election into the Royal Society (1788)." *Notes and Records of the Royal Society of London* (1950), 7:264–67.

——— "Relations between Fellows of the Royal Society and French Men of Science When France and Britain Were at War." *Notes and Records of the Royal Society of London* (1952), 9:244–300.

——— *The Sciences Were Never At War.* London: Thomas Nelson, 1960.

——— "Voltaire, F.R.S." *Notes and Records of the Royal Society of London* (1951), 8:247–52.

de la Blancherie, Pahin. *Correspondance générale sur les sciences et les arts.* Paris: Au bureau de la Correspondance, 1779.

Dejob, Charles. "De l'établissement connu sous le nom de Lycée et d'Athénée et de quelques établissements analogues." *Revue Internationale de l'Enseignement* (1889), 18:4–38.

Delandine, François. *Couronnes académiques ou Recueil des prix proposés par les sociétés savants,* etc. Paris: Cuchet, 1787.

Delaunay, Paul. *La Vie médicale aux XVIe, XVIIe et XVIIIe siècles.* Paris: Éditions Hippocrate, 1935.

Delorme, Suzanne. "L'Académie Royale des Sciences: Ses Correspondants en Suisse." *Revue d'Histoire des Sciences* (1951), 4:159–70.

——— "Une Famille de grands Commis de l'État, amis des sciences au XVIIIe siècle: Les Trudaine." *Revue d'Histoire des Sciences* (1950), 3:101–09.

Denina, Abbé [Carlo]. "De l'influence qu'a eue l'Académie de Berlin sur d'autres grands établissemens [sic] de la même nature." *Mémoires de l'Académie de Berlin (Depuis Frédéric II)* (1792–1798), 45:562–73.

Desaive, J. P., J. P. Goubert, et al. *Médecins, climat et épidémies à la fin du XVIIIe siècle.* Paris: C.N.R.S., 1972.

Desplat, Christian. *Un Milieu socio-culturel provincial: L'Académie royale de Pau au XVIIIe siècle.* Pau: Béarn-Adour, 1971.

De Terra, Helmut. *Humboldt, The Life and Times.* New York: Knopf, 1955.

Dorn, Harold. "The Art of Building and the Science of Mechanics." Ph.D. dissertation, Princeton University, 1970.

Drake, Stillman. *Discoveries and Opinions of Galileo.* Garden City, N.Y.: Doubleday Anchor Books, 1957.

——— *Galileo at Work: His Scientific Biography.* Chicago: University of Chicago Press, 1978.

——— *Galileo Studies: Personality, Tradition, and Revolution.* Ann Arbor: University of Michigan Press, 1970.

Du Hamel, Jean Baptiste. *Regiae Scientiarum Academiae Historia.* Paris: S. Michellet, 1698.

Dulieu, Louis. "La Contribution montpelliéraine aux recueils de l'Académie royale des sciences." *Revue d'Histoire des Sciences* (1958), 11:250–62.

—— "Jean Astruc." *Revue d'Histoire des Sciences* (1973), 26:113–35.

—— "Le Mouvement scientifique montpelliérain aux XVIIIe siècle." *Revue d'Histoire des Sciences* (1958), 11:227–49.

Dumas, J. B. *Histoire de L'Académie royale des sciences, belles-lettres et arts de Lyon.* 2 vols. Lyon: Giberton, 1839.

Durant, Will and Ariel. *The Story of Civilization.* Vol. 9: *The Age of Voltaire.* New York: Simon and Schuster, 1965.

Duveen, Denis I. and Roger Hahn. "Laplace's Succession to Bézout's Post of *Examinateur des Elèves de l'Artillerie:* A Case History in the 'Lobbying' for Scientific Appointments in France during the Period Preceding the French Revolution." *Isis* (1957), 48:416–27.

Dvoichenko-Markov, Eufrosina. "The American Philosophical Society and Early Russian-American Relations." *Proceedings of the American Philosophical Society* (1950), 94:549–610.

—— "Benjamin Franklin, the American Philosophical Society, and the Russian Academy of Science." *Proceedings of the American Philosophical Society* (1947), 91:250–57.

Dyson, Sir Frank W. "Histoire de l'observatoire de Greenwich." *Revue Scientifique* (1931), 69:1–13.

Eamon, William and Françoise Paheau. "The Accademia Segreta of Girolamo Ruscelli: A Sixteenth-Century Italian Scientific Society." *Isis* (1984), 75:327–42.

Eisenstein, Elizabeth. *The Printing Revolution in Early Modern Europe.* Cambridge: Cambridge University Press, 1983.

Elliott, Philip. *The Sociology of the Professions.* London: Macmillan, 1972.

Emerson, Roger L. "The Edinburgh Society for the Importation of Foreign Seeds and Plants, 1764–1773." *Eighteenth-Century Life* (1982), 7:73–95.

—— "The Philosophical Society of Edinburgh, 1737–1747." *British Journal for the History of Science* (1979), 12:154–91.

—— "The Philosphical Society of Edinburgh, 1748–1768." *British Journal for the History of Science* (1981), 14:133–76.

Encyclopédie ou Dictionnaire Raisonné des Sciences, des Arts et des Métiers, [Diderot and d'Alembert, eds.]. 10 vols. Paris 1751–1765. *Supplément à l'Encyclopédie.* 4 vols. Amsterdam, 1776–1777. Microprint Reprint, New York: Redex Microprint Corp., 1969.

Farrington, Benjamin. *Francis Bacon.* New York: Henry Schuman, 1949.

Fäy, Bernard. "Learned Societies in Europe and America in the Eighteenth Century." *American Historical Review* (1932), 37:255–67.

Fayet, Joseph. *La Révolution française et la science: 1789–1795.* Paris: Librairie Marcel Rivière, 1960.

Ferrâo, António. *A Academia des Sciências de Lisboa.* Coimbra: Imprensa da Universidada, 1923. [Separata do "Boletim da Classe de Letras," 15.]

Fersman, A. *The Pacific: Russian Scientific Investigation.* New York: Greenwood Press, 1969.

Feyl, Othmar. *Beiträge zur Geschichte der slawischen Verbindung und internationalen Kontake der Universität Jena.* Jena: Gustav Fischer, 1960.

Fleur, M. E. "Quelques Documents sur la Société royale des sciences et arts de Metz." *Mémoires de l'Académie nationale de Metz* (1935–36), 26:99–105.

Fleure, H. J. "The Manchester Literary and Philosophical Society." *Endeavour* (1947), 6:147–52.

Forbes, [Edward]. "Address by Principal Forbes on the History of the Society (1862)." Royal Society of Edinburgh, *General Index*, 1890.

Forbes, Eric G. *The Euler-Mayer Correspondence, 1751–1755: A New Perspective on Eighteenth-Century Advances in the Lunar Theory.* New York: American Elsevier, 1971.

Forbes, Robert James. "The Hollandsche Maatschappij der Wetenschappen (Dutch Society of Science)." *Journal of the Royal Society of Arts* (1963), 111:897–900.

Forbes, Robert James, ed. *Martinus van Marum, Life and Works.* 6 vols. Haarlem: Hollandsche Maatschappij der Wetenschappen, 1969–1976.

Foucault, Michel. *Naissance de la clinique.* Paris: Presses Universitaires de France, 1965.

Fouchon, C. "Les Sociétés Savantes d'Orléans devant le décret de la Convention du 8 août 1793." *Mémoires de la Société d'Agriculture, sciences, belles-lettres et arts d'Orléans* (1912), 12:64–111.

Fox, Robert. "Science, the University, and the State in Nineteenth-Century France." In Geison, ed., *Professions and the French State*, pp. 66–145.

Fox, Robert and George Weisz, eds. *The Organization of Science and Technology in France, 1808–1914.* Cambridge: Cambridge University Press; Paris: Éditions de la Maison des Sciences de l'Homme, 1980.

Franzoja, Matteo. "Introduzione storica." Accademia di Scienze, Lettere ed Arti di Padua, *Saggi scientifici e letterarj* (1786), 1:i–xii.

Frémy, Edouard. *L'Académie des derniers Valois: Académie de poésie et de musique 1570–1576: Académie du Palais 1576–1585.* Paris: Ernest Leroux, s.d.

Fries, Robert E. *On the Origin and Foundation of the Royal Swedish Academy of Science.* Stockholm: Almqvist & Wiksells, 1940.

Frijhoff, W. Th. M. *La Société Néerlandaise et ses gradués, 1575–1814.* Amsterdam: APA—Holland University Press, 1981.

Frisinger, H. Howard. *The History of Meteorology to 1800.* New York: Science History Publications, 1977.

Frängsmyr, Tore. "Swedish Science in the Eighteenth Century." *History of Science* (1974), 12:29–42.

Fueter, Eduard. *Geschichte der exakten Wissenschaften in der Schweizerischen Aufklärung (1680–1780).* Arrau and Leipzig: Sauerländer, 1941.

Furet, François, ed. *Livre et société dans la France du XVIIIe siècle.* The Hague: Mouton, 1965.

Fuss, Paul-Henri, ed. *Correspondance mathématique et physique de quelques célèbres géomètres du XVIIIe siècle, précédée d'une notice sur la vie et sur les écrits de Léonard Euler.* 2 vols. Saint Petersburg: Académie Impériale des Sciences, 1843; reprint, New York: Johnson Reprint, 1968.

Gage, A. T. *A History of the Linnean Society of London.* London: Printed for the Linnean Society, 1938.

Gallois, L. "L'Académie des sciences et les origines de la carte de Cassini." *Annales de Géographie* (1909), 17:193–204 and 289–310.

Garrison, Fielding, H. "The Medical and Scientific Periodicals of the Seventeenth and Eighteenth Centuries." *Bulletin of the Institute of the History of Medicine* (1934), 2:285–343.

Gaston, Jerry. "Sociology of Science and Technology." In P. Durbin, ed., *A Guide to the Culture of Science, Technology, and Medicine,* pp. 465–526. New York: Free Press, 1980.

Gauja, Pierre. *L'Académie des sciences de l'Institut de France.* Paris: Gauthier-Villars, 1934.

——— "Les Origines de l'Académie des sciences de Paris." In Académie des Sciences, L'Institut de France, *Troisième centenaire* (1967), pp. 1–51.

Gautier, M. "Notice historique sur l'Académie delphinale." *Bulletin de l'Académie Delphinale* (1876), 11:lxxvi–xciv.

Gaxotte, Pierre. *L'Académie française.* Paris: Hachette, 1965.

Gay, Peter. *The Enlightenment: An Interpretation.* 2 vols. New York: Knopf, 1966–69.

Geison, Gerald L., ed. *Professions and the French State, 1700–1900.* Philadelphia: University of Pennsylvania Press, 1984.

Gelfand, Toby. *Professionalizing Modern Medicine: Paris Surgeons and Medical Science and Institutions in the 18th Century.* Westport, Conn.: Greenwood Press, 1980.

——— "A 'Monarchical Profession' in the Old Regime: Surgeons, Ordinary Practitioners, and Medical Professionalization in Eighteenth-Century France." In Geison, ed., *Professions and the French State,* pp. 149–80.

Génique, Gaston. "Académie de Châlons, ses origines et son caractère au XVIIIe siècle." *Mémoires de la Société d'Agriculture, Commerce, Sciences et Arts du Département de la Marne* (1947–1952), 27:8–19.

Gertz, Otto. *Kungl. Fysiografiska Sällskapet i Lund, 1772–1940: Historisk Överblick och Personförteckningar.* Lund: Håkan Ohlsson, 1940.

Gesellschaft Naturforschender Freunde zu Berlin. "Festschrift zum 200jährigen Bestehen der Gesellschaft," Konrad Herter and Kurt Becker, eds., *Sitzungsberichte der Gesellschaft Naturforschender Freunde zu Berlin* (1973), p. 13.

Gillispie, Charles Coulston. "The *Encyclopédie* and the Jacobin Philosophy of Science." In Marshall Calgett, ed., *Critical Problems in the History of Science,* pp. 255–90. Madison: University of Wisconsin Press, 1959.

——— *The Montgolfier Brothers and the Invention of Aviation.* Princeton: Princeton University Press, 1983.

——— *Science and Polity in France at the End of the Old Regime.* Princeton: Princeton University Press, 1980.

——— "Science in the French Revolution." In Bernard Barber and Walter Hirsch, eds., *The Sociology of Science.* New York: Free Press of Glenco, 1962.

Gillispie, Charles Coulston, ed. in chief. *Dictionary of Scientific Biography.* 16 vols. New York: Scribner's: 1970–1980.

Gillispie, Charles Coulston and A. P. Youschkevitch. *Lazare Carnot, Savant.* Princeton: Princeton University Press, 1971.

Gilpin, Robert. *France in the Age of the Scientific State.* Princeton: Princeton University Press, 1968.

Giuliani, Renzo. "I Georgofili e l'agricoltura Toscana." *Atti e Memorie della Accademia Petrarca di Lettere, Arti, ed Scienze* (1952–1957), 36:175–93.

Gnuchev. V. F. *Materiali gla Istorii Ekspeditsii Akademii Nauk v XVIII i XIX vekax.* Moscow, 1940.

Golder, F. A. *Russian Expansion on the Pacific: 1641–1850.* Cleveland: Arthur H. Clark, 1914.

Goubert, Pierre. *The Ancien Régime: French Society, 1600–1750.* Steve Cox, tr. New York: Harper & Row, 1974.

Gray, Lewis Cecil. *History of Agriculture in the Southern United States to 1860.* 2 vols. Washington, D.C.: Carnegie Institute, 1933.

Greene, John C. "Science, Learning and Utility: Patterns of Organization in the Early American Republic." In Oleson and Brown, eds., *The Pursuit of Knowledge in the Early American Republic,* pp. 1–20.

Griffiths, David M. Review [Kopelevich, *The Rise of Scientific Academies* (1974)]. *Isis* (1977), 68:154–56.

Guerlac, Henry. *Essays and Papers in the History of Modern Science.* Baltimore: Johns Hopkins University Press, 1977.

——— *Lavoisier, The Crucial Year.* Ithaca, N.Y.: Cornell University Press, 1961.

Guillon, M. "L'Ancienne Académie royale d'Orléans." *Mémoires de la Société d'Agriculture, sciences, belles-lettres et arts d'Orléans* (1908), 8:369–444.

Gunther, R. T. *Early Science in Oxford.* 14 vols. Oxford, 1923–45; reprint, London: Dawsons, 1968.

Hahn, Roger. *Anatomy of a Scientific Institution: The Paris Academy of Sciences, 1666–1803.* Berkeley, Los Angeles, and London: University of California Press, 1971.

——— "The Application of Science to Society: The Societies of Arts." *Studies on Voltaire and the Eighteenth Century* (1963), 24/27:829–36.

——— "Scientific Careers in Eighteenth-Century France." In Crosland, ed., *The Emergence of Science in Western Europe,* pp. 127–38.

——— "Scientific Research as an Occupation in Eighteenth–Century Paris." *Minerva* (1975), 13:501–13.

Hall, A. Rupert. *The Cambridge Philosophical Society: A History 1819–1969.* Cambridge: Cambridge Philosophical Society, 1969.

_____ "Merton Revisited." *History of Science* (1963), 2:1–15.

_____ *Philosophers at War: The Quarrel between Newton and Leibniz*. Cambridge: Cambridge University Press, 1980.

_____ *The Scientific Revolution 1500–1800: The Formation of the Modern Scientific Attitude*. 2d ed. Boston: Beacon Press, 1966.

Hall, A. Rupert, ed. [Accademia del Cimento], *Essays of Natural Experiments*. Richard Waller tr. New York: Johnson Reprint, 1964.

Hall, A. Rupert and Marie Boas Hall. "Les Liens publics et privés dans les relations franco-anglaises (1660–1720): I. D'après la correspondance d'Oldenburg. II. D'après la correspondance de Newton." *Revue de Synthèse* (1976), 97:51–75.

Hall, Marie Boas. "Oldenburg and the Art of Scientific Communication." *British Journal for the History of Science* (1965), 2:277–90.

_____ "Science in the Early Royal Society." In Crosland, ed., *The Emergence of Science in Western Europe*, pp. 57–77.

_____ *The Scientific Renaissance, 1450–1630*. New York: Harper Torchbooks, 1962.

_____ "Sources for the Early History of the Royal Society." In Birch, *History* (1968), pp. xxvi–xxx.

Hammer, Karl and Jürgen Voss, eds. *Historische Forschung im 18. Jahrhundert: Organisation, Zielsetzung, Ergebnisse*. Bonn: Röhrscheid, 1976.

Hammermayer, Ludwig. "Akademiebewegung und Wissenschaftsorganisation: Formen, Tendenzen und Wandel in Europa während der zweiten hälfte des 18. Jahrhunderts." In Erik Amburger, ed., *Wissenschaftspolitik in Mittel- und Osteuropa*, pp. 1–84. Berlin: Ulrich Camen, 1976.

_____ "Die Benediktiner und die Akademiebewegung im Katholischen Deutschland 1720 bis 1770." *Studien und Mitteilungen zur Geschichte des Benediktinerordens* (1960), 70:45–146.

_____ "Europäische Akademiebewegung und italienische Aufklärung." *Historisches Jahrbuch* (1961), 81:247–63.

_____ *Grüdungs- und Frühgeschichte der bayerischen Akademie der Wissenschaften*. [Münchener Historische Studien. Abteilung Bayerische Geschichte, 4: Max Spindler, ed.] Kallmünz/Opf: Verlag Michael Lassleben, 1959.

_____ "Süddeutsch-russische Wissenschaftsbeziehungen im 18. Jahrhundert: Korrespondenzen der Bayerischen Akademie der Wissenschaften mit Gelehrten der Kaiserlichen Akademie und der Freien Ökonomischen Gesellschaft zu St. Petersburg, 1761–1806." In Albrecht, Kraus, & Reindel, eds., *Festschrift für Max Spindler* (Munich, 1969), pp. 503–38.

Handlon, C. Rollins. "The Decline and Fall of Scientific Societies." *Surgery* (1959), 46:1–8.

Hankins, Thomas. *Jean d'Alembert: Science and the Enlightenment*. Oxford: Clarendon Press, 1970.

Hannaway, Caroline C. F. "Medicine, Public Welfare and the State in 18th-Century France: The Société Royale de Médecine of Paris: 1776–1793." Ph.D. dissertation; Johns Hopkins University, 1974.

—— "The Société Royale de Médecine and Epidemics in the *Ancien Régime*." *Bulletin of the History of Medicine* (1972), 46:257–73.

Hans, Nicholas A. "Academies." *Encyclopedia Britannica*, (14th ed., 1967 version), 1:57–60.

—— *New Trends in Education in the Eighteenth Century*. London: Routledge & Kegan Paul, 1951.

Hansen, James Roger. "Scientific Fellowship in a Swiss Community Enlightenment: A History of Zurich's Physical Society, 1746–1798." Ph.D. dissertation, Ohio State University, 1981.

Harnack, Adolf von. *Geschichte der königlich preussischen Akademie der Wissenschaften zu Berlin*. 3 vols. Berlin: Reichsdruckerie, 1900; 1 vol., Berlin: G. Stilke, 1901.

Harnack, Axel von. "Die Akademien der Wissenschaften." Fritz Milkau, ed., *Handbuch der Bibliotheckswissenschaften* (1931), 1:850–76.

Hartley, Harold, ed. *The Royal Society: Its Origins and Founders*. London, 1960.

Hartmann, Fritz and Rudolf Vierhaus, eds. *Der Akademiegedank in 17. und 18. Jahrhundert*. Bremen & Wolfenbüttel: Jacobi Verlag, 1977.

Hartung, Frank E. "Science as an Institution." *Philosophy of Science* (1951), 18:35–54.

Haumant, Émile. *La Culture française en Russie (1700–1900)*. Paris: Librairie Hachette, 1910.

Hebbe, P. M. "Les Relations scientifiques russe-suédoises pendant la deuxième moitié du XVIIIe siècle." *Lychnos* (1938), p. 392.

Heilbron, J. L. *Electricity in the 17th and 18th Centuries: A Study of Early Modern Physics*. Berkeley: University of California Press, 1979.

—— *Elements of Early Modern Physics*. Berkeley: University of California Press, 1982.

Hildebrand, Bengt. *Kungl. Svenska Vetenskapsakademien Förhistoria, Grundläggning och Första Organisation*. Stockholm, 1939.

Hill, Christopher. *Intellectual Origins of the English Revolution*. Oxford: Oxford University Press, 1965.

Hindle, Brooke. *The Pursuit of Science in Revolutionary America: 1735–1789*. Chapel Hill: University of North Carolina Press, 1956.

Holmberg, Arne. *Some Notes on the Royal Swedish Academy of Science*. Stockholm: Almqvist & Wiksells, 1939.

Home, Roderick W. "Science as a Career in 18th-Century Russia: The Case of F. U. T. Aepinus." *Slavonic and East European Review* (1973), 51:75–94.

Hoppen, K. Theodore. *The Common Scientist in the Seventeenth Century: A Study of the Dublin Philosophical Society 1683–1708*. London: Routlege & Kegan Paul, 1970.

—— "The Nature of the Early Royal Society." *British Journal for the History of Science* (1976), 9:1–24 and 243–73.

—— *The Papers of the Dublin Philosophical Society, 1683–1708*. Dublin: Irish Manuscripts Commission, 1983.

—— "The Royal Society and Ireland I." *Notes and Records of the Royal Society* (1963), 18:125–35.

—— "The Royal Society and Ireland II." *Notes and Records of the Royal Society* (1965) 20:78–99.

Horn, D. B. *A Short History of the University of Edinburgh 1556–1889*. Edinburgh: University Press, 1967.

Hubrig, Hans. *Die patriotischen Gesellschaften des 18. Jahrhunderts*. Berlin: Verlag Julius Beltz, 1957.

Hudson, Derec and Kenneth W. Luckhurst. *The Royal Society of Arts (London) 1754–1954*. London: John Murray, 1954.

Hufbauer, Karl. *The Formation of the German Chemical Community (1720–1795)*. Berkeley: University of California Press, 1982.

Huggins, Sir William. *The Royal Society, or Science in the State and in the Schools*. London and New York: Stechert, 1906.

Hume, Abraham. *The Learned Societies and Printing Clubs of the United Kingdom*. London: G. Willis, 1853.

Hunter, Michael. "The Social Basis and Changing Fortunes of an Early Scientific Institution: An Analysis of the Membership of the Royal Society, 1660–1685." *Notes and Records of the Royal Society* (1976), 31:9–114.

Imperial Academy of Sciences, Saint Petersburg. [Imperatorskaia Akademiia Nauk.] *Materiali gla Istorii Imperatorskoi Akademii Nauk*. 10 vols. Saint Petersburg: Academic Press, 1885–1900.

—— *Protokoli Zasidanii Konferensii Imperatorskoi Akademii Nauk c 1725 po 1803 goda (Procès verbaux des séances de l'Académie impériale des sciences depuis sa foundation jusqu'à 1803)*. 4 vols. Saint Petersburg: Academy Press, 1897–1911.

—— *Zapiski Imperatorskoi Akademii*. Vol. 6. Saint Petersburg, 1865.

Imperiale Reale Accademia de Scienze, Lettere ed Arti degli Agiate in Rovereto. *Memorie pubblicate per commenorare il suo centocinquantesimo anno di vita*. Rovereto: 1901.

Institut de France, Académie des Sciences. *Index Biographique de l'Académie des Sciences*. Paris: Gauthier-Villars, 1979.

Itard, Jean. "Lagrange." C. C. Gillispie, ed., *Dictionary of Scientific Biography*, 7:559–73.

Jacob, Margaret C. *The Newtonians and the English Revolution 1689–1720*. Ithaca, N.Y.: Cornell University Press, 1976.

Jacquot, Jenn. "Sir Hans Sloane and French Men of Science." *Notes and Records of the Royal Society* (1953), 10:85–98.

Jaeggli, Alvin E. "Recueil des pièces qui ont remporté le prix de l'Académie Royale des Sciences depuis 1720 jusqu'en 1772." *Gesnerus* (1977), 34:408–14.

Jecht, Richard. "Kurzer Wegweiser durch die Geschichte der Oberlausitzischen Gesellschaft der Wissenschaften von 1779 bis 1928." *Neues Lausitzisches Magazin* (1929), 105:1–59.

Joachim, Johannes. "Die Anfänge der Königlichen Sozietät der Wissenschaften zu Göttingen." *Abhandlungen der Gesellschaft der Wissenschaften zu Göttingen*, Philologisch-Historische Klasse, Dritte Folge (1936), 19:2–30.

Johnson, Francis R. "Gresham College, Precursor of the Royal Society." In Weiner and Noland, eds., *Roots of Scientific Thought* (New York: Basic Books, 1957), pp. 328–53.

Johnson, Terrence J. *Professions and Power.* London: Macmillan, 1972.

Johnson, William A. *Christopher Polhem, The Father of Swedish Technology.* Hartford, Conn.: Trinity College, 1963.

Joravsky, David. Review [*Istoria Akademii Nauk SSSR*, K. V. Ostrovitianov, ed.] *Isis* (1961), 52:591–92.

Jorpes, J. Erik. *Jac. Berzelius, His Life and Work*, Barbara Steele, tr. Stockholm: Almqvist & Wiksells, 1966.

Justin, E. *Les Sociétés royales d'agriculture aux XVIIIe siècle, 1757–1793.* Saint-Lô, 1935.

Kalousek, Joseph. *Geschichte der königlich Böhmischen Gesellschaft der Wissenschaften.* 2 vols. Prague: Verlag der Gesellschaft, 1884–85.

Karlberg, Gun. *Royal Society of Sciences of Uppsala: Catalogue of Publications: 1716–1967.* Uppsala: Almqvist & Wiksells International, 1977.

Kastler, Alfred. "Évolution de l'Âge moyen des membres de l'Académie des sciences depuis la foundation de l'Académie." *Comptes rendus de l'académie des sciences de Paris* (February 12, 1973), 276:65–66.

Kearney, Hugh. *Science and Change 1500–1700.* New York and Toronto: McGraw-Hill, World University Library, 1971.

—— "Puritanism & Science: Problems of Definition." *Past and Present* (1965), 31:104–10.

Kellner, J. "Alexander von Humbolt and the History of International Scientific Collaboration." *Scientia (Revista di Scienza)* (1960), 95:252–6.

Kendall, James. "The Royal Society of Edinburgh." *Endeavour* (1946), p. 5.

Khrgian, A. Kh. *Meteorology: A Historical Study.* Jerusalem: Israel Program for Scientific Translations, 1970.

Kindelberger, Elizabeth R. "The *Société royale des sciences de Montpellier*." Ph.D. dissertation, Johns Hopkins University, 1979.

Kington, J. A. "Meteorological Observing in Scandinavia and Iceland during the Eighteenth Century." *Weather* (June 1972), pp. 222–33.

Kister, Adolf. *Die Pflege der Naturwissenschaften in Mannheim zur Zeit Karl Theodors: Geschichte der Kurpfälzischen Akademie der Wissenschaft in Mannheim.* Mannheim: Altertumsverein, 1930.

Kopelevich, Iudif' Khaimovna. "The Petersburg Astronomy Contest in 1751," *Soviet Astronomy* (1966), 9:653–66.

—— *Vozniknovenie nauchnykh akademii (seredina XVII—seredina XVIII v.)* [The Rise of Scientific Academies from the Middle of the 17th to the Middle of the 18th Century]. Leningrad: Nauka, 1974.

Krassovsky, Dimitry and Robert Vosper. "The Structure of the Russian Academy of Science from its Beginning to 1945." University of Kentucky Libraries, *Occasional Contributions* #39. Lexington: 1952.

Kraus, Andreas. "Die Bedeutung des deutschen Akademien des 18. Jahrhunderts für die historische und naturwissenschaftliche Forschung." In Hartmann & Vierhaus, eds., *Der Akademiegedank* (1977), pp. 139–70.

_____ *Vernunft und Geschichte: Die Bedeutung der deutschen akademien für die Entwicklung der Geschichtswissenschaft in späten 18. Jahrhundert.* Freiberg: Herder, 1963.

Krauss, Werner. *Studien zur deutschen und französischen Aufklärung* [Neue Beiträge zur Literaturwissenschaft, 16]. Berlin: Butten & Loening, 1963.

Kronick, David A. "The Fielding H. Garrison List of Medical and Scientific Periodicals of the 17th and 18th Centuries: Addenda et Corrigenda." *Bulletin of the History of Medicine* (1958), 32:456–74.

_____ *A History of Scientific Periodicals: The Origins and Development of the Scientific and Technical Press, 1665–1790,* 2d ed., Metuchen, N.J.: Scarecrow Press, 1976; 1st ed., New York: Scarecrow Press, 1962.

_____ "Scientific Journal Publication in the Eighteenth Century," *Biographical Society Papers* (1965), 59:28–44.

Kuenen, J. P. *Gedenkboek van het Bataafsch Genootschap der Proefondervindelijke Wijsbegeerte te Rotterdam, 1769–1919.* Rotterdam, 1919.

Kuhn, Thomas S. *The Essential Tension: Selected Studies in Scientific Tradition and Change.* Chicago: University of Chicago Press, 1977.

_____ "History of Science." *International Encyclopedia of the Social Sciences* (New York: Crowell Collier and Macmillan, 1968), 14:74–83. [Reprinted in Kuhn, *The Essential Tension,* pp. 105–26.]

_____ "Mathematical versus Experimental Traditions in the Development of Physical Science." *Journal of Interdisciplinary History* (1976), 7:1–31. [Reprinted in Kuhn, *The Essential Tension,* pp. 31–65.]

_____ "Scientific Growth: Reflections on Ben-David's 'Scientific Role.'" *Minerva* (1972), 10:166–78.

_____ *The Structure of Scientific Revolutions,* 2d ed. enlarged. [International Encyclopedia of Unified Sciences, II, #2.] Chicago: University of Chicago Press, 1970; 1st ed., Chicago: University of Chicago Press, 1962.

Kungl. Svenska Vetenskapsakademie. *Bicentenaire de l'Académie royale des sciences de Suède Année 1939.* Uppsala: Almqvist & Wiksells, 1941.

Kunik, A. A. *Sbronik materialof gla Istorii Imperatorskoi Akademii Nauk v XVIII veka.* 2 vols. Saint-Petersburg, 1865.

Lautard, J. B. *Histoire de l'Académie de Marseille.* 3 vols. Marseille: 1826–1843.

Lavalleye, Jacques. *L'Académie royale des sciences, des lettres et des beaux-arts de Belgique.* Brussels, 1973.

Lefèvre-Pontalis, Eugène. *Bibliographie des Sociétés Savantes de la France.* Paris: Imprimerie nationale, 1887.

LeGrand, Homer E. "Chemistry in a Provincial Context: The Montpellier Société Royale des Sciences in the 18th Century." *Ambix* (1982), 29:88–105.

Leleu, M. "L'Académie d'Amiens, son histoire avant 1750." *Mémoires de l'Académie d'Amiens* (1901), 48:133–89.

Lenoble, Robert. *Mersenne ou la naissance du mécanisme.* Paris: Librairie Philosophique J. Vrin, 1943.

Letteron, Abbé. "Les Sociétés Savantes à Bastia: Notes pour servir à leur histoire." *Bulletin de la Société des sciences historiques et naturelles de la Corse* (1916–17), 3–4:99–205.

Ligou, Daniel. *Montauban à la fin de l'Ancien Régime et aux débuts de la Révolution, 1787–1794.* Paris: Marcel Rivière, 1958.

Liljencrantz, Axel. "Christopher Polem und die Gründung der ersten Schwedischen naturwissenschaftlichen Gesellschaft; nebst anderen Bermerkungen über des *Collegium Curiosorum.*" *Lychnos* (1940), pp. 53–54.

Lilley, Samuel. "Social Aspects of the History of Science." *Archives Internationales d'Histoire des Sciences* (1949), 2:376–443.

Lindroth, Sten. *Kungl. Svenska Vetenskapsakademiens Historia: 1739–1818.* 2 vols.· in 3. Stockholm: Kungl. Vetenskapsakademien, 1967.

—— "Science and Enlightenment in Sweden: 1740–1765." *Lynchnos* (1957–58), 192–93.

—— *Swedish Men of Science.* Burnett Anderson, tr. Uppsala: Almqvist & Wiksells, 1952.

Lingelbach, William E. "B. Franklin and the Scientific Societies." *Journal of the Franklin Institute* (1956), 261:9–31.

Lipski, Alexander. "The Foundation of the Russian Academy of Sciences." *Isis* (1953), 44:349–55.

Lloyd, G. E. R. *Early Greek Science: Thales to Aristotle.* New York: Norton, 1970.

—— *Greek Science after Aristotle.* New York: Norton, 1973.

Lomholt, Asger. *Det Kongelige Danske Videnskabernes Selskab 1742–1942: Samlinger till Selskabets Historie.* 4 vols. Copenhagen: Ejnar Munskgaard, 1942–1961.

—— "Outline of the History of the Academy." In Royal Danish Academy of Sciences and Letters, *The Royal Danish Academy of Sciences and Letters,* pp. 9–27. Copenhagen: Royal Danish Academy, 1981.

Longardi, Piero and Piero Galdi. *Le Accademie in Italia.* Turin: Edizioni Radio Italiana, 1956.

Lönnberg, Einar. *Kungl. Vetenskapsakademie Naturhistoriska Riksmusetts Historia.* Stockholm: Almqvist & Wiksells, 1916.

Lubimenko, Inna I. "Un Académicien russe à Paris." *Revue d'Histoire Moderne* (1935), 10:414–47.

Lubimenko, Inna I., compiler and ed. *Outchenaia Korrespondentsia Akademii Nauk XVIII veka, Nauchinoe opisanie 1766–1782 [Correspondance scientifique de l'Académie des Sciences au XVIIIe siècle: Description scientifique sous la haute direction de l'académicien D. S. Rojdestvensky, rédigée par I. I. Lubimenko avec la collaboration de*

G. A. Kniazev et L. B. Modzalevsky: L'Introduction d'Inna Lubimenko est suivie de sa traduction en français, pp. 35–48]. Moscow and Leningrad: 1937.

Luck, James Murray. *Science in Switzerland.* New York: Columbia University Press, 1967.

Lyons, Sir Henry. *The Royal Society, 1660–1940: A History of its Administration under its Charters.* Cambridge: Cambridge University Press, 1944; reprint New York: Greenwood Press, 1968.

—— "The Society's Finances: 1662–1830." *Notes and Records of the Royal Society* (1938), 2:73–87.

McClellan, James E. III. "The Académie Royale des Sciences, 1699–1793: A Statistical Portrait." *Isis* (1981), 72:541–67.

—— "The International Organization of Science and Learned Societies in the 18th Century." Ph.D. dissertation, Princeton University, 1975.

—— "Un Manuscrit inédit de Condorcet: 'Sur l'utilité des académies.'" *Revue d'Histoire des Sciences* (1977), 30:241–53.

—— "Scientific Associations." *Academic American Encyclopedia.* 17:144–46. Danbury, Conn.: Grolier, 1982.

—— "The Scientific Press in Transition: Rozier's Journal and the Scientific Societies." *Annals of Science* (1979), 36:425–49.

McCormick, Richard P. *Experiment in Independence: New Jersey in the Critical Period, 1781–1789.* New Brunswick, N.J.: Rutgers University Press, 1950.

McKie, Douglas. "Fontenelle et la Société Royale de Londres." *Revue d'Histoire des Sciences* (1958), 10:334–38.

—— "The *Observations* of the Abbé François Rozier." *Annals of Science* (1957), 13:73–89.

—— "The Origins and Foundation of the Royal Society of London." *Notes and Records of the Royal Society* (1960), 15:1–37.

—— "The Scientific Periodical from 1665 to 1798." *Philosophical Magazine* (1948), 150th Anniversary Commemorative Number, pp. 122–32.

—— "Scientific Societies to the End of the Eighteenth Century." *Philosophical Magazine* (1948), 150th Anniversary Commemorative Number: pp. 133–43.

McLachlan, H. *Warrington Academy: Its History and Influence.* Manchester: Chetham Society, 1943.

McLachlan, Patricia Petruschke. "Scientific Professionals in the Seventeenth Century." Ph.D. dissertation, Yale University, 1969.

MacLeod, Roy and Peter Collins, eds. *The Parliament of Science: The British Association for the Advancement of Science 1831–1981.* Northwood, G.B.: Science Reviews, 1981.

Mahaffy, J. P. "On the Origins of Learned Academies in Modern Europe." *Proceedings of the Royal Irish Academy* (1913), 30C:429–44.

Maindron, Ernest. *L'Académie des sciences.* Paris: Felix Alcan, 1888.

—— *Les Fondations de prix à l'Académie des sciences: Les Lauréats de l'Académie, 1714–1880.* Paris: Gauthier-Villars, 1881.

Manuel, Frank E. *A Portrait of Isaac Newton*. Cambridge, Mass.: Belknap Press, 1968; Washington, D. C.: New Republic Books, 1979.

Marguet, F. "La 'Connaissance des Temps' et son histoire." *Revue Générale des Sciences Pures et Appliquées* (1912), 23:133–40.

Marsak, Leonard. "Bernard de Fontenelle: The Idea of Science in the French Enlightenment." *American Philosophical Society Transactions* (1959), vol. 49, part 7.

Martin, Thomas. "The Origins of the Royal Institution." *British Journal for the History of Science* (1962), 1:49–63.

Mason, Stephen F. *A History of the Sciences*. New York: Collier, 1962.

Maugain, Gabriel. *Étude sur l'évolution intellectuelle de l'Italie de 1657 à 1750 environ*. Paris: Hachette, 1909.

Maurel, Blanche. "Une Société de pensée á Saint-Domingue: Le 'Cercle des philadelphes', au Cap Français." *Revue Française d'Historie d'Outre Mer* (1961), 177:234–66.

Maury, L.-F. Alfred. *Les académies d'autrefois: L'Ancienne Académie des Sciences*, 2d ed. Paris: Didier—Librarie Académique, 1864.

Maylender, Michele. *La Storia delle Accademie in Italia*. 5 vols. Bologna: L. Cappelli, 1926–30; reprint Rome: Arnaldo Forni, s.d. [1970s].

Medici, Michele. *Memorie storiche intorno le Accademie scientifiche e letterairie della cità di Bologna*. Bologna: Spaderie, 1852.

Meding, Henri L. *L'Académie Impériale Leopoldino-Carolina des Naturalistes (Curieux de la Nature)*. Paris: Victor Masson, 1854.

Meister, Richard. *Geschichte der Akademie der Wissenschaften in Wien: 1847–1947*. Vienna: Adolf Holzhausens, 1947.

Mendelsohn, Everett. "The Emergence of Science as a Profession in Nineteenth-Century Europe." In Karl Hill, ed., *The Management of Scientists*, pp. 3–48. Boston: Beacon Press, 1963.

Menshutkin, Boris N. *Russia's Lomonosov: Chemist, Courtier, Physicist, Poet*. Jeannette Eyre Thal and Edward J. Webster, trs. Princeton: Princeton University Press, 1952.

[Mersenne, Marin.] *Correspondance du P. Marin Mersenne, Religieux Minime*. 12 vols. Mme. Paul Tannery and Cornelis de Waard, eds. Paris: C.N.R.S., 1933–72.

Merton, Robert K. *Science, Technology and Society in Seventeenth-Century England*. Rev. ed. New York: Harper Torchbooks, 1970.

Merz, John Theodore. *A History of European Thought in the Nineteenth Century*. 4 vols. London: Blackwood, 1904–1912; vols. 1 & 2 reprinted as *History of European Scientific Thought* (New York: Dover, 1965).

Midbøe, Hans. *Det Kongelige norske videnskabers selskabs historie, 1760–1960*. 2 vols. Trondheim, 1960.

Middleton, W. E. Knowles. *The Experimentors: A Study of the Accademia del Cimento*. Baltimore: Johns Hopkins University Press, 1971.

_____ *Invention of Meteorological Instruments.* Baltimore: Johns Hopkins University Press, 1969.

_____ "Science in Rome, 1675–1700, and the Accademia Fisicomathematica of Giovanni Giustino Ciampini." *British Journal for the History of Science* (1975), 8:138–54.

Miller, David Philip. "Between Hostile Camps: Sir Humphry Davy's Presidency of the Royal Society of London: 1820–1827." *British Journal for the History of Science* (1983), 16:1–47.

_____ "Sir Joseph Banks: An Historiographical Perspective." *History of Science* (1981), 19:284–92.

Misland, Ph. *Notes et documents pour servir à l'histoire de l'Académie des sciences, arts et belles-lettres de Dijon.* Paris: Aubry, 1871.

Modzalevskii, B. L. *Spisok Chlenov Imperatorskoi Akademii Nauk 1725–1907.* Saint Petersburg, 1908.

Montandon, Cléopâtre. "Sciences et société à Genève au XVIIIe siècle." *Gesnerus* (1975), 32:16–34.

Montagne, Roland. "L'Expédition hydrographique de Chabert au Canada." *Revue d'Histoire des Sciences* (1964), 17:155–59.

Mornet, Daniel. *Les Origines intellectuelles de la Révolution française, 1715–1781.* 6th ed. Paris, 1967; 1st ed., 1933.

Morrell, Jack and Arnold Thackray. *Gentlemen of Science: Early Years of the British Association for the Advancement of Science.* Oxford: Clarendon Press, 1981.

Moschetti, Andrea. "La R. Accademia di Scienze, Lettere ed Arti di Padoua." *Atti e Memorie della Regia Accademia di Scienze, Lettere ed Arti di Padoua* (1934–35), 51:1–25.

Murua y Valerdi, Augustin. "Historia de la Real Academia de Ciencias y Artes." In Real Academia de Ciencias y Artes de Barcelona, *Fiestas cientificas celebradas con motivo del CL aniversario de su fundación* (1914), pp. 143–83.

Nachmanson, Ernst. "Forteckning över Göteborgs Kungl. Vetenskaps -och Vitterhets-Samhällets, Ledamöter, 1774–1927." In Göteborgs Kungl. Vetenskaps -och Vitterhets-Samhällets, *Handlingar*, Fjärde Följden (1928), 32(2):1–90.

Neave, E. W. J. "Chemistry in Rozier's Journal." *Annals of Science* (1950), 6:416–21; (1951), 7:101–06, 284–99, 393–400; (1952), 8:28–45.

Newton, Isaac. *Principia Mathematica.* Motte-Cajori tr. 2 vols. Berkeley: University of California Press, 1966.

Nevskaja, Nina I. "Joseph-Nicolas Delisle, 1688–1768." *Revue d'Histoire des Sciences* (1973), 26:289–313.

Nicholson, Francis. "The Literary & Philosophical Society, 1781–1851." *Memoirs and Proceedings of the Manchester Literary & Philosophical Society* (1923–24), 68:97–148.

Nicholson, Marjorie H. *Science and Imagination.* Ithaca, N.Y.: Cornell University Press, 1962.

Noel, M. "Notice historique sur la Société Académique de Cherbourg." *Mémoires de la Société [Impériale] Académique de Cherbourg* (1856), 7:1–29.

Nordenmark, N. V. E. *Anders Celsius, Professor i Uppsala.* Uppsala: Almqvist & Wiksells, 1936.

—— *Pehr Wilhelm Wargentin, Kungl. Vetenskaps-akademiens Sekreterare och Astronom 1749–1783.* Uppsala: Almqvist & Wiksells, 1939.

Nordin-Pettersson, B. S. "K. Svenska Vetenskapsakademiens äldre prisfrågor och belöningar 1739–1820." K. Svenska Vetenskapsakademie *Årsbok* (1959), pp. 435–516.

Oberhummer, Wilfrid. "Die Akademien der Wissenschaften." In Werner Schuder, ed., *Universitas Litterarum, Handbuch der Wissenschaftskunde,* pp. 700–708. Berlin: Walter de Gruyter, 1955.

Oergel, D. "Die Akademie nützlicher Wissenschaften zu Erfurt von ihrer wieder Belebund durch Dalberg bis zu ihrer endgültigen Anerkennung durch die Krone Preussen, 1776–1816." *Jährbucher der königlichen Akademie gemein-nütziger Wissenschaften zu Erfurt, Festschrift* (1940), 30:139–224.

[Oldenburg, Henry]. *The Correspondence of Henry Oldenburg.* 11 vols. A. R. Hall and M. B. Hall, eds. Vols. 1–9: Madison: University of Wisconsin Press, 1965–1973; vols. 10–11: London: Mansell Press, 1975–1977. (Vols. 12–13 projected.)

Oleson, Alexandra and Sandborn C. Brown, eds. *The Pursuit of Knowledge in the Early American Republic: American Scientific and Learned Societies from Colonial Times to the Civil War.* Baltimore: Johns Hopkins University Press, 1976.

Ollier, M. "Les Deux Premiers Siècles de l'Académie de Lyon." In Académie national des sciences, belles-lettres et arts de Lyon, *Le deuxième centenaire, 1700–1900,* 2:1–23. Lyon: A. Rey, 1900.

Ornstein, Martha [Bronfenbrenner]. *The Rôle of Scientific Societies in the Seventeenth Century.* New York: 1913; 2d ed. Chicago: University of Chicago Press, 1928; 3d ed. (New York Academy of Medicine, History of Medicine Series #6), 1938; reprints London: Archon, 1963; New York: Arno Press, 1975.

Outram, Dorinda. "The Ordeal of Vocation: The Paris Academy of Sciences and the Terror." *History of Science* (1983), 21:251–95.

Pace, Antonio. *Benjamin Franklin and Italy.* Philadelphia: American Philosophical Society, 1958.

Parker, Harold T. "French Administrators and French Scientists during the Old Regime and the Early Years of the Revolution." *Ideas in History* (1965), pp. 85–109.

Parker, Irene. *Dissenting Academies in England.* Cambridge: Cambridge University Press, 1914.

Pasquier, L. Gustave du. *Léonard Euler et ses amis.* Paris: J. Hermann, 1927.

Paul, Charles B. *Science and Immortality: The Éloges of the Paris Academy of Sciences (1699–1791).* Berkeley: University of California Press, 1980.

Pavlova, Galina E. "Lalande and the Saint Petersburg Academy of Sciences."

Proceedings of the 10th International Congress of the History of Science (Ithaca, N.Y., 1967), 2:743–46.

Pingaud, Léonce. "Documents pour servir à l'histoire de l'Académie de Besançon, 1752–1789." In Académie des sciences belles-lettres et arts de Besançon, *Procès Verbaux et Mémoires* (1893), pp. 234–334.

Polanyi, Michael. "The Republic of Science: Its Political and Economic Theory." *Minerva* (1962), 1:54–73.

Ponceau, Stephen Du. *An Historical Account of the Origin and Formation of the American Philosophical Society* [1840]. Philadelphia: American Philosophical Society, 1914.

Porter, Roy and Mikulas Teich, eds. *The Enlightenment in National Context*. Cambridge: Cambridge University Press, 1981.

Prescott, James Arthur. "The Russian Free Economic Society: Foundation-Years." *Agricultural History* (1977), 51:503–12.

Price, Derek J. da Solla. *Little Science, Big Science*. New York: Columbia University Press, 1963.

——— *Science Since Babylon*. Enlarged ed. New Haven: Yale University Press, 1975.

Proust, Jacques. *L'Encyclopédisme dans le Bas-Languedoc au XVIIIe siècle*. Montpellier: Faculté des Lettres et Sciences Humaines de Montpellier, 1968.

Purver, Margery. *The Royal Society: Concept and Creation*. Cambridge, Mass.: M.I.T. Press, 1967.

Quetelet, Adolphe. "Premier Siècle de l'Académie royale de Belgique." *Centième Anniversaire de fondation de l'Académie royale des sciences, belles-lettres et des beaux-arts de Belgique*, 1:1–174. Brussels, 1872.

Rabb, Theodore K. "Religion and the Rise of Modern Science." *Past and Present* (1965), 31:111–26.

Ramsbottom, J. "The Royal Swedish Academy of Science." *Nature* (1939), 144:270–71.

Rance, A.-J. "Histoire de l'Académie d'Arles." *Revue de Marseille et de Provence* (1885), 32:49–58, 145–56.

Rappaport, Rhoda. "Government Patronage of Science in Eighteenth-Century France." *History of Science* (1969), 8:119–36.

——— "The Liberties of the Paris Academy of Sciences, 1716–1785." In Harry Woolf, ed., *The Analytic Spirit, Essays in the History of Science in Honor of Henry Guerlac*, pp. 225–53. Ithaca, N.Y., and London: Cornell University Press, 1981.

Reichenbach, E. and Uschmann, G., eds. *Beiträge zur Geschichte der Präsidenten der deutschen Akademie der Naturforscher Leopoldiana*. Leipzig: J. A. Barth, 1970.

Ribeiro, José Silvestre. *Historia dos estabelecimentos scientificos, litterarios e artisticos de Portugal*. 5 vols. Lisbon: Academia Real das Sciencias, 1871–76.

Rickett, Harold William. "The Royal Botanical Expedition to New Spain, 1788–1820." *Chronica Botanica*. (1947/1948), 11:1–86.

Robertson, D. Maclaren. *A History of the French Academy 1635–1910.* New York: G. W. Dillingham, 1910.

Robinet, André. "La Vocation académicienne de Malebranche." *Revue d'Histoire des Sciences* (1959), 12:1–18.

Robinson, Eric. "The Derby Philosophical Society." *Annals of Science* (1953), 9:359–68.

Roche, Daniel. "L'Histoire dans les activités des académies provinciales en France au XVIIIe siècle." In Hammer and Voss, eds., *Historische Forschung im 18. Jahrhundert,* (Bonn, 1976), pp. 260–93.

———— "Milieux académiques provinciaux et société des lumières." In François Furet, ed., *Livre et société dans la France du XVIIIe siècle.* pp. 94–156.

———— "Sciences et pouvoirs dans la France du XVIIIe siècle: 1666–1803." *Annales* (1974), 3:738–39.

———— *Le siècle des lumières en province: Académies et académiciens provinciaux, 1680–1789.* 2 vols. Paris and The Hague: Mouton, 1978.

La Rochelle [town], Bibliothèque municipale. *Réaumur et l'Académie de la Rochelle au XVIIIe siècle* [Exhibition catalogue]. 1958.

Roddier, Henri. "Pour une histoire des académies." *Actes du IIIe Congrès national de la Société française de littérature comparée,* Dijon, 1959, pp. 45–55. Paris: Didier, 1960.

Ronan, Colin A. *Astronomers Royal.* Garden City, N.Y.: Doubleday, 1969.

Rosen, George. *From Medical Police to Social Medicine: Essays on the History of Health Care.* New York: Science History Publications, 1974.

Rosen, Richard L. "The Academy of Sciences of the Institute of Bologna." Ph.D. dissertation, Case Western Reserve University, 1971.

Ross, Sydney. "Scientist: The Story of a Word." *Annals of Science* (1962), 18:65–85.

Rossiter, Margaret. "The Organization of Agricultural Improvement in the United States, 1785–1865." In Oleson and Brown, eds., *The Pursuit of Knowledge in the Early American Republic,* pp. 279–98.

Rousseau, G. S. and Roy Porter, eds. *The Ferment of Knowledge: Studies in the Historiography of Eighteenth-Century Science.* Cambridge: Cambridge University Press, 1980.

Royal Dublin Society [of Arts]. *Bicentenary Souvenir: 1731–1931.* Dublin, 1931.

Royal Society of Edinburgh. *General Index to the First Thirty-Four Volumes of Transactions (1783–1888) with a History of the Institution.* Edinburgh: Neill, 1890.

Royal Society of London. *Book Catalogue of the Library of the Royal Society.* Alan J. Clark, compiler. Frederick, MD: University Publications of America, 1982.

———— *The Record of the Royal Society of London.* 4th ed. London: 1940.

Rozier, François. *Nouvelles Tables des articles contenus dans les volumes de l'Académie royale des sciences de Paris, depuis 1666 jusqu'en 1770.* 4 vols. Paris: Ruault, 1775–76.

Rudwick, M. J. S. "The Foundation of the Geological Society of London: Its Scheme for Co-operative Research and Its Struggle for Independence." *British Journal for the History of Science* (1963), 2:325–55.

Rübel, Eduard. "Geschichte der naturforschenden Gesellschaft in Zürich." *Neujahrsblatt der Naturforschenden Gesellschaft in Zürich* (1947), vol. 149.

Russo, François. *Éléments de bibliographie de l'histoire des sciences et des techniques.* 2d ed. Paris: Hermann, 1969.

Rydberg, Sven. *Svenska Studieresor till England under Frihetstiden.* Uppsala: Almqvist & Wiksells, 1951.

Salomon-Bayet, Claire. *L'Institution de la science et l'expérience du vivant: Méthode et expérience à l'Académie royale des sciences 1666–1793.* Paris: Flammarion, 1978.

Sampolo, Luigi. "Su la origine, le vicende e il rinnovamento della Accademia." *Atti della Accademia di Scienze, Lettere e Arti di Palermo* (1971–73), 32(2):25–75.

Sarton, George. "Vindication of Father Hell." *Isis* (1944), 35:98–99.

Savarit, C.-M. "Les Académies de province au travail." *Revue des Deux Mondes* (1930), 58:464–69.

Schmieder, Carl Christoph. *Geschichte der Entstehung und neuern Einrichtung der Naturforschenden Gesellschaft in Halle.* Halle: Hendels Verlag, 1809.

Schmitt, C. B. "Science in the Italian Universities in the 16th and early 17th Centuries." In M. P. Crosland, ed., *The Emergence of Science in Western Europe,* pp. 35–56.

Schnabel, Franz. *Deutsche Geschichte in neunzehnten Jahrhundert.* Vol. 3: *Erfahrungswissenschaft und Technik.* Freiburg im Breisgau: Herder, 1934.

Schneiders, W. "Gottesreich und gelehrte Gesellschaft: Zwei politische Modelle bei G. W. Leibniz." In Hartman and Vierhaus, eds., *Der Akademiegedank* (1977), pp. 47–62.

Schofield, Robert E. "Histories of Scientific Societies: Needs and Opportunities for Research." *History of Science* (1963), 2:70–84.

—— *The Lunar Society of Birmingham: A Social History of Provincial Science and Industry in Eighteenth Century England.* Oxford: Clarendon Press, 1963.

—— *Mechanism and Materialism: British Natural Philosophy in an Age of Reason.* Princeton: Princeton University Press, 1970.

Schroeder-Gudehus, Brigitte. "Division of Labor and the Common Good: The International Association of Academies, 1899–1914." In Carl Gustaf Bernhard et al., eds., *Science, Technology and Society in the Time of Alfred Nobel,* pp. 3–20. Oxford: Pergamon Press/Nobel Foundation, 1982.

Schumann, E. "Geschichte der Naturforschenden Gesellschaft in Danzig, 1743–1892." [Festschrift zur Feier des 150 Jährigen Bestehens der naturforschenden Gesellschaft in Danzig am 2 Januar 1893.] *Schriften der naturforschenden Gesellschaft in Danzig* (1893), vol. 8, pt. 2.

Scudder, Samuel H. *Catalogue of Scientific Serials, 1633–1876.* [Library of Harvard University, Special Publications #1.] Cambridge, Mass.: Library of Harvard University, 1879.

Shafer, Robert Jones. *The Economic Societies in the Spanish World, 1763–1821.* Syracuse, N.Y.: Syracuse University Press, 1958.

Shapin, Steven. "Property, Patronage and the Politics of Science: The Founding of the Royal Society of Edinburgh." *British Journal for the History of Science* (1974), 7:1–41.

—— "The Royal Society of Edinburgh: A Study of the Social Context of Hanoverian Science." Ph.D. dissertation, University of Pennsylvania, 1971.

—— "Social Uses of Science." In Rousseau and Porter, eds., *The Ferment of Knowledge*, pp. 93–139.

Shapin, Steven and Arnold Thackray. "Prosopography as a Research Tool in History of Science: The British Scientific Community 1700–1900." *History of Science* (1974), 12:1–28.

Sharlin, Harold I. *The Convergent Century: The Unification of Science in the Nineteenth Century.* London and New York: Abelard-Schuman, 1966.

Sheenan, Donal. "The Manchester Literary and Philosophical Society." *Isis* (1941), 33:519–23.

Shinn, Terry. *L'École Polytechnique 1794–1914.* Paris: Presses de la Fondation Nationale des Sciences Politiques, 1980.

Simon, Joseph. "Documents inédits pour servir à l'histoire de l'Académie de Nîmes." *Mémoires de l'Académie de Nîmes* (1897), 20:1–12.

Singels, N. J. *Vit de Geschiedenis van het Provinciaal Utrechtsch Genootschap, 1773–1923.* Utrecht: A. Ossthoek, 1923.

Smeaton, W. A. "The Early Years of the Lycée and the Lycée des Arts." *Annals of Science* (1955), 11:257–67; 309–19.

—— "Louis-Bernard Guyton de Morveau F.R.S. and his Relations with British Scientists." *Notes and Records of the Royal Society* (1967), 22:113–30.

Smend, Rudolf. "Die Göttinger Gesellschaft der Wissenschaften." *Festschrift zur Feier des zweihundertjährigen Bestehens der Akademie der Wissenschaften in Göttingen* (1951), pp. v–xix.

Smith, Alan G. R. *Science and Society in the 16th and 17th Centuries.* New York: Science History Publications, 1972.

Solomon, Howard M. *Public Welfare, Science and Propaganda in Seventeenth-Century France: The Innovations of Théophraste Renaudot.* Princeton: Princeton University Press, 1972.

Sonntag, Otto. "Albrecht von Haller on Academies and the Advancement of Science: The Case of Göttingen." *Annals of Science* (1975), 32:379–91.

Spence, Jonathan. *To Change China: Western Advisers in China 1620–1960.* Boston: Little, Brown, 1969.

Sprat, Thomas, *The History of the Royal Society for the Improving of Natural Knowledge.* London: 1667; 2d ed. 1702. Johnson reprint, 1958; with Introduction by Jackson I. Cope and H. W. Jones.

Stackelberg, J. von. "Die Akademie Française." In Hartman & Vierhaus, eds., *Der Akademiegedank* (1977), pp. 26–46.

Staszewski, Jacek. "Die ersten wissenschaftlichen Geselleschaften in Polen und Ihre Bedeutung für die Entwicklung der Aufklärung." In Eric Amberger et al., eds., *Wissenschaftspolitik in Mittel- und Osteuropa*, pp. 309–20. Berlin: Verlag Uhrich Camer, 1976.

Stearn, William T. "A Royal Society Appointment with Venus in 1769: The Voyage of Cook and Banks in the Endeavor in 1768–1771 and its Botanical Results." *Notes and Records of the Royal Society* (1969), 24:64–90.

Stearns, Raymond Phineas. "Colonial Fellows of the Royal Society of London." *Osiris* (1948), 8:73–121.

—— "Fellows of the Royal Society in North Africa and the Levant." *Notes and Records of the Royal Society* (1954), 11:75–90.

—— *Science in the British Colonies of America.* Urbana: University of Illinois Press, 1970.

Steward, Dugald. *Account of the Life and Writing of William Robertson.* London: Cadell, Davies & Balfour, 1801.

Stillwell, Margaret Bingham. *The Awakening Interest in Science during the First Century of Printing.* New York: Bibliographical Society of America, 1970.

Stimson, Dorothy. *Scientists and Amateurs: A History of the Royal Society.* New York: Henry Schuman, 1948.

Stone, Bruce Winchester. "The Role of the Learned Societies in the Growth of Scientific Boston: 1780–1848." Ph.D. dissertation, Boston University, 1974.

Stroup, Alice. "Wilhelm Homberg and the Search for the Constituents of Plants at the 17th-Century Académie Royale des Sciences." *Ambix* (1979), 26:184–201.

Tannery, Paul. "Les Sociétés savantes et l'histoire des sciences [1904]." In *Mémoires scientifiques*, 10:183–90. Toulouse: Privat, 1930.

Taton, René. *Les Origines de l'Académie royale des sciences.* Paris: Palais de la Découverte, 1966.

—— "Le Rôle des correspondances scientifiques dans la diffusion de la science aux XVIIe et XVIIIe siècles." *Proceedings of XIV International Congress of the History of Science* (1974), pt. 2, pp. 214–30.

Taton, René, ed. *Enseignement et diffusion des sciences en France au XVIIIe siècle.* Paris: Hermann, 1964.

Taylor, E. G. R. *The Haven-Finding Art.* London: Hollis and Carter, 1951.

Taylor, Kenneth L. "Cotte." In C. C. Gillispie, ed., *Dictionary of Scientific Biography*, 3:435–6.

Teich, Mikulas. "Bohemia: From Darkness into Light." In Roy Porter and Mikulas Teich, eds., *Enlightenment in National Context*, pp. 141–63. Cambridge: Cambridge University Press, 1981.

—— "Ignaz von Born als Organisator wissenschaftlicher Bestrebungen in der Habsburger Monarchie." In E. Amburger, ed. *Wissenschaftspolitik*, pp. 195–205.

—— "The Royal Bohemian Society of Sciences and the First Phase of Organized Scientific Advance in Bohemia." *Historica* (1960), 2:161–82.

Terrin, A. "Histoire de l'Académie de Nîmes." *Mémoires de l'Académie de Nîmes* (1933–35), pp. xvi–lxix.

Thackray, Arnold. *Atoms and Powers: An Essay on Newtonian Matter Theory and the Development of Chemistry*. Cambridge, Mass.: Harvard University Press, 1970.

—— "Natural Knowledge in Cultural Context: The Manchester Model." *American Historical Review* (1974), 79:672–709.

Théodoridès, Jean. "Humboldt and England." *British Journal for the History of Science* (1966), 3:39–55.

Thiele, Richard. "Die Gründung der Akademie nützlicher Wissenschaften zu Erfurt und die Sicksale derselben bis zu ihrer Wiederbelebung durch Dalberg, 1754–1776." *Jährbucher der königlichen Akademie gemeinnütziger Wissenschaften zu Erfurt, Festschrift* (1940), 30:1–138.

Thompson, Thomas. *History of the Royal Society from Its Institution to the End of the Eighteenth Century*. London: Robert Baldwin, 1812.

Thomson, Sir George. "Anglo-Portuguese Scientific Relations." *Nature* (1943), 152:618–19.

Thorndike, Lynn. *A History of Magic and Experimental Science*. 8 vols. New York: Columbia University Press, 1929–1958.

Thornton, John L. and R. I. J. Tully. *Scientific Books, Libraries & Collectors: A Study of Bibliography and the Book Trade in Relation to Science*, 3d ed. revised. London: The Library Association, 1971.

Tisserand, Roger. *Au Temps de l'Encyclopédie: L'Académie de Dijon de 1740 à 1793*. Paris: Boivin, 1936.

—— *Les Concurrents de J. J. Rousseau à l'Académie de Dijon pour le prix de 1754*. Paris: Boivin, 1935.

Toellner, R. "Entstehung und Programm der Göttinger Gelehrten Gesellschaft unter besonderer Berücksichtingung des Hallerschen Wissenschaftsbegriffes." In Hartmann and Vierhaus, eds., *Der Akademiegedank*, pp. 97–116.

Torlais, Jean. "L'Académie de la Rochelle et la diffusion des sciences au XVIIIe siècle." *Revue d'Histoire des Sciences* (1959), 12:111–25.

Trénard, Louis. "L'Académie de Lyon et ses relations étrangères au XVIIIe siècle." *Actes du IIIe Congrès national de la Société française de littérature comparée*, Dijon, 1959, pp. 65–86. Paris: Didier, 1960.

Truesdell, C. A. *Essays in the History of Mechanics*. Berlin and New York: Springer-Verlag, 1968.

Turnbull, H. W. "Early Scottish Relations with the Royal Society." *Notes and Records of the Royal Society* (1940), 3:22–38.

Turner, Steven R. "The Prussian Universities and the Research Imperative, 1806–1848." Ph.D. dissertation, Princeton University, 1972.

Uggla, Aruid Hj. "Early Relations between the Royal Society and Sweden." *Lychnos* (1940), p. 324.

Varnhagen, Francisco Adolpho. *História Geral do Brasil*. 7th ed. Sâo Paulo: Melhoramentos, 1962.

Vávra, Jaroslav. "Die Olmützer Societas incognitorum und die Petersburger Akademie der Wissenschaften." In W. Steinitz et al., eds., *Ost und West in der Geschichte des Denken und der Kulturellen Beziehungen*, [Quellen und Studien zur Geschichte Osteuropas, 15], pp. 278–89. Berlin: Akademie Verlag, 1966.

Verdoorn, Frans. "The Development of Scientific Publications and Their Importance in the Promotion of International Scientific Relations." *Science* (1948), 107:492–97.

Vierhaus, Rudolf, ed. *Deutsche patriotische und gemeinnützige Gesellschaften* [Wolfenbütteler Forschungen, 8]. Munich: Kraus International Publications, 1980.

Vivie, Aurelin. "Les Lauréats de l'Académie, 1713–1893." *Actes de l'Académie nationale des sciences, belles-lettres et arts de Bordeaux* (1893), 55:423–91.

von Sydow, Carl-Otto. "The Society of Sciences and Henric Benzelius' Journey to Lapland in 1711." *Lychnos* (1962), pp. 162–63.

Voss, Jürgen. "Die Akademien als Organisationsträger der Wissenschaften im 18. Jahrhundert." *Historische Zeitschrift* (1980), 231:43–74.

―― "Die Societe [sic] Patriotique de Hesse-Hombourg (1775–1781)." In Vierhaus, ed., *Deutsche patriotische und gemeinnützige Gesellschaften*, pp. 195–221.

Vucinich, Alexander. Review [*Istoria Akademii Nauk SSSR*, K. V. Ostrovitianov, ed.]. *Science* (1960), 131:151–52.

―― *Science in Russian Culture: A History to 1860*. Stanford, Calif.: Stanford University Press, 1963.

Waff, Craig B. "Universal Gravitation in the Motion of the Moon's Apogee: The Establishment and Reception of Newton's Inverse Square Law, 1687–1749." Ph.D. dissertation, Johns Hopkins University, 1976.

Walter, Friedrich. *Geschichte Mannheims von den ersten Anfängen bis zum Übergang an Baden*. Mannheim: Verlag der Stadtgemeinde, 1907.

Waters, David F. "Nautical Astronomy and the Problem of Longitude." In John G. Burke, ed., *The Uses of Science in the Age of Newton*, pp. 143–69. Berkeley: University of California Press, 1983.

Watson, Robert Spence. *The History of the Literary and Philosophical Society of Newcastle-upon-Tyne*. London: Walter Scott, 1897.

Weld, Charles Richard. *A History of the Royal Society with Memoirs of the Presidents*. 2 vols. London: John Parker, 1848; reprint: New York: Arno Press, 1975.

Westenrieder, Lorenz. *Geschichte der Bayerschen Akademie der Wissenschaften*. 2 vols. Munich: 1784, 1807.

Westfall, Richard S. *The Construction of Modern Science: Mechanisms & Mechanics*. Cambridge: Cambridge University Press, 1971.

―― *Never at Rest: A Biography of Isaac Newton*. Cambridge: Cambridge University Press, 1980.

Whitehill, Walter Muir. "Early Learned Societies in Boston and Vicinity." In

Oleson and Brown, eds. *The Pursuit of Knowledge in the Early American Republic*, pp. 151–73.

Williams, L. P. "Science, Education, and the French Revolution." *Isis* (1953), 44:311–30.

—— "Science, Education, and Napoleon I." *Isis* (1956), 47:369–82.

Williams, L. P. ed., *Album of Science: The Nineteenth Century.* New York: Scribner, 1978.

Wilson, Iris Higbie. "Scientists in New Spain: The Eighteenth Century Expeditions." *Journal of the West* (1962), 1:24–44.

Winau R. "Zur Frühgeschichte der Akademia Curiosorum." In Hartman and Vierhaus, eds., *Der Akademiegedank* (1977), pp. 117–38.

Winter, Eduard. *Die Registres der Berliner Akademie der Wissenschaften, 1746–1766: Documente für Wirken Leonard Eulers in Berlin.* Berlin: Akademie-Verlag, 1957.

Winter, H. J. J. "Scientific Notes from the Early Minutes of the Peterborough Society." *Isis* (1939), 31:51–60.

Wood, Henry Trueman. *A History of the Royal Society of Arts.* London: John Murray, 1913.

Woodward, Horace B. *The History of the Geological Society of London.* London: Geological Society, 1907; reprint, New York: Arno Press, 1978.

Woodward, Stephen Wyndham. "Paper Read to the Society of Antiquaries of London [re Spalding Gentlemen's Society] 13 October 1960." Mimeo typescript, 1960.

Woolf, Harry. *The Transits of Venus, A Study of Eighteenth-Century Science.* Princeton: Princeton University Press, 1959.

Wooley, Sir Richard. "Captain Cook and the Transit of Venus in 1769." *Notes and Records of the Royal Society* (1969), 24:19–32.

Youschkevitch, A. P. "Euler." In C. C. Gillispie, ed., *Dictionary of Scientific Biography* 4:467–84.

Youschkevitch, A. P. and E. Winter. *Die Berliner und die Petersburger Akademie der Wissenschaften in Briefwechsel Leonard Eulers.* 3 vols. Berlin: Akademie Verlag, 1959–1976. [Cited as *Briefwechsel.*]

Zacek, Joseph F. "The Virtuosi of Bohemia: The Royal Bohemian Society of Sciences." *East European Quarterly* (1968), 2:147–59.

Zupko, R. E. *French Weights and Measures before the Revolution: A Dictionary of Provincial and Local Units.* Bloomington and London: Indiana University Press, 1978.

Index